W9-DFC-517

ECOLOGICAL ASSESSMENTS OF EFFLUENT IMPACTS ON COMMUNITIES OF INDIGENOUS AQUATIC ORGANISMS

A symposium
sponsored by ASTM
Committee D-19 on Water
AMERICAN SOCIETY FOR
TESTING AND MATERIALS
Ft. Lauderdale, Fla.
29–30 Jan. 1979

ASTM SPECIAL TECHNICAL PUBLICATION 730
J. M. Bates, Ecological Consultants, Inc.,
and C. I. Weber, U.S. Environmental
Protection Agency, editors

ASTM Publication Code Number (PCN)
04-730000-16

AMERICAN SOCIETY FOR TESTING AND MATERIALS
1916 Race Street, Philadelphia, Pa. 19103

Library of Congress Catalog Card Number: 80-69061

NOTE

The Society is not responsible, as a body,
for the statements and opinions
advanced in this publication.

Printed in Baltimore, Md.
May 1981

Foreword

The symposium on Ecological Assessments of Effluent Impacts on Communities of Indigenous Aquatic Organisms was held on 29–30 Jan. 1979 in Ft. Lauderdale, Fla. The American Society for Testing and Materials, through its Committee D-19 on Water, sponsored the event. J. M. Bates, Ecological Consultants, Inc., and C. I. Weber, U.S. Environmental Protection Agency, presided as cochairmen of the symposium and served as editors of this publication.

Related ASTM Publications

Manual on Water, STP 442A (1978), $28.50, 04-442010-16

Aquatic Toxicology and Hazard Evaluation, STP 634 (1977), $30.75, 04-634000-16

Aquatic Toxicology: Second Conference, STP 667 (1979), $37.75, 04-667000-16

Aquatic Toxicology: Third Conference, STP 707 (1980), $39.50, 04-707000-16

Biological Monitoring of Water and Effluent Quality, STP 607 (1976), $24.25, 04-607000-16

Biological Methods for the Assessment of Water Quality, STP 528 (1972), $16.25, 04-528000-16

Estimating the Hazard of Chemical Substances to Aquatic Life, STP 657 (1978), $19.50, 04-657000-16

Biological Data in Water Pollution Assessment: Quantitative and Statistical Analyses, STP 652 (1978), $17.50, 04-652000-16

Measurement of Organic Pollutants in Water and Wastewater, STP 686 (1979), $36.50, 04-686000-16

Bacterial Indicators/Health Hazards Associated with Water, STP 635 (1977), $34.75, 04-635000-16

Native Aquatic Bacteria: Enumeration, Activity, and Ecology, STP 695 (1979), $25.00, 04-695000-16

A Note of Appreciation to Reviewers

This publication is made possible by the authors and, also, the unheralded efforts of the reviewers. This body of technical experts whose dedication, sacrifice of time and effort, and collective wisdom in reviewing the papers must be acknowledged. The quality level of ASTM publications is a direct function of their respected opinions. On behalf of ASTM we acknowledge with appreciation their contribution.

ASTM Committee on Publications

Editorial Staff

Contents

Introduction

Interest in environmental impact assessment has greatly increased in recent years. While significant advances have been made in bioassays, biomonitoring, and other toxicological measurements, advancements in the development of methodologies for assessment of effluent impacts on the biological integrity of receiving waters has lagged.

The Ecosystems Section of ASTM Subcommittee D19.23 on Biological Field Methods, which was instituted in 1977, has attempted to address problems attendant to in-stream assessments. Consequently, a request for this symposium was prepared in January of 1978, and the symposium was held 29-30 Jan. 1979 in Ft. Lauderdale, Fla.

The purpose of the symposium was to provide a forum for scientists engaged in the development and application of methods of evaluating the effects of effluents on the biological integrity of surface waters. Emphasis was to be placed on methods of evaluating the effects of effluents on standing crop, community structure, and community function. The call for papers emphasized the importance of the feasibility of application and the priority ordering of community parameters to be investigated within the constraints of available time, personnel, and budget.

The seventeen papers presented in this volume will hopefully be useful to those who are presently engaged in studies dealing with aquatic ecological assessments. A brief critique of these papers is presented in the summary of this volume.

J. M. Bates

Ecological Consultants, Inc., Ann Arbor, Mich. 48103; symposium chairman and editor.

C. I. Weber[1]

Evaluation of the Effects of Effluents on Aquatic Life in Receiving Waters—An Overview

REFERENCE: Weber, C. I., **"Evaluation of the Effects of Effluents on Aquatic Life in Receiving Waters—An Overview,"** *Ecological Assessments of Effluent Impacts on Communities of Indigenous Aquatic Organisms, ASTM STP 730,* J. M. Bates and C. I. Weber, Eds., American Society for Testing and Materials, 1981, pp. 3-13.

ABSTRACT: Federal water pollution control legislation strongly emphasizes the need to protect aquatic life from adverse effects of effluents. Effluent biomonitoring, therefore, plays an important role in the federal water pollution control program, and includes two basic functions: (1) measurement of the toxicity and biostimulatory properties of effluents, and (2) measurement of the effects of effluents on the biological integrity of receiving waters, which includes the abundance, species composition, metabolism, and condition of indigenous aquatic organisms. Standardized methods for sample collection and preparation are widely available. However, additional tools are urgently needed to assist biologists in sample analysis and data interpretation. The tools needed include species identification manuals, information on the life histories of aquatic organisms, and descriptive models with which to evaluate field data. The models would describe the standing crop, species composition, and community metabolism expected at a given sampling station in the absence of pollution, taking into account the season of the year, climate, ambient meteorological and hydrological conditions, water chemistry, and the nature of the substrate. Effective data interpretation also requires utilization of available parametric and nonparametric methods for statistical analyses of the field data in order to determine the significance of observed differences in the properties of aquatic communities at "control" and "polluted" sampling stations.

KEY WORDS: water pollution, biomonitoring, effluents, biological integrity, toxicity tests, aquatic organisms, ecology

Objectives of the Water Pollution Control Program

The protection of aquatic life and human health from adverse effects of point and nonpoint sources of pollution is the central theme of current

[1] Chief, Aquatic Biology Section, Biological Methods Branch, Environmental Monitoring and Support Laboratory, U.S. Environmental Protection Agency, Cincinnati, Ohio, 45268.

federal water pollution control legislation. This theme permeates the Federal Water Pollution Control Act Amendments of 1972 (Public Law 92-500) and 1977 (Public Law 95-217, or the Clean Water Act of 1977), which state that the objective of the legislation is to "restore and maintain the physical, chemical, and biological integrity of the Nation's waters." The legislation also states that it is national policy that the discharge of toxic substances in toxic amounts be prohibited. The goals of the legislation are to achieve a degree of water quality that provides for the protection and propagation of aquatic life by mid-1983 and the elimination of discharges of pollutants to surface waters by 1985. Other federal legislation relating to water pollution control includes the Resource Conservation and Recovery Act (Public Law 94-580), the Marine Protection and Sanctuaries Act (Public Law 92-532), and the Toxic Substances Control Act (Public Law 94-469).

The threat to human health and welfare posed by surface water pollution has two principle facets: (1) direct effects, which may result from the consumption of contaminated water supplies and food, and (2) indirect effects, which may result from the impact of pollutants on the quantity and quality of aquatic organisms used for human food, the recreational use of water, the aesthetic quality of the aquatic environment, and the integrity of the biosphere (Fig. 1).

The definition of toxicity provided in Paragraph 502(13) of the Clean Water Act of 1977 (CWA) is very explicit and encompasses essentially all the possible adverse effects on *all* types of organisms. These effects include death, disease, behavior abnormalities, physiological malfunctions, physical deformities (including birth defects), mutations, and cancer.

The concept of "biological integrity" is not as clearly defined as that of "toxicity" in the CWA, but can be inferred from the frequent use of terminology such as "the protection and propagation of shellfish, fish, and wildlife," "balanced indigenous populations of shellfish, fish, and wildlife," and "ecosystem diversity, productivity, and stability, and species and community (structure)." Further clarification of the intent of the Congress in using the term "biological integrity" is found in Paragraph 502(15) of the CWA, where the definition of "biological monitoring" includes the determination of the effects (of pollution) on (all types of) aquatic life, which, together with terminology used in other sections of the CWA, can only be interpreted to include all communities of aquatic organisms, such as the phytoplankton, zooplankton, periphyton, macrophyton, macroinvertebrates, and fish.

The concept of biological integrity is easily defined in terms of the basic properties of communities of aquatic organisms, which are (1) the standing crop or abundance (expressed in terms of numbers of organisms, weight, size, or biomass), (2) community structure (the kinds of organisms present and the relative abundance of each kind), and (3) community metabolism and condition (rates of physiological processes, such as photosynthesis and

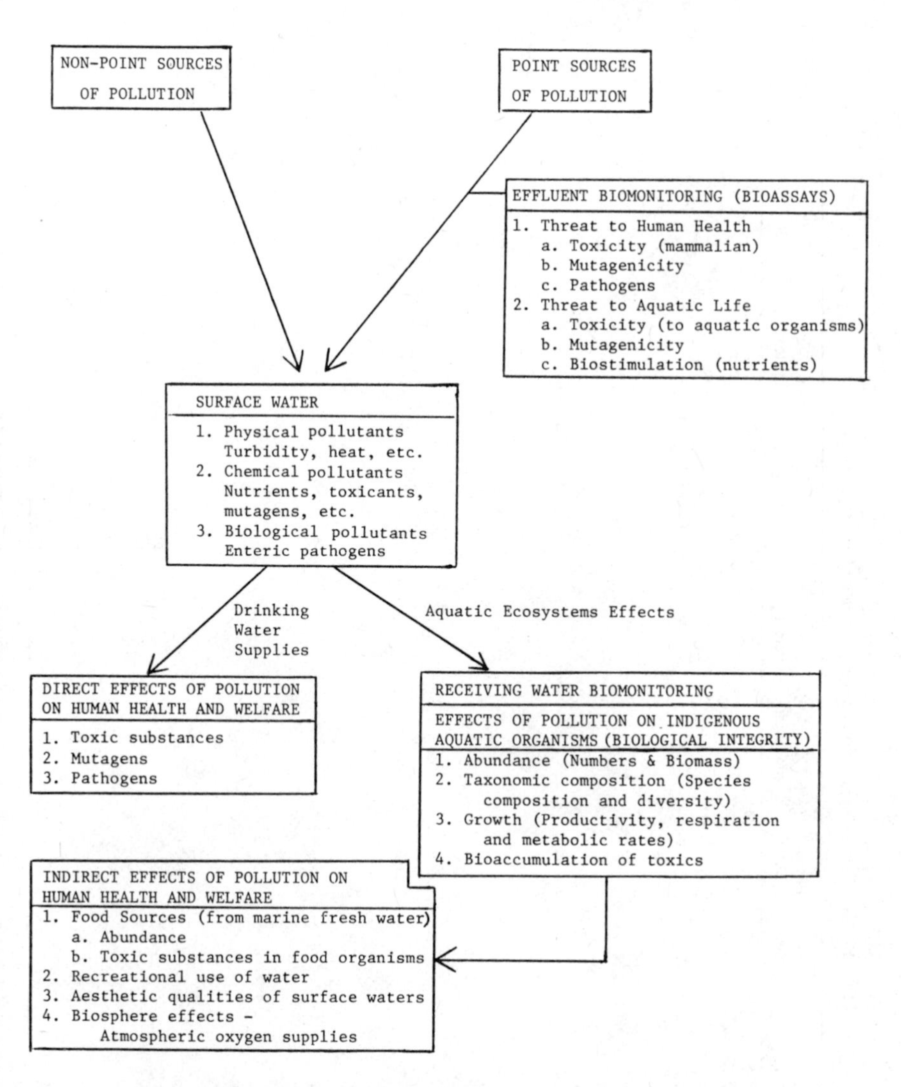

FIG. 1—*Biological effects of pollution on human health and aquatic life.*

nitrogen fixation, accumulation of toxic substances, disease, histopathological conditions, parasitism, and flesh tainting).

Effluent Biomonitoring

Effluent biomonitoring includes two basic activities: (1) measurement of the biological properties of effluents, using captive organisms exposed to the effluent in the laboratory or in the environment (Table 1), and (2) mea-

TABLE 1—*Use of captive organisms in biomonitoring and toxicity tests.*

Type of Test	Phyto-plankton	Zoo-plankton	Periphyton	Macro-phyton	Macroinver-tebrates	Fish
In situ tests						
Bioaccumulation						
Toxic metals	×	×	×	×	×	×
Pesticides (organics)	×	×	×	×	×	×
Flesh tainting					×	×
Toxicity tests						
Acute toxicity					×	×
Histopathology					×	×
Histochemistry			×	×	×	×
Cholinesterase						×
In-plant tests (effluents)						
Bioaccumulation						
Toxic metals			×	×	×	×
Pesticides (organics)			×	×	×	×
Flesh tainting					×	×
Toxicity tests						
Acute toxicity	×	×			×	×
Low-level responses						×
Behavioral responses					×	×
Histopathology					×	×
Biostimulatory tests						
Algal growth response (AGP)	×		×			
Laboratory tests						
Bioaccumulation						
Toxic metals	×	×	×	×	×	×
Pesticides (organics)	×	×	×	×	×	×
Flesh tainting					×	×
Toxicity tests						
Acute toxicity	×	×			×	×
Chronic toxicity	×	×	×	×	×	×
Histopathology					×	×
Low-level responses						×
Behavioral responses					×	×
Biostimulatory tests						
Algal growth response (AGP)	×		×			

surement of the effects of effluents on the biological integrity of the receiving waters (Table 2). Effluents may adversely affect the receiving water if they are biostimulatory or toxic. Biostimulation may result from the discharge of dissolved or particulate inorganic or degradable organic macro- and micro-nutrients, which accelerate the growth of heterotrophic bacteria, algae, aquatic weeds, and animals, and generally cause a reduction in species diversity and an increase in standing crop and community metabolism. In contrast, the discharge of toxic substances generally results in a reduction in standing crop and community metabolism as well as in species diversity. However, because species vary widely in their response to different types, combinations, and concentrations of pollutants, it is not possible to construct generalizations that are universally valid.

Direct measurements of the toxic or biostimulatory properties of effluents prior to discharge are generally carried out in a base or mobile laboratory, using one or a few species of organisms that are selected because of their relatively greater sensitivity to toxicants or their ease of culture or both, or because they are considered representative of the indigenous organisms in the receiving water. Within the American Society for Testing and Materials (ASTM), methodology for these tests has been addressed in five previous Special Technical Publications [*1–5*][2] and is being considered by task groups within ASTM Committees D-19 on Water and E-35 on Pesticides. Methods for effluent toxicity tests have also been published by the U.S. Environmental Protection Agency (EPA) [*6*] and are included in *Standard Methods for the Examination of Water and Wastewater* [*7*]. Some of these methods may also be applied in or with receiving waters, but, because they cannot be used to determine biological integrity, they are excluded from the discussion that follows.

Although the relative toxicity and biostimulatory properties of effluents can be measured prior to discharge, the data are of very limited use in predicting the effects of effluents on the biological integrity (standing crop, taxonomic composition, and condition) of communities of indigenous aquatic organisms in receiving waters for several reasons:

1. Only a small number of species have been used in bioassays.
2. Most toxicity data have been derived from tests with pure compounds.
3. Knowledge of the relationship between the effects of toxicants and the environmental conditions existing in receiving waters is extremely limited.
4. Communities of aquatic organisms in receiving waters are complex and highly dynamic, and the life histories and species–species and species–environment relationships are very poorly understood.

Because of technical and economic constraints, laboratory bioassays can not be carried out on every species present (or expected to be present in the

[2] The italic numbers in brackets refer to the list of references appended to this paper.

TABLE 2—*Properties of indigenous communities of aquatic organisms used in determining the biological integrity of surface waters.*

Parameter	Phyto-plankton	Zoo-plankton	Periphyton	Macro-phyton	Macroin-vertebrates	Fish
Standing crop						
Count	×	×	×	×	×	×
Volume	×	×	×		×	
Wet weight	×	×	×	×	×	×
Dry weight	×	×	×	×	×	
Ash-free weight	×	×	×	×	×	
DNA content	×					
ATP content	×	×	×			
Chlorophyll *a* content	×		×	×		
Taxonomic composition						
Species identification	×	×	×	×	×	×
Indicator species	×	×	×	×	×	×
Number of individuals/species	×	×	×	×	×	×
Total number of species	×	×	×	×	×	×
Diversity index	×	×	×	×	×	×
Pigment composition						
Biomass/chlorophyll *a*	×		×			
Chlorophyll *a*/chlorophyll *b*	×		×			
Chlorophyll *a*/chlorophyll *c*	×		×			
Pheophyton content	×		×			
Nitrogen fixation	×		×			
Metabolic activity or condition						
Primary productivity						
Carbon-14 uptake	×		×	×		
Oxygen evolution	×		×	×		
Respiration rate						
Plankton dark-bottle oxygen uptake	×					
Electron transport	×					
Benthic respirometer oxygen uptake			×	×	×	
Nitrogen fixation	×		×			
Chemical composition				×		
Macronutrient content				×		
Enzyme content						
Acetylcholinesterase					×	×
Phosphatase	×			×		
Nitrate reductase	×					
Toxic organics and metals content	×	×	×	×	×	×
Flesh tainting					×	×
Histopathology					×	×
Condition factor						×

absence of pollution) in the receiving water. Therefore, toxicity data are available on only a relatively small number of species, which in turn have been tested with only a limited number of compounds. Even within the same genus, species are known to vary greatly in their sensitivity to the same pollutants. Thus, it is not possible to predict accurately the effects of pollutants on organisms other than the ones used in the test.

Effluents commonly contain many substances that, in combination, may have synergistic or antagonistic effects. Most toxicity data, however, have been derived from tests using pure compounds, and the data cannot be used to predict the effects of these substances in the presence of other materials.

Depending upon the climate and day-to-day weather, environmental conditions may undergo large, natural, cyclic, seasonal changes, as well as brief but large excursions from seasonal norms, which place stress on aquatic organisms and make them more susceptible to pollutants. In contrast, environmental conditions employed in laboratory bioassays are generally constant and optimum, and provide no information on biological effects over the normal range of conditions obtaining in the receiving waters.

Because of the prevailing competition between organisms in natural communities, relatively small changes in behavior, rates of growth or reproduction, or other life processes may result in competitive advantages or disadvantages that result in major shifts in community biomass, taxonomic composition, and metabolism. However, too little is known about these processes to predict the effects of pollutants on community stability.

Evaluating Biological Integrity

From the preceding discussion, it is evident that the only valid method of determining the status of or trends in the integrity of communities of aquatic organisms is to go into the field and examine the properties of such communities directly. To determine the biological integrity at a given location, the communities of indigenous organisms are sampled, and the properties of the communities are compared with those expected in the absence of pollution. The data collected are used to answer the following questions: Are the expected species present in the expected abundance? Are they carrying out life functions (such as growth, reproduction, etc.) at expected rates, and are they free of pollution-related disease, chronic (sublethal) effects [as defined in Paragraph 502(13) of the Act], and harmful concentrations of toxic substances? The parameters commonly measured in characterizing biological integrity are listed in Table 2.

The effectiveness of studies of the integrity of communities of aquatic organisms depends on factors such as the efficacy of the field and laboratory methods employed, the training and experience (competence) of the staff, the technical and physical resources available to them, and attention given to quality assurance. Valid methods are required for sample collection, preservation, and preparation for analysis; enumeration and identification of organisms; measurements of biomass, metabolic rates, and bioaccumulation of toxics; histopathological examination; and data processing, analysis, and interpretation. Standardized methods for sample collection and analysis are already available from a variety of sources and are being developed within ASTM by Subcommittee D19.23 on Biological Field Methods [*7–10*]. It will be necessary, of course, to update these methods periodically to include technological advances. There is also still a need for adequate taxonomic references (species identification manuals) for some important groups of organisms. Guidelines for staffing, parameter selection, and quality assurance for biological field studies are available from the EPA [*11,12*].

Although the interpretation of data on the biological integrity of surface waters is ultimately a matter of professional judgement, tools and guidelines are urgently needed to assist the biologist in making judgements about the significance of field data (that is, are conditions satisfactory or unsatisfactory? Are they improving or deteriorating? What is the degree of change from normal?). Thus, "norms" or "standards" (models) are needed in order to compare field data on standing crop, species composition, and community metabolism, and limits must be established for the magnitude of pollution-caused changes to be permitted in these properties. It is essential that these models take into account normal seasonal variations in the properties of aquatic communities due to climate, weather, hydrology, and other environmental factors not related to pollution.

Some guidelines are available for the interpretation of biomass data [*11*], but much additional effort is needed in this area. A very extensive literature exists on indicator organisms, including species associated with different levels of organic pollution, trophic levels, and metals [*8,13–19*]. Our program has funded the compilation and summarization of data and the preparation of reports on the environmental requirements and pollution tolerance of several groups of organisms, including freshwater diatoms [*20*], blue-green algae [*21*], midges [*22*], mayflies [*23*], caddis flies [*24*], stone flies [*25*], mussels [*26*], and zooplankton [*27*]. We have also developed a coded master list of common species of aquatic organisms [*28*] to ensure uniformity in the use of taxonomic nomenclature and, therefore, comparability in species identifications. This list also plays a key role in the EPA's computerized biological data management system (BIO-STORET) [*29*] and has been adopted by many other federal, state, and private agencies.

A wide variety of mathematical expressions are available to describe community structure, including log-normal distributions of the numbers of individuals within species, information theory equations, and equitability, and to compare community structure at different stations, such as control and impacted stations in the same body of water, or at the same station over time [*30–32*]. Wilhm's D-BAR [*30*], which is probably the most widely used species diversity index, was originally restricted largely to macroinvertebrate data, but its use is now expanding to include other groups of organisms [*33*].

Aside from measurements of photosynthesis and respiration, field studies of the effects of pollution at metabolic and cellular levels have been neglected. An exception to this is the measurement of acetylcholinesterase activity in the central nervous systems of fish and invertebrates to detect sublethal effects of pesticides [*34*]. Metabolic rates respond much more quickly to the effects of pollution than biomass and community structure and often require much less effort to measure. Histopathological effects should also be considered, but histological examinations of aquatic organisms have been restricted largely to marine specimens [*35,36*].

In the past, biological data interpretation has been highly subjective, relying heavily on qualitative information, such as species lists, with insufficient regard for nonpollution-related variations in biological parameters. In the future, more emphasis must be placed on the use of parametric and nonparametric statistical analyses to support conclusions on the differences in biological conditions at control and impacted stations and changes within stations over extended periods of time [*32*].

References

[*1*] *Biological Methods for Assessment of Water Quality, ASTM STP 528,* American Society for Testing and Materials, Philadelphia, 1973.

[*2*] *Biological Monitoring of Water and Effluent Quality, ASTM STP 607,* American Society for Testing and Materials, Philadelphia, 1976.

[*3*] *Aquatic Toxicology and Hazard Evaluation, ASTM STP 634,* American Society for Testing and Materials, Philadelphia, 1977.

[*4*] *Estimating the Hazards of Chemical Substances to Aquatic Life, ASTM STP 657,* American Society for Testing and Materials, Philadelphia, 1978.

[*5*] *Aquatic Toxicology, ASTM STP 667,* American Society for Testing and Materials, Philadelphia, 1979.

[*6*] Peltier, W., *Methods for Measuring the Acute Toxicity of Effluents to Aquatic Life,* EPA 660/4-78-012, Environmental Monitoring and Support Laboratory, U.S. Environmental Protection Agency, Cincinnati, Ohio, 1978.

[*7*] *Standard Methods for the Examination of Water and Wastewater,* 14th ed., American Public Health Association, Washington, D.C., 1976.

[*8*] *Biological Field and Laboratory Methods for Measuring the Quality of Surface Waters and Effluents,* EPA-670/4-73-001, Weber, C. I., Ed., Environmental Monitoring and Support Laboratory, U.S. Environmental Protection Agency, Cincinnati, Ohio, 1973.

[*9*] "Standard Practices for Measurement of Chlorophyll Content of Algae in Surface Waters (D 3731-79)," *1979 Annual Book of ASTM Standards,* Part 31, *Water,* American Society for Testing and Materials, Philadelphia, 1979, pp. 971–975.

[*10*] "Standard Practice for Evaluating an Effluent for Flavor Impairment to Fish Flesh (D 3696-78)," *1979 Annual Book of ASTM Standards,* Part 31, *Water,* American Society for Testing and Materials, Philadelphia, 1979, pp. 976–984.

[*11*] *Model State Water Monitoring Program,* Monitoring and Data Support Division, Office of Water and Hazardous Materials, U.S. Environmental Protection Agency, Washington, D.C., 1975.

[12] *Basic Water Monitoring Program,* Monitoring and Data Support Division, Office of Water and Hazardous Materials, U.S. Environmental Protection Agency, Washington, D.C., 1977.
[13] Weber, C. I. in *Bioassay Techniques and Environmental Chemistry,* G. E. Glass, Ed., Ann Arbor Science Publishers, Ann Arbor, Mich., 1973, pp. 119–138.
[14] Stoermer, E. F., *Transactions of the American Microscopical Society,* Vol. 97, 1978, pp. 2–16.
[15] Gannon, J. E. and Stembegger, R. S., *Transactions of the American Microscopical Society,* Vol. 97, 1978, pp. 16–35.
[16] Collins, G. B. and Weber, C. I., *Transactions of the American Microscopical Society,* Vol. 97, 1978, pp. 36–43.
[17] Cairns, J., Jr., *Transactions of the American Microscopical Society,* Vol. 97, 1978, pp. 44–49.
[18] Hart, C. W., Jr. and Fuller, S. L. H., *Pollution Ecology of Freshwater Invertebrates,* Academic Press, New York, 1974.
[19] Hart, C. W., Jr. and Fuller, S. L. H., *Pollution Ecology of Estuarine Invertebrates,* Academic Press, New York, 1979.
[20] Lowe, R. L., *Environmental Requirements and Pollution Tolerance of Freshwater Diatoms,* EPA-670/4-74-005, Environmental Monitoring and Support Laboratory, U.S. Environmental Protection Agency, Cincinnati, Ohio, 1974.
[21] Vanlandingham, S., *A Guide to the Identification and Environments of the Bluegreen Algae (Cyanophyta),* Environmental Monitoring and Support Laboratory, U.S. Environmental Protection Agency, Cincinnati, Ohio, in press.
[22] Beck, W. M., Jr., *Environmental Requirements and Pollution Tolerance of Chironomidae,* EPA-600/4-77-024, Environmental Monitoring and Support Laboratory, U.S. Environmental Protection Agency, Cincinnati, Ohio, 1977.
[23] Hubbard, M. D. and Peters, W. L., *Environmental Requirements and Pollution Tolerance of Ephemeroptera,* EPA-600/4-78-061, Environmental Monitoring and Support Laboratory, U.S. Environmental Protection Agency, Cincinnati, Ohio, 1978.
[24] Harris, T. L. and Lawrence, T. M., *Environmental Requirements and Pollution Tolerance of Trichoptera,* EPA-600/4-78-063, Environmental Monitoring and Support Laboratory, U.S. Environmental Protection Agency, Cincinnati, Ohio, 1978.
[25] Surdick, R. F. and Gaufin, A. R., *Environmental Requirements and Pollution Tolerance of Plecoptera,* EPA-600/4-78-063, Environmental Monitoring and Support Laboratory, U.S. Environmental Protection Agency, Cincinnati, Ohio, 1978.
[26] Bates, J., Dennis, S., van der Schalie, H., and Isom, B., *Taxonomy and Ecology of the Freshwater Mussels of the United States,* Environmental Monitoring and Support Laboratory, U.S. Environmental Protection Agency, Cincinnati, Ohio, in press.
[27] Gannon, J. L., *Environmental Requirements and Pollution Tolerance of Zooplankton,* Environmental Monitoring and Support Laboratory, U.S. Environmental Protection Agency, Cincinnati, Ohio, in press.
[28] Weber, C. I., *BIO-STORET Master Species List,* Environmental Monitoring and Support Laboratory, U.S. Environmental Protection Agency, Cincinnati, Ohio, 1976.
[29] Nacht, L. and Weber, C. I., *Final Design Specification, Biological Data Storage and Retrieval System (BIO-STORET),* Environmental Monitoring and Support Laboratory, U.S. Environmental Protection Agency, Cincinnati, Ohio, 1976.
[30] Wilhm, J. L., *Journal of the Water Pollution Control Federation,* Vol. 65, 1970, pp. R221–R224.
[31] MacArthur, R. H., *Proceedings of the National Academy of Sciences of the United States of America,* Washington, D.C., Vol. 43, 1957, pp. 293–295.
[32] *Biological Data in Water Pollution Assessment: Quantitative and Statistical Analyses, ASTM STP 652,* American Society for Testing and Materials, Philadelphia, 1978.
[33] Weber, C. I. and McFarland, B., "The Effects of Copper on the Periphyton of a Small Calcareous Stream," this volume, pp. 101–131.
[34] Coppage, D. L. and Matthews, E., *Bulletin of Environmental Contamination and Toxicology,* Vol. 11, 1974, pp. 483–488.
[35] Yevich, P. P. and Barszcz, C. A., *Preparation of Aquatic Animals for Histopathological*

Examination, Environmental Monitoring and Support Laboratory, U.S. Environmental Protection Agency, Cincinnati, Ohio, 1979.

[*36*] Goldberg, E. K., Bowen, A. T., Farrington, J. W., Harvey, G., Martin, J. H., Parker, P. L., Risebrough, R. W., Robertson, W., Schneider, E., and Gamble, E., *Environmental Conservation*, Vol. 5, 1978, pp. 101-126.

E. E. Herricks,[1] *M. J. Sale,*[2] *and E. D. Smith*[3]

Environmental Impact Analysis of Aquatic Ecosystems Using Rational Threshold Value Methodologies

---

REFERENCE: Herricks, E. E., Sale, M. J., and Smith, E. D., **"Environmental Impact Analysis of Aquatic Ecosystems Using Rational Threshold Value Methodologies,"** *Ecological Assessments of Effluent Impacts on Communities of Indigenous Aquatic Organisms, ASTM STP 730,* J. M. Bates and C. I. Weber, Eds., American Society for Testing and Materials, 1981, pp. 14–31.

ABSTRACT: The techniques generally applicable to environmental impact analysis fall short of effective predictive capacity because of incomplete data bases and the lack of procedures for systematic data analysis to maximize data utility. To overcome these difficulties, the Construction Engineering Research Laboratory of the Department of the Army has developed a computer-based environmental impact analysis system. As part of this continuing activity, the Environmental Impact Computer System (EICS) has been used as a basis for development of a detailed subprogram for predicting impacts on aquatic ecosystems. The EICS system is based on the development of methodology for determination of rational threshold values (RTV). RTV includes indexes of both structural and functional community conditions that are integrated into a decision algorithm. Data for community parameters are obtained from readily available resources [STORET, National Pollution Discharge Elimination System (NPDES) permit reporting, and other state and federal water-quality monitoring efforts], system subprograms that provide population predictions based on state-of-the-art population models, and actual field data. The protocols described include interactive, real-time manipulation of data based on site-specific information. The application of these protocols provides a new dimension in impact analysis methodology. Ecosystem-specific threshold levels can be set, and impacts caused by alternatives can be evaluated efficiently and economically. This paper summarizes aquatic ecosystem RTV analysis procedures, discusses feasibility of application, and provides a sound quantitative approach to priority ordering of community parameters for aquatic ecosystem impact analysis.

KEY WORDS: impact analysis, aquatic ecosystems, population modeling, monitoring, impact mitigation, ecology, effluents, aquatic organisms

[1]Associate professor, Department of Civil Engineering, University of Illinois, Urbana, Ill., 61801.

[2]Graduate research assistant, Department of Civil Engineering, University of Illinois, Urbana, Ill. 61801; present address: Oak Ridge National Laboratory, Division of Environmental Sciences, Oak Ridge, Tenn. 37830.

[3]Environmental engineer, Construction Engineering Research U.S. Army Corps of Engineers, Champaign, Ill. 61820.

The assessment of effluent impacts on receiving waters has traditionally involved two distinct, yet highly integrated, activities—data collection and data interpretation. Sample collection, processing, organism identification, and initial data synthesis are generally considered data collection activities. The results of data collection are most often expressed in terms of community structure (kinds and numbers of organisms), while initial data synthesis may include calculation of diversity indexes or correlative relationships (cluster analysis, ordination). Recently, measures of functional condition [adenosine tiphosphate (ATP) activity, productivity, respiration, degradation, and behavior] have been included in initial data collection activities. Thus, initial impact assessment quantifies the structure and function of the ecosystem.

After quantification of the ecosystem parameters, it is necessary to interpret the results. This second assessment activity is often unquantified, depending on the experience of the ecologist making the interpretations. Unfortunately, a hierarchy of interpretative quality exists. The biologist who collected the data can make the most valid ecological interpretations, based on his intimate knowledge of the data collection strengths and weaknesses. As data interpretation is integrated into management or regulatory processes, this intimate knowledge of the ecological system is often lost. The final result is the use of only synthetic or interpretative information that has lost a direct association with the data collection and thus with an essential validation component. Fortunately, as data collection is improved and standardized, the reliability of interpretation is also improved. This interpretative improvement occurs through sound statistical design of data collection efforts, which bring about added costs in time and money. Unfortunately, the requirements of existing data resources and time or fiscal limitations may preclude additional data collection.

Recognizing the limitations to data interpretation, ecologists have been faced with even greater challenges. Recent legislation calls for major improvement in both interpretative and predictive analysis of ecosystems. For example, the revision of regulations for implementation of the National Environmental Policy Act (NEPA) [*1*][4] and recent water quality legislation (specifically the Clean Water Act of 1977) call for predictive analysis to assure mitigation of impacts and maintenance of ecosystem integrity. Progress toward the development of predictive ecosystem analysis has been slow. One cause is that biologists are reluctant to make ecological predictions because of a lack of detailed and quantitative data. Typically, the inherent variability of ecosystems and the limitations to the collection of statistically valid data constrain both interpretation and prediction. New approaches are necessary. Assumptions that account for biological variability, yet simplify analysis and interpretation, must be made to facilitate federally mandated environmental impact analysis (EIA) activities.

[4]The italic numbers in brackets refer to the list of references appended to this paper.

The development of quantitative, interpretable data for the EIA process has been one of the major foci of the environmental impact monitoring, management, assessment, and planning activities of the Construction Engineering Research Laboratory of the Department of the Army. The requirements of the Army for speedy, accurate, and interpretative EIA activity has led to the development of a number of computer-based systems of analysis, the Environmental Impact Computer System (EICS) [*2*]. The sophistication of the EICS system has evolved with the EIA process and clarification of the EICS content and format. The result has been the development of a second generation of predictive analysis, based on the concept of the rational threshold value (RTV). An RTV is variably defined based on the scientific or technical discipline involved. Based on significance, an RTV quantifies environmental impact through identification of levels (or thresholds) that cause damage. Generally defined, an RTV is a logically derived quantitative value that denotes the onset of effect. An RTV establishes a quantitative base for the EIA process. The use of RTVs is dependent on the validity of the data base, as well as on consistent evaluation of the information available.

The protocols proposed in this paper incorporate numerous assumptions that simplify complex ecological data. RTVs (and corresponding impacts) are evaluated in the context of a hierarchial organization of ecosystems. Population level effects are used as the primary indicator of impact; single-species toxicity testing data is used to predict population trends. Analysis and prediction of impact effects on two or more populations provide a basis for extrapolation to community or ecosystem effects. Supplementing population trend predictions, commonly used indexes of community or ecosystem conditions (saprobic indexes, algal growth potential, etc.) are also incorporated into the predictive community or ecosystem analysis.

Information Flow

To achieve the goals set forth in NEPA and recent water-quality legislation, predictive capabilities are developed from either available data or additional sampling efforts. In the decision process, the normal flow of information begins with data collection, analysis, and interpretation. This information can be used or processed in two ways. Monitoring, which relies on interpretation of change, is identified as an error control process (Fig. 1). When data is processed through a prediction algorithm, the system is termed "prediction controlled." A prediction-controlled system processes changes of attributes (individual environmental parameters such as air, water quality, etc.) through an algorithm derived from environmental setting or monitoring data to develop impact predictions. Using a prediction control system, alternative plans can be generated and tested to evaluate relative impacts of the

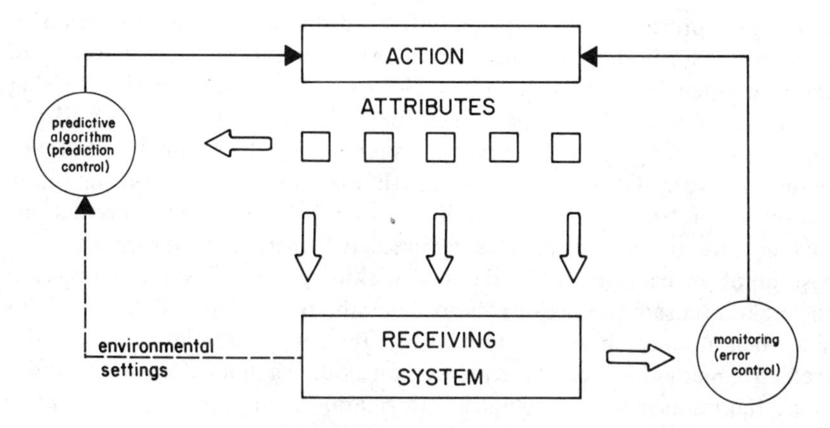

FIG. 1—*Environmental management information flow.*

selected alternatives. Although prediction control methodologies are used in many disciplines, they have not been widely applied to ecosystem analysis.

It will be useful to discuss briefly the relative information flows in prediction control and error control systems and the limitations of both. In error control systems a major limitation is that information gathering always occurs after the fact. The action is initiated, and the impact on some attribute is observed. If possible, this information is fed back to change the action, resulting in improvement of the receiving system. The result of error control is potential damage unless the feedback loop has a short time delay between data acquisition and the initiation and completion of control activity. The development of rapid biological monitoring for aquatic ecosystems has produced response times in seconds [*3*], rather than the days to months required for the typical sampling and analysis programs, but the predictive capacity of monitoring is limited. The limitations of prediction control are different. The predictive capacity is based on data (generally a change in an attribute due to a planned action) being processed through an algorithm for initiation of control processes. Two factors control prediction success—the validity of the algorithm and the quality of the data upon which the prediction is based. Even if both the algorithm validity and data quality are assured, if only prediction control systems are used, no feedback is available to verify the success of the control process. Thus, prediction control systems used alone are unreliable and may damage the receiving system.

This brief discussion supports the agrument that monitoring alone may not prevent damage; similarly, predictive algorithms used alone may lead to alteration and damage of the receiving system. Unless monitoring (or at least detailed assessment) and predictive analysis are combined in information processing, prevention of impacts will be difficult. The incorporation of mechanisms for both prediction control and error control forms the basis of the second-generation EIA methodologies embodied in RTV analysis. Using monitoring or assessment data, a protocol has been developed to take advantage of a wide range of specific models, such as models of trophic condition, dissolved oxygen (DO), population growth, and environmental setting information, taken from monitoring data. It uses this information to predict impact on aquatic ecosystems. Using predictive methodologies changes the position of the ecologist in the decision-making process. If an ecologist cannot make accurate predictions, he is generally forced into a defensive position when attempting to qualify data or subjective evaluations. Using valid predictive methodologies, the ecologist can assume a more offensive position. Thus, the importance of ecological information is improved when social or political decisions are made that affect the environment.

Analytical Protocols

The protocols of analysis for determining the threshold of effect and its corresponding impact on aquatic ecosystems incorporate state-of-the-art analysis and monitoring methods coupled with models which make predictions about population size through time. Protocols for aquatic ecosystem evaluation are hierarchically arranged on systematic and trophic relationships. Species-specific thresholds of effect for each attribute are identified from toxicity testing. Even though limited by the quality and availability of data (relatively few toxicants have been evaluated, with even fewer species, and test environmental conditions are generally variable), short-term testing at nominally high concentrations of toxicants provides initial threshold estimates. The results of identification of the threshold for higher levels of ecological organization are mixed [*4*], in part due to the complexity of the systems involved and the variable concentration and duration of stresses. At the community level, the most common approach to monitoring involves stress–response relationships indicated by changes in the numbers of organisms or species composition of the community. Impact can be evaluated through time (following removal of stress) or distance (evaluation of recovery downstream due to modification or removal of stress) [*5*]. Thresholds can be established using a variety of techniques that include both structural [*6*] and functional [*7*] parameters. The RTV analysis protocol incorporates several indicators of community condition for which thresholds can be developed. Ecosystem response also varies with stress duration and intensity. For example, a short-term stress usually does not affect all components of the ecosystem. Although damage may occur at the species or population level, the ecosystem's structural changes may be minor, and function can be maintained. For chronic stress the ecosystem often accommodates the insult unless the intensity of the stress has eliminated major functional groups. Because the structural and functional components of ecosystems constantly change, it has been argued that ecosystem response curves are linear and exhibit no threshold response [*8*].

RTV analysis protocols are based on a simplified hierarchial organization of ecosystems. Organism-specific response thresholds established from toxicity testing are extrapolated to population growth models. Two or more population response estimates are combined in developing a community response prediction. Finally, by integrating a community-level prediction with a variety of interactive components (physical, chemical, and biological parameters), a prediction is possible of ecosystem response. Even though the response may be linear, describing characteristics such as ecosystem inertia, elasticity, or resiliency [*9*] provides a basis for threshold analysis [*10*]. The successful application of these protocols depends on three assumptions: (1) that the environmental setting data (including water quality and biological data) accurately describe the receiving system; (2) that the models and the application of the models, based on analytical or experimental studies of the environmental setting, are accurate; and (3) that the environmental setting information (largely habitat potential) can be coupled with information about the receiving-system biota (biotic potential) through the use of population prediction models. In the development of these protocols, the major difficulties arise in coupling habitat and biotic potential. Although quantitative approaches exist, this coupling is largely subjective.

The most important component of the analysis protocols is consistent application of analysis procedures (either objective or subjective) through an interactive computer-based system. Each potential impact is quantified in terms of an RTV, with output in the form of impact significance. This output is used in a mitigation loop to test alternatives. When project design variables (for example, effluent concentrations) are incorporated into RTV functions, new constraints or objectives for project management can be formulated and analyzed with mathematical programming routines. Management objectives (for example, minimizing cost or maximizing equity) can be clearly stated, eliminating many difficulties inherent in joining the dissimilar metrics of engineering or economics and of the environment or aesthetics. This procedure is cost-effective. Available data can be used to identify gaps, which can be filled by selective monitoring. Similarly, predictive methodologies allow "experiments" to be conducted in the computer, reducing overall environmental costs. The end result is more efficient data collection and interpretation.

Models and Indexes Available for Aquatic Ecosystem RTV Analysis

The protocol for aquatic ecosystem RTV analysis calls for use of state-of-the-art models and indexes in an interactive, computer-based system to reduce environmental setting information to a format suitable for threshold analysis. Figure 2 illustrates a simplified impact analysis procedure showing intermediate RTV analysis steps, including attribute modification algorithms and population models. The attribute modification algorithms can be used directly to predict impacts or provide data concerning the alteration of habitats that provide input to the predictive population models.

As previously stated, the validity of this simplified cause-effect impact analysis is dependent on the quality of the input data and the accuracy of the modification algorithms. The quality of the input data is based on the study design and analysis accuracy, both factors that can be associated with high costs. The modification algorithms are derived from models or indexes that simplify or reduce the complexity of environmental setting data. Table 1 summarizes the impacts, input, and output of models suitable for RTV analysis. These models have many limitations if viewed and used individually, but considerable predictive strength is developed when one is combined with one or more other models and placed in the hands of an experienced ecologist. The protocol for RTV analysis incorporates subjective evaluations, but we feel that the protocol structure constrains this subjectivity or, at a minimum, forces consistent, though subjective, evaluation during the RTV analysis process. A second advantage is integration of subjective analysis with various models to provide procedural redundancy and quality control.

Procedures of Analysis for Aquatic Ecosystem RTV

The initial input of environmental setting information to the RTV analysis is made in the attribute modification algorithms identified in Fig. 2. The fundamental assumption of this analysis is that it is possible to determine attribute characteristics at specific points in space and time from available data or from manipulation of environmental setting and effluent data through modification algorithms. For example, if data are available for N and P, a calculation of a trophic status index (TSI) is possible directly from available data. If modification algorithms are used, attribute characteristics are determined and used to build the vector $\mathbf{WQ}_{ij}$, which is defined as the set of modified attributes at the ith point in space and jth point in time. These

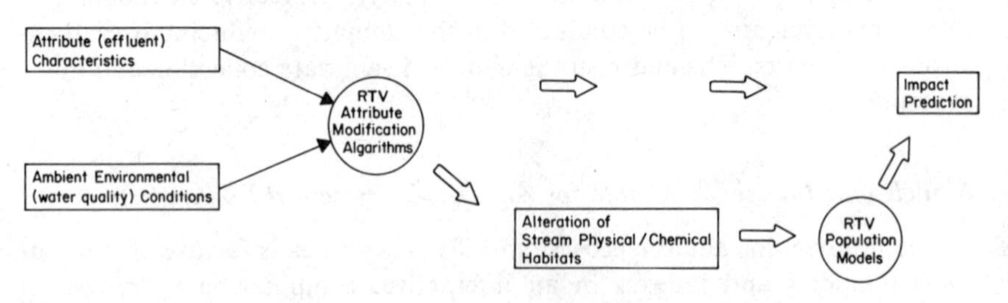

FIG. 2—*Impact analysis procedure.*

TABLE 1—*Summary of possible RTV models.*

RTV Model	Impacts Addressed	Inputs	Outputs	References
Water quality index	overall water quality	DO, fecal coliforms, pH, nitrate nitrogen, phosphate phosphorus, BOD_5, temperature, total solids, turbidity	relative condition of overall water quality	*11—14*
BOD/DO models	organic, point-source pollution—mainly in rivers and streams (may include NOD and SOD)	stream channel morphometry, temperature, BOD, biological rate constants, point-source discharges, stream discharge	oxygen deficits, instream concentration of NH_3, NO_3, BOD, DO, etc.	*15*
Saprobic index (SI)	organic, point-source pollution—mainly in rivers and streams	BOD_5	saprobian classification of biological communities	*16, 17*
Trophic state index (TSI)	eutrophication—mainly in lakes and reservoirs	transparency (Secchi disk), [Chlorophyll *a*], [total phosphorus]	indication of lake trophic condition	*18*
Nutrient loading models	eutrophication—mainly in lakes and reservoirs	basin morphometry, phosphorus inflow and outflow, stream flow, land use	projected lake trophic condition	*19, 20–22*
Simulated autotrophic index (SAI)	eutrophication—in both lentic and lotic environments	nutrient concentration, carbonate system, light, temperature	relative dominance of autotrophic component of microbial community	*23*
Relative algal growth potential (RAGP)	eutrophication—in both lentic and lotic environments	nutrient concentration, carbonate system, light, temperature	percent of maximum growth rate for components of algal community, limiting environmental parameters	*24*
Toxicity unit (TU)	environmental toxicity	indicator-species-specific LC_{50}, modifying factors (that is, hardness, etc.)	overall accute toxicity of environment	*25, 26*
Population growth index (PGI)	impacts on reproduction and survival	age-specific fecundity and survival functions	net population reproductive rate per generation (idealized)	*20, 27*
Population simulations	cumulative and long-term effects on higher levels of tropic structure	population parameters such as fecundity and survivorship and age structure of initial standing crop	projected levels, stability, recovery rates from short-term impacts	*28–30*

calculations require the use of environmental setting information, which results in the complex set of procedures illustrated in Fig. 3, an analysis protocol for an introduction of a typical domestic and industrial wastewater to a stream system.

The attributes of the effluent are separated into conservative and nonconservative components using steady-state, one-dimensional models. The development of predictive models for conservative components follows Eq 1, a typical mass balance based on dilution (the terms for diffusion or advective mixing are not included)

$$C = \frac{C_e Q_e + C_a Q_a}{Q_e + Q_a} \tag{1}$$

where

C_e and Q_e = the effluent attribute concentration and discharge, and
C_a and Q_a = the ambient attribute concentration and discharge.

Nonconservative components are modeled on the basis of a first-order decay equation, which is subject to sedimentation, decay, chemical transformation, and biological uptake.

$$\frac{dC}{dt} = kC \pm \text{terms} \tag{2}$$

The application of these equations and the use of other predictive models or indexes define the elements of vector $\mathbf{WQ}_{ij}$. For example, it is necessary to derive (or actually sample) in-stream values for the effluent as well as environmental setting attributes, especially those that might modify toxicity, such as hardness, pH, temperature, or dissolved oxygen. Once derived, the elements of $\mathbf{WQ}_{ij}$ can be applied directly to a water quality index (WQI) (see Table 1) to measure cumulative water quality alteration. Similarly, a single parameter, such as biological oxygen demand (BOD), can be modified by the BOD/DO model and used with the saprobic index to assess the system trophic status (Table 1). This information can be used directly to determine impact or provide input to habitat modification algorithms to assist in population impact predictions.

The structure of the habitat modification routines is illustrated through the use of toxicity data. In the simplest example, an addition of a conservative toxicant to a stream system (zinc in Fig. 3) leads to toxicity prediction based on calculations of the toxicant concentration and the annual survival of the target species. A more accurate prediction of toxicity is possible by using modification of the toxic effect by environmental conditions (hardness, temperature, etc.). For example, the discharge of a waste containing BOD,

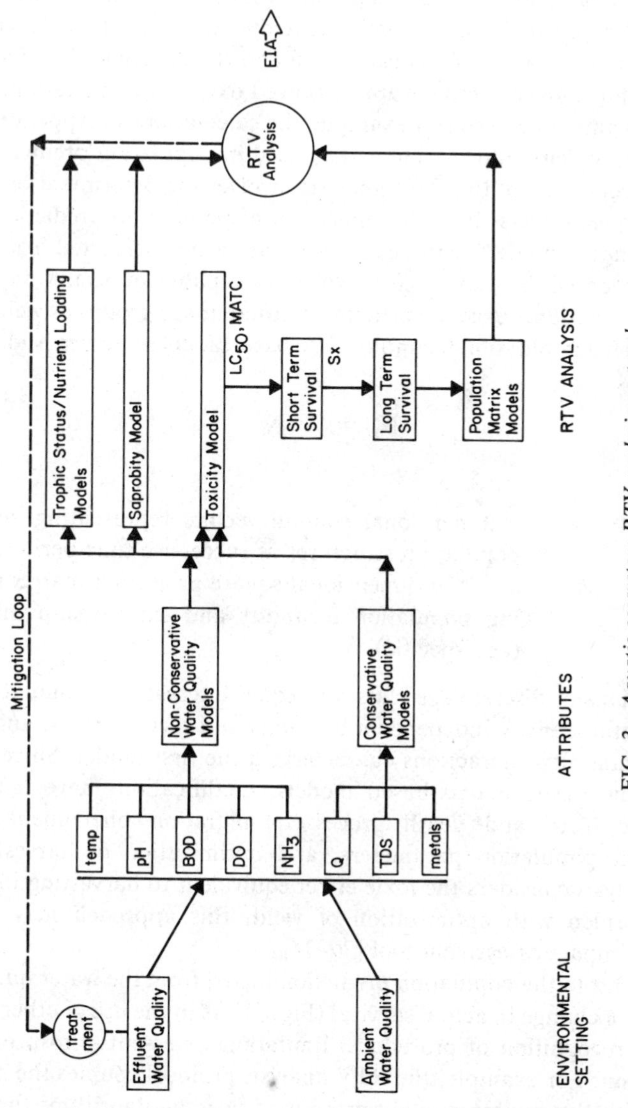

FIG. 3—*Aquatic ecosystem RTV analysis protocol.*

ammonia, and heavy metals would necessarily incorporate conservative and nonconservative predictive analysis, in addition to the possible modification due to environmental conditions (Fig. 4). The RTV procedure includes analysis of dissolved oxygen with time and distance determination of the nitrification of ammonia to develop time/distance concentrations (both using BOD/DO models). From the ambient temperature and pH, the concentration of unionized ammonia is calculated and the differential toxicity of zinc specified for a given temperature, dissolved oxygen concentration, pH, and unionized ammonia level (for example, the calculations in Appendix I). This short-term toxicity information is then used in population predictions.

The second step of the RTV analysis involves the determination of potential biological effects through application of population prediction models. The simulation model for these analyses has been constructed from a matrix model proposed by Leslie [*28*], which is capable of examining impacts resulting in variable excess mortality on different age groups as well as effects of decreasing population fecundity. The basic model structure is identified in Eq 3.

$$\mathbf{N}_t = \mathbf{A}^t \cdot \mathbf{N}_0 \tag{3}$$

where

$\mathbf{N}_t$ and $\mathbf{N}_{t+1}$ = n-dimensional column vectors representing age-specific population structures at successive time periods, and

$\mathbf{A}$ = a $n \times n$ dimensional square projection matrix representing population fecundity and survivorship information (age-specific).

Assumptions of discrete age classes of equal interval in a population with a 1:1 sex ratio, density-independent fecundity and survival rates, and no effect from community interactions characterized the first model. Since introduction of the Leslie matrix-based models, modifications have incorporated stochastic birth and death processes, migration phenomena, density-dependent population parameters, and optimization of harvests. (*Note*: RTV analysis considers the toxic effect equivalent to harvesting individuals; when coupled with optimization of yield, this approach may provide a valuable impact assessment tool [*30–34*].)

The input to the population prediction model from the water quality information is a change in actual survival (Fig. 5). As in the use of other models or indexes, recognition of procedural limitations is essential to proper model application. For example, the RTV analysis protocol couples the population model with the environmental setting modification algorithms through age-specific survivorship rates (s_i) used in developing the modified **A** matrices (see Appendix II). These age-specific survivorship rates are defined in terms of single-time steps equal to the age class interval of the population; thus, the impact prediction will not truly reflect the discontinuous or variable condi-

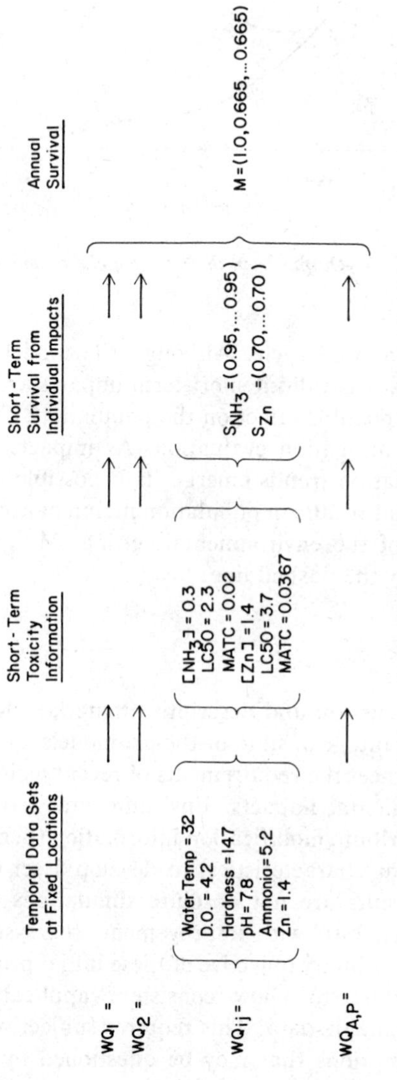

FIG. 4—*Calculations for the RTV population model.*

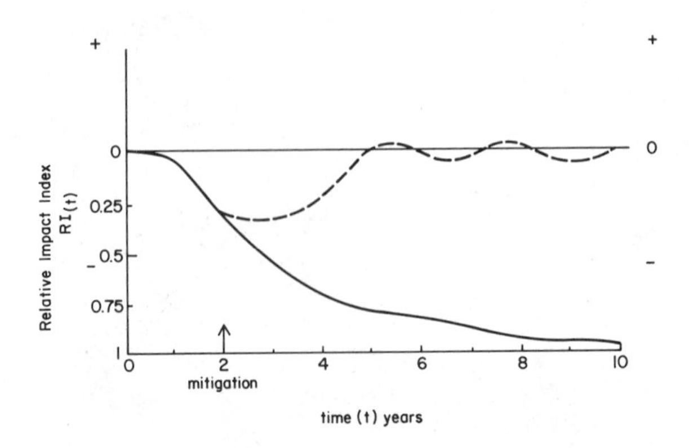

FIG. 5—*Output from the RTV population model.*

tions typical of an annual cycle. Although the model lacks this detailed predictive capability, it is valid if short-term impacts are evaluated to determine the maximum possible effect on the population. This conservative approach is useful in mitigation evaluation. As impacts are mitigated, new predictions of population trends emerge. It is possible from this process to define a threshold that results in population maintenance at levels acceptable within the context of the environmental setting. Mitigation effects are illustrated in Fig. 5 by the dashed line.

Summary

The foregoing discussion and descriptive materials describe a set of protocols that takes advantage of state-of-the-art models and the experience of a trained ecologist to meet the requirements of recent legislation for predictive analysis of environmental impacts. Environmental setting data is used in conjunction with attribute modification information (derived from suspected or monitored effluent characteristics) to develop a set of population trend predictions. The results are not absolute simulations of water quality or ecosystem conditions, but, with improvement, the system may eventually reach that goal. The primary objective of these initial protocols is to organize the analysis procedure to allow consistent application of interpretive methodologies to available data. This requires subjective evaluations and is dependent on assumptions that may be questioned by critics who require statistically valid data before any decision is made. Although this procedure is often not quantitative, the protocols used assure consistent application of the subjective methods through an interactive computer-based analysis system. This provides redundancy in analysis and allows application of quality control procedures while providing direction and guidance for more efficient data collection efforts in ecosystem monitoring. Although many assumptions are made as complex attributes are modified, these assumptions are clearly stated and allow review of analysis procedures as well as evaluation of input data quality.

The proposed protocols are not the only solution to questions raised about impact analysis and the determination of ecosystem integrity, but provide an operation system that will lead to advances in analysis and interpretation of ecological data. The protocol structure will allow incorporation of these new methodologies with little difficulty. These RTV protocols take maximum advantage of both error control and prediction control systems for the assessment of the fate and effect of environmental pollutants.

Acknowledgments

This work was supported in part by Contracts DACA 88-878-0-008 from the Department of the Army, Construction Engineering Research Laboratory, Corps of Engineers, Champaign, Ill. The authors are grateful to Drs. R. K. Jain and E. D. Brill for stimulating discussions and review of many aspects of this work.

APPENDIX I

Sample Calculations of Toxicity Modification

During summer low-flow conditions, the following water quality is either observed or predicted from available models or simulations.

Water temperature = 32°C
Dissolved oxygen (at critical deficit) = 4.2 ml/litre
Hardness = 147 mg/litre as calcium carbonate
pH = 7.8
Total ammonia (at point of critical DO deficit) = 5.2 mg/litre
Zinc concentration = 7.1 mg/litre
Calculate unionized ammonia

$$[NH_3] = \frac{[NH_3 + NH_4{}^+]}{1 + 10^{[(-0.0323\ \mathrm{Temp}) + 10.0533 - \mathrm{pH}]}}$$

Designate MATC for ammonia

$$96 = \mathrm{h}\ LC_{50} \text{ for bluegills}$$
$$NH_3 = 2.3 \text{ mg/litre}$$
$$\mathrm{MATC} = 0.02$$

Calculate LC_{50} for zinc (Zn) as modified by DO and hardness

$$[\mathrm{Zn}] = 7.1 \text{ mg/litre}$$
$$96 = \mathrm{h}\ LC_{50} \text{ for bluegills}$$
$$\mathrm{Zn} = 18.356(\log_{10}\mathrm{DO}) + 0.042\ (\text{hardness}) - 13.94 = 3.67$$

Designate MATC for zinc

$$\text{MATC} = 0.01 \times 96 = \text{h}$$
$$\text{LC}_{50} = 0.0367$$

APPENDIX II

In order to simulate the dynamics of populations affected by specific environmental impacts, a new projection matrix is derived (Eq 3)

$$\mathbf{A}_{\text{m}} = \mathbf{E}\cdot\mathbf{I}\cdot\mathbf{A} \tag{3}$$

where

$\mathbf{A}_{\text{m}}$ = the $n \times n$ matrix of survivorship and fecundity values modified by the effects of environmental impacts,
$\mathbf{E}$ = an n-dimensional row vector containing the relative impacts on population fecundity and age-specific survivorship,
$\mathbf{I}$ = an $n \times n$ identity matrix, and
$\mathbf{A}$ = the natural projection matrix.

$$\mathbf{A} = \begin{bmatrix} f_0 & f_i & f_2 & f_3 \\ p_0 & 0 & 0 & 0 \\ 0 & p_i & 0 & 0 \\ 0 & 0 & p_2 & 0 \end{bmatrix}$$

where

f_i = the age-specific natality of the ith age group (for example, the average number of eggs produced by a female of the ith age group during one time interval);
p_i = the age-specific survivorship rate for individuals in the ith age group, the proportion of individuals of the ith age category which survive to the $(i + 1)$th age category; and
$\mathbf{E}$ = impact modification vector.

$$\mathbf{E} = (m, s_0, s_1, s_2, s_i, s_n)$$

where

m = the proportional reduction of the population fecundity by impacts (Note: not age-specific), and
s_i = the age-specific reduction of the survivorship rate of individuals in the ith age group by environmental impacts.

Several RTV indicators are available in the application of population models. Relative impact (RI) can be measured as the ratio of the stressed population to the unimpacted population. A property of the projection matrices is important for this analysis. That is, the age-specific population vector at time t (that is, t years after impact) can be calculated as the projection matrix to the tth power times the original age-specific population vector at time zero (Eq 3). For all impact assessments, the original age structure used in these calculations is the stable age distribution (determined by the eigenvector associated with the dominant latent root of the unmodified projection matrix) with a value of 1 in the oldest age class [35]. The relative impact index can be expressed

$$\mathrm{RI}(t) = \frac{\mathbf{I_r} \cdot \mathbf{A_m}^t \cdot \mathbf{N_0} - \mathbf{I_r} \cdot \mathbf{A}^t \cdot \mathbf{N_0}}{\mathbf{I_r} \cdot \mathbf{A}^t \cdot \mathbf{N_0}}$$

where

$\mathrm{RI}(t)$ = the relative impact at time t after an activity is initiated,
$\mathbf{I_r}$ = an identity row vector, which acts to add up the individuals in each age class,
$\mathbf{A_m}$, $\mathbf{A}$ = the modified and unmodified projection matrices, and
$\mathbf{N_0}$ = the stable age structure just described.

The values of $\mathrm{RI}(t)$ can vary between -1 and $+\infty$, designating negative and positive impacts, respectively. Thresholds are placed as absolute values of $\mathrm{RI}(t)$ for example, $\mathrm{RI}(t)$ less than 0.5 is a threshold). Not all properties of the projection matrix models have been used in this brief example. Additional properties can be used to evaluate population stability, elasticity, and so forth, which are beyond the scope of this paper.

The utility of $\mathrm{RI}(t)$ can be demonstrated in the context of the example used previously. The following question might be generated from this example: "What is the significance in terms of in-stream population maintenance if, during low-flow conditions for short periods in the summer months, measurably toxic concentrations are present in portions of the receiving stream of a wastewater effluent?" Given a specific target population, $\mathrm{RI}(t)$ can be calculated to answer this question. From available data [36,37], the matrix **A** is developed

$$\mathbf{A} = \begin{bmatrix} 0 & 0 & 100 & 9\,400 & 16\,900 & 31\,200 \\ 0.0001 & 0 & 0 & 0 & 0 & 0 \\ 0 & 0.33 & 0 & 0 & 0 & 0 \\ 0 & 0 & 0.87 & 0 & 0 & 0 \\ 0 & 0 & 0 & 0.86 & 0 & 0 \\ 0 & 0 & 0 & 0 & 0.6 & 0 \end{bmatrix}$$

$$\mathbf{N_o} = \begin{bmatrix} 67\,522 \\ 7 \\ 2 \\ 2 \\ 2 \\ 1 \end{bmatrix}$$

From toxicity data, **E** = (1.0, 0.665, 0.665, 0.0665), producing Table 2.

TABLE 2—*Example results from impact prediction protocol.*

Year	$\mathbf{I} \cdot \mathbf{A}^t \cdot \mathbf{N}_0$	$\mathbf{I} \cdot \mathbf{A}_m{}^t \cdot \mathbf{N}_0$	RI(t)
0	67 536	67 536	0.0000
1	84 014	84 009	−0.00006
2	83 110	55 268	−0.335
3	76 617	33 881	−0.55779
4	75 737	22 313	−0.70538
5	83 489	18 587	−0.77737
6	88 954	15 350	−0.82743
7	94 438	12 281	−0.86995
8	91 099	7 885	−0.91345
9	89 841	5 355	−0.94039
10	94 162	4 094	−0.95652

References

[1] Council of Environmental Quality, *Federal Register*, Vol. 43, No. 230, 29 Nov. 1978.

[2] Jain, R. K. and Webster, R. D., *Journal of Water Research Planning and Management*, Vol. 103, No. WR2, Nov. 1977, pp. 257–270.

[3] Cairns, J., Jr, Sparks, R. E., and Waller, W. T., *Biological Methods for the Assessment of Water Quality, ASTM STP 528*, American Society for Testing and Materials, Philadelphia, 1973, pp. 127–143.

[4] Sharma, R. K., Buffington, J. D., and McFadden, J. T., Eds., *Biological Significance of Environmental Impacts*, NR-CONF-002, U.S. Nuclear Regulatory Commission, Washington, D.C. 1976, pp. 1–327.

[5] Herricks, E. E., *The Restoration and Recovery of Damaged Ecosystems*, University of Virginia Press, Charlottesville, Va., 1977, pp. 43–71.

[6] *Biological Data in Water Pollution Assessment: Quantitative and Statistical Analyses, ASTM STP 652*, American Society for Testing and Materials, Philadelphia, 1978.

[7] *Biological Monitoring of Water and Effluent Quality, ASTM STP 607*, American Society for Testing and Materials, Philadelphia, 1973.

[8] Woodwell, G. M. in *Ecosystem Analysis and Prediction*, S. A. Levin, Ed., Society of Industrial and Applied Mathematics, July 1974, Philadelphia, Pa., pp. 9–21.

[9] Cairns, John, Jr., *Integrity of Water*, GPO 055-001-01068-1, U.S. Environmental Protection Agency, Washington, D.C., 1975, pp. 171–188.

[10] Stauffer, J. R., Jr., Hocutt, C. H., Hendricks, M. L., and Markham, S. L. in *Surface Mining and Fish/Wildlife Needs in the Eastern United States*, D. E. Samuel, J. R. Stauffer, C. H. Hocutt, and W. T. Mason, Jr., Eds., FWS/OBS-78-81, U.S. Department of the Interior, Washington, D.C. Dec. 1978, pp. 105–118.

[11] Nossa, G. A., *The Computation and Graphical Display of the NSF Water Quality Index from the STORET Data Base Using the Integrated Plotting Package*, EPAWQ1001, U.S. Environmental Protection Agency, Washington, D.C., 1976.

[12] Donigian, A. S., Jr., and Linsley, R. K., *The Use of Continuous Simulation in the Evaluation of Water Quality Management Plans*, Contract No. 14-31-0001-5215, Office of Water Research and Technology, U.S. Department of the Interior, Washington, D.C., Aug. 1976.

[13] Dunnette, D. A., *Journal of the Water Pollution Control Federation*, Vol. 51, No. 1, 1979, pp. 53–61.

[14] Ott, W. R., *Water Quality Indices: A Survey of Indices Used in the United States*, EPA-600/4-78-005, U.S. Environmental Agency, Washington, D.C., 1978.

[15] Streeter, H. W. and Phelps, E. B., *Public Health Bulletin* 146, U.S. Public Health Service, Washington, D.C., 1925.

[16] Sladecek, V., *Internationale Vereinigung Für Theoretische und Angewandte Limnologie*, Vol. 17, 1969, pp. 546–559.

[*17*] Sladeck, V., *Erg Limnologie*, Vol. 7, 1973, pp. 1-218.
[*18*] Carlson, R. E., *Limnology and Oceanography*, Vol. 22, No. 2, 1977, pp. 361-369.
[*19*] Gakstatter, J. H., Allum, M. O., and Omernik, J. M. in *Water Quality Criteria Research of the U.S. EPA*, EPA-600/3-76-079, U.S. Environmental Protection Agency, July 1976, Washington, D.C., pp. 185-205.
[*20*] Tapp, J. S., *Journal of the Water Pollution Control Federation*, Vol. 50 No. 3, Part 1, 1978, pp. 484-492.
[*21*] Hormack, P. D. and Hutchins, M. L., "Input/Output Models as Decision Criteria for Lake Restoration," Technical Compliance Report Project C-7232, Water Research Center, University of Maine, Orono, Maine, 1978.
[*22*] Bradford, W. L. and Maiero, D. J., *Journal of the Environmental Engineering Division, ASCE*, Vol. 104, No. EE5, 1978, pp. 981-998.
[*23*] Weber, C. I. in *Bioassay Techniques and Environmental Chemistry*, G. Glass, Ed., Ann Arbor Science Publishers, Ann Arbor, Mich., 1973, pp. 119-138.
[*24*] Herricks, E. E. and Sale, M. J., "Development of Rational Threshold Values for Aquatic Ecosystems," Final Report, Contract No. DACA 88-77-M-D170, U.S. Army Construction Engineering Research Laboratory, Champaign, Ill., 1978, pp. 1-72.
[*25*] Lubinski, K. S., Sparks, R. E., and Jahn, L. A., "Development of Toxicity Indices for Assessing the Quality of the Illinois River," Report No. 96, UILU-WRC-74-0096, University of Illinois at Urbana-Champaign, 1974, pp. 1-46.
[*26*] Sparks, R. E., *Report of the Stream and Lake Classification Project to the Illinois Institute for Environmental Studies*, Institute for Environmental Studies, University of Illinois at Urbana-Champaign, 1977, Urbana, Ill., pp. 34-40.
[*27*] Lotka, A. J., *Science*, Vol. 26, 1907, pp. 21-22.
[*28*] Leslie, P. H., *Biometrika*, Vol. 35, 1945, pp. 185-212.
[*29*] Usher, M. B. in *Mathematical Models in Ecology*, J. N. R. Jeffers, Ed., Blackwell, 1972, London, pp. 29-60.
[*30*] Pollard, J. H., *Biometrika*, Vol. 33, No. 3, 1966, pp. 397-415.
[*31*] Usher, M. B., *Biometrics*, Vol. 26, 1970, pp. 1-12.
[*32*] Mendelssohn, R., *American Naturalist*, Vol. 110, 1970, pp. 339-349.
[*33*] Doubleday, W. G., *Biometrics*, Vol. 51, 1975, pp. 189-200.
[*34*] VanWinkle, W., DeAngeles, D. L., and Blum, S. R., *Transactions of the American Fisheries Society*, Vol. 107, No. 3, 1978, pp. 395-401.
[*35*] Saila, S. B. and Dorda, E. in *Assessing the Effects of Power-Plant-Induced Mortality on Fish Populations*, W. VanWinkle, Ed., Pergamon, New York, 1977, pp. 311-332.
[*36*] Mayhew, J., *Transactions of the Iowa Academy of Sciences*, Vol. 63, 1956, pp. 705-713.
[*37*] Carlander, K. D., *Handbook of Freshwater Fishery Biology*, Vol. 2, 1977, University of Iowa Press, Awes, Iowa, pp. 73-118.

P. P. Russell,[1] *A. J. Horne,*[2] *and J. F. Thomas*[3]

Application of Laboratory Scale Model Streams Toward Assessing Effluent Impacts in Freshwater Lotic Environments

REFERENCE: Russell, P. P., Horne, A. J., and Thomas, J. F., **"Application of Laboratory Scale Model Streams Toward Assessing Effluent Impacts in Freshwater Lotic Environments,"** *Ecological Assessments of Effluent Impacts on Communities of Indigenous Aquatic Organisms, ASTM STP 730,* J. M. Bates and C. I. Weber, Eds., American Society for Testing and Materials, 1981, pp. 32–48.

ABSTRACT: Laboratory scale model streams were developed to assess the effects of oil-shale-related effluents on aufwuchs. The model stream design allowed variation of water temperature, chemical composition, flow rate, and illumination intensity and periodicity. Chemical constancy of the stream water was provided by continuously metering new water to the streams with doses of test effluents. Omega-9 water, the experimental retort water used, was studied at dilutions between 0.013 and 2.12 percent for 9 days. Day 3 samples generally showed stimulated growth at all dilutions. Measurements on Days 6 and 9 however, had lowered levels of growth at 1 percent or more Omega-9 water concentration. The more dilute effluent loads stimulated growth throughout the test period. The most marked effect of Omega-9 water was to lower the proportional contribution of diatoms to the aufwuchs biomass. Measurements of control model streams suggest a species composition heterogeneity of 10 to 15 percent with a total unloaded stream spatial/temporal variability of no greater than about 25 percent. The biomass measurements were very reproducible, but chlorophyll *a* and respiration rate assays were hampered by excessive analytic error. Field confirmation of these tentative results is strongly recommended.

KEY WORDS: ecology, effluents, aquatic organisms, model streams, artificial substrates, aufwuchs, periphyton, oil shale, effluent impacts, water pollution, synfuels

The United States Government and private industry are increasingly directing attention toward domestic resources to meet our projected energy

[1]Director, Environmental Engineering, Fireman's Fund Insurance Companies, San Francisco, Calif. 94720.

[2]Professor of Civil Engineering, University of California, Berkeley, Calif. 94720.

[3]Professor of Civil Engineering and chairman of the Division of Sanitary, Environmental, Coastal, and Hydraulic Engineering, University of California, Berkeley, Calif. 94720.

needs. Oil shale reserves, particularly the vast deposits underlying portions of Colorado, Utah, and Wyoming, constitute an enormous fossil fuel energy source. It is estimated that in excess of 160×10^9 m^3 (10^{12} bbl) of oil equivalent are contained in the oil shale deposits of the Piceance Creek Basin in northwestern Colorado and the immediate vicinity [*1,2*].[4]

The extent and timing of development of a shale oil industry depend upon refinement of adequate technology for extracting the synthetic petroleum within imposed environmental constraints and the competitive position of this synfuel in the domestic energy market. There is promising evidence that environmentally sound oil shale usage will be economical [*3,4*]. Extensive research must be conducted to substantiate this optimism by assessing what environmental impacts can be expected from a commercial shale oil industry. The results presented here are an estimate of the effect of release of one of the industry effluents on biota in the creeks and rivers of the area. Specifically, the study centers on the influence of retort water dilutions on lotic aufwuchs. Current industry development plans provide for zero discharge of retort water, but, nevertheless, unintentional release through spills and leakage is a real possibility.

Aufwuchs, the microbial community attached to submerged surfaces, is an ecologically important biological group whose environmental requirements can be reproduced quite readily in the laboratory. Thus, this community was selected as the subject for the bioassay study. Colonization and development of aufwuchs on bare artificial substrates in model streams were monitored for 9 days. Control stream growth and responses to dilutions of an experimentally produced oil shale retort water were simultaneously observed in parallel streams. Measurements of aufwuchs biomass, chlorophyll *a*, and respiration rate were used to quantify growth and metabolism on the artificial substrates. The purpose of this research was to evaluate both the effects of oil-shale-related effluents on aufwuchs and the adequacy of the laboratory model stream technique as a monitoring tool for receiving water responses to wastewater discharges. The algal assay bottle test was developed to standardize the responses of planktonic algae to nutrients and toxic substances [*5*]. This laboratory model stream approach has great potential for providing similar information on lotic habitats.

Materials and Methods

Aufwuchs bioassays were conducted in four model streams. A schematic of the apparatus with one of the four streams shown is presented in Fig. 1. Each stream consisted of a riffle reach 122 cm long bounded by a pool at each end. The width of the riffle and pool reaches was 9.5 cm, and the capacity of each stream was 12 dm^3. The inner surfaces of the streams were coated with epoxy

[4] The italic numbers in brackets refer to the list of references appended to this paper.

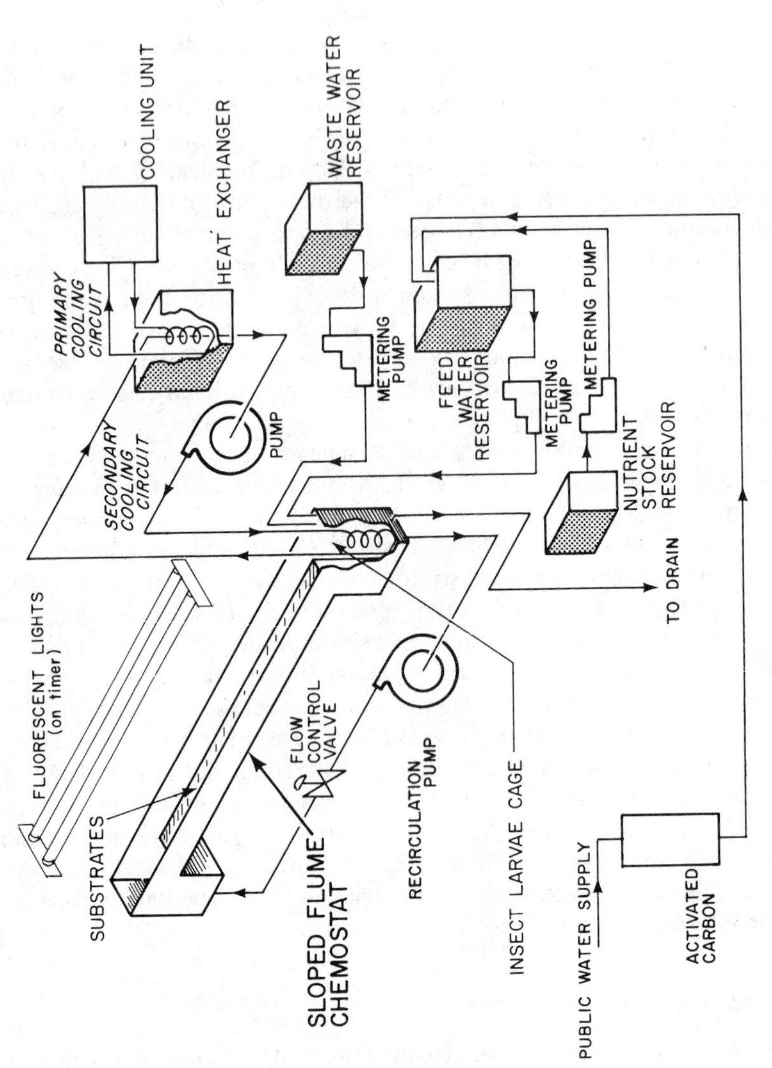

FIG. 1—*Model stream schematic.*

paint. The other materials in contact with the stream water were polyethylene, Teflon, Plexiglas, polyvinyl chloride, and the steel pump impeller and casing. Flow was produced in the streams by pumping water from the lower pools to the respective upper pools. A nominal flow rate of 45 cm/s and water depth of 2 to 3 cm were obtained by simultaneously adjusting the slope of the stream beds and the throttle valves on the discharge of the centrifugal pumps used for recirculation. Fluorescent tubes suspended over the riffle reaches provided illumination. Each stream received approximately equal exposure to a standard fluorescent tube, Sylvania Lifeline F40D, and a plant growth fluorescent tube, Sylvania Lifeline F40-GRO, suspended 30 cm above the water surface. A 15-h:9-h light:dark photoperiod was employed. Temperature control was effected by means of a cooling coil in the lower pool of each stream. The stream temperature was monitored daily, and the cooling system maintained the mean temperature of the four streams between 20.9 and 25.2°C. The temperature difference between streams on any given day never exceeded ±0.3°C of the mean, and, over the course of any of the 9-day experimental runs, the range of the daily stream temperature means was never greater than 2.8°C.

The chemical constancy of the stream waters was accomplished by metering fresh water to the streams on a continuous basis. Fresh water was drawn from the Berkeley (East Bay Municipal Utilities District) water supply and passed through a column of activated carbon to remove residual chlorine. The tap water was modified chemically to produce a nutrient solution similar to that suggested by Guillard but with reduced nitrate content [*6*]. Wastewater was added to the freshwater solution before introducing it into the streams. Incoming fresh water displaced stream water, which exited via overflow ports in stilling wells connected to each stream. The rate of water delivery to the streams was adjusted to provide a mean hydraulic residence time of 4 h, a turnover of six stream volumes per day.

Each experimental run consisted of monitoring the development of aufwuchs on initially bare standardized substrates over 9 days. Seed organisms were obtained by collecting stones (5 cm diameter) from the Stanislaus River in California (1300-m elevation) and transporting them in chilled containers to the laboratory. The stones were distributed to the lower reservoirs of the four streams. Subsequent experimental runs used the seed culture growing in the control stream of the previous run for aufwuchs recruitment.

The standard substrates were fabricated of 2.5-cm lengths of 1.59-cm (5/8-in.) outside diameter Tygon tubing. The outer surface of the substrate was roughened with sandpaper to promote aufwuchs attachment. To assure that aufwuchs colonization would occur only on the outer surface, the substrates were slipped on snugly fitting Teflon rods. Three substrates were mounted, abutting each other, on each rod, and short collars of the same tubing were slipped over the ends of the rod so that growth on the ends of the substrates was precluded. When removing a sample substrate, the end collar was first discarded.

Aufwuchs growth and metabolism were measured at 3-day intervals from the beginning of each run. Triplicate measurements of biomass, chlorophyll *a*, and respiration were performed on each sampling day for the four streams. The biomass sample was secured by scraping aufwuchs onto a glass fiber filter disk, Watman GF/C, using a toothbrush and squeeze bottle. Total solids were estimated by oven drying the sample at 103°C, weighing it to the nearest 0.1 mg, and correcting for the weight loss of blank filter disks carried through the procedure. Volatile solids measurements were obtained by heating the total solids samples in a muffle furnace at 550°C for 30 min, cooling them, reweighing them, and adjusting for the blanks. The volatile solids were equal to the weight difference between samples heated to 103 and 550°C.

Chlorophyll *a* in the aufwuchs was estimated by spectrophotometric measurement of methanol extracts prepared following the method of Talling [7]. Substrates were immersed in centrifuge tubes of boiling methanol for 30 s then removed. After centrifugation, absorbance was measured at 665 and 750 nm. The chlorophyll *a* on the substrates was computed by

$$\text{chl}\,a = 13.9 \times (A_{665} - A_{750}) \times V/L$$

where chl *a* is in milligrams of chlorophyll *a* per substrate; A_{665} and A_{750} are the absorbance readings at 665 and 750 nm, respectively; V is the sample volume in cubic decimetres, and L is the spectrophotometer cell path length in centimetres. The constant 13.9 was suggested by Talling for use with methanol extracts of phytoplankton pigment.

The respiration rates were measured by placing substrates in 60-cm^3 biochemical oxygen demand (BOD) bottles filled with water from the stream sampled. Aluminum foil was wrapped around the bottles, and they were immersed in a water bath maintained at the temperature of the stream. The dissolved oxygen concentration in the bottles was measured with a YSI dissolved-oxygen meter (Yellow Springs Instrument Co. Model 57). After a 2-h quiescent incubation, the specimen substrate was replaced with a magnetic stirring bar which circulated the water during measurement. The sample dissolved oxygen content was read after 1-min probe equilibration. The oxygen consumption due to the aufwuchs (DO) was computed by

$$\text{DO} = (\overline{\text{DOB}} - \text{DOS}) \times 57.4/2$$

where DO is the oxygen uptake in micrograms of oxygen per hour per substrate, $\overline{\text{DOB}}$ is the mean oxygen concentration in grams per cubic metre in the blank bottles containing only stream water, DOS is the dissolved oxygen concentration in bottles with substrates in grams per cubic metre, 57.4 cm^3 is the capacity of a BOD bottle with a substrate inside, and the constant 2 converts the incubation period oxygen consumption to a per hour basis.

Six experimental runs were performed. In each run, clean substrates, mounted on Teflon rods with the main axis horizontal and perpendicular to the flow, were positioned in the riffle reaches of the streams. No wastewater was tested in Run 1, the purpose being to gather baseline data on the streams in the unloaded condition. Runs 2 through 6 examined retort waters, with one stream of each run left unloaded as a control check and the other three streams dosed with three dilutions of effluent. The stream used as the control was varied from one run to the next. Between runs, all four streams were operated for at least 1 week on freshwater solution without any test effluent addition. Prior to commencing each experimental run, the riffle reaches were wiped clean of aufwuchs and nine substrates from the control stream of the immediately preceding run were positioned in each stream as a seed source for aufwuchs colonization.

The wastewater examined in Runs 2 and 3 was a retort water sample obtained from the U.S. Department of Energy's Laramie Energy Technology Center as produced during the Rock Springs Site 9 experimental *in situ* oil shale processing project near Rock Springs, Wyoming [8]. Omega-9 water is the designation given this effluent. It was produced, prepared, and distributed to interested researchers as a standard stock material for interlaboratory comparison. The preparation of Omega-9 water for distribution included filtration to effect a nominal 0.4-μm exclusion of suspended matter. Table 1 describes the chemical characteristics of Omega-9 water. It must be pointed out that this water is not necessarily representative of waters that will be produced during *in situ* oil shale processing. Therefore, results presented here are applicable only to Omega-9 water. Runs 4 through 6 examined dilutions of other retort waters. Data were used from the control streams only of these runs, the intent being to assess the inherent variability of the measurements under the no-load condition.

Results

The effects of Omega-9 water on the total solids, volatile solids, percent volatile solids, chlorophyll *a*, and respiration rate of aufwuchs are summarized in Figs. 2 through 6, respectively. The parameter values are given on a per substrate basis rather than per unit area because, for the purposes of comparing these data with field stream measurements, it is not now clear whether the appropriate areal denominator is the entire outer surface of the substrate, the upper half of the outer surface, or the area of the upper half projected onto a plane. It is clear from visual observation that the community on the upper side of the substrates was distinctly different from that on the under side. Thus, by expressing the amount of growth or metabolism on a per substrate basis, a weighted average of the contribution of the two aufwuchs types is used. The foremost graph in Figs. 2 through 6 represents the mean values computed from all the measurements on unloaded streams. The

TABLE 1—*Water Quality characterization of Omega-9 water.* [a]

Characteristic	Value, g/m^3 unless otherwise noted
Alkalinity (as $CaCO_3$)	16 200 ± 480
Biochemical oxygen demand, 5-day	740
Carbon	
Bicarbonate (as HCO_3^-)	15 940
Carbonate (as CO_3^{-2})	500
Inorganic (as C)	3 340 ± 390
Organic (as C)	1 003 ± 192
Chemical oxygen demand	8 100 ± 5 700
Conductivity	20 400 ± 3 840 μmhos/cm
Cyanide (as CN^-)	0.42 to 2.9
Hardness, total (as $CaCO_3$)	110
Nitrogen	
Ammonia [b] (as NH_3)	3 795 ± 390
Ammonium (as NH_4^+)	3 470 ± 830
Kjeldahl (as N)	3 420 ± 420
Nitrate (as NO_3^-)	0.17
Organic (as N)	148 to 630
Oil and grease	580
pH	8.65 ± 0.26
Phenols	60 ± 30
Phosphorus, orthophosphate (as PO_4^{-3})	0.08 to 24.6
Solids	
Fixed	13 430 ± 415
Total	14 210 ± 120
Total dissolved	14 210 ± 193
Sulfur	
Sulfate (as SO_4^{-2})	1 990 ± 250
Sulfide (as S)	0.0
Sulfite (as S)	<20
Tetrathionate (as $S_4O_6^{-2}$)	280
Thiosulfate (as $S_2O_3^{-2}$)	2 740 ± 730
Thiocyanate (as SCN^-)	123 ± 18

[a] From Fox et al [9].
[b] This value is the sum of NH_3 and NH_4^+.

means plotted on this zero percent effluent plate are presented in Table 2 along with the range and coefficient of variation for each variable. Results of Runs 2 and 3 are shown together in these figures. Run 2 examined 0.013, 0.13, and 0.53 percent Omega-9 water dilutions, while the 0.27, 1.06, and 2.12 percent dilutions were studied in Run 3.

An analysis of variance was performed on the data plotted in each of Figs. 2 through 6. Three factors were used in the analyses—sampling time, stream identity, and Omega-9 water dilution. This three-factor model was very significant ($P = 0.001$) in explaining the variance observed in measurements of each of the variables. The t-test significance of the differences between the mean of all the measurements of the control streams and the mean for each

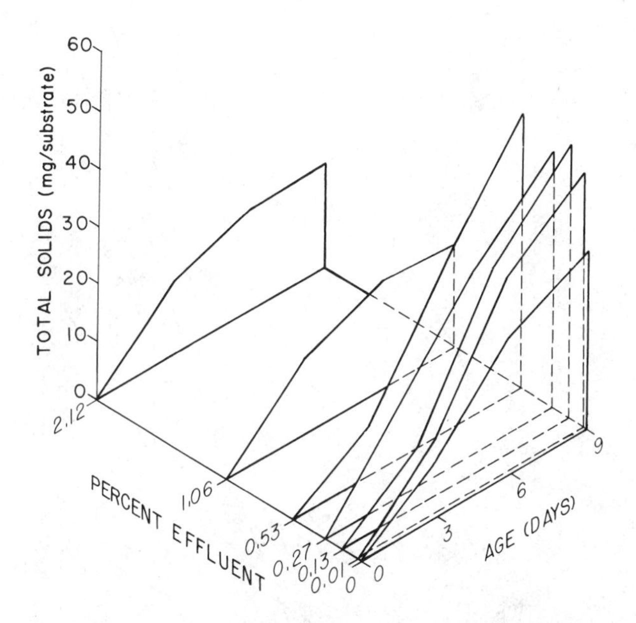

FIG. 2—*Aufwuchs total solids versus Omega-9 water dilution and sample time.*

of the Omega-9 water dilution streams was computed. Table 3 summarizes these results for each variable.

The factors in the analysis of variance together accounted for 74.2 percent of the variance in the total solids data. Individually, the sampling time and Omega-9 water dilution were very significant ($P = 0.001$), while the stream identity was not. The interaction between the sampling time and Omega-9 water dilution was also very important ($P = 0.001$), indicating that the effect of the effluent changed as development of the aufwuchs progressed. All of the dilutions in the Day 3 samples increased the mean total solids accumulation over the mean of the control streams (Fig. 2). Thereafter, total solids stimulation was observed at concentrations of 0.53 percent or less. The 1.06 and 2.12 percent sample means on Days 6 and 9 were depressed compared with the control mean.

The significance of the factors and two-way interactions, as shown by the analysis of variance of the volatile solids data, is the same as that evaluated for the total solids data. With this variable, the three-factor model explained 77.9 percent of the total volatile solids variance. All of the volatile solids means were elevated in relation to the control streams' mean, except those from the two highest concentrations of Omega-9 water on Day 9 (Fig. 3). These two depressed sample means were not statistically different from those of the controls, however. The volatile solids stimulation was typically very significant.

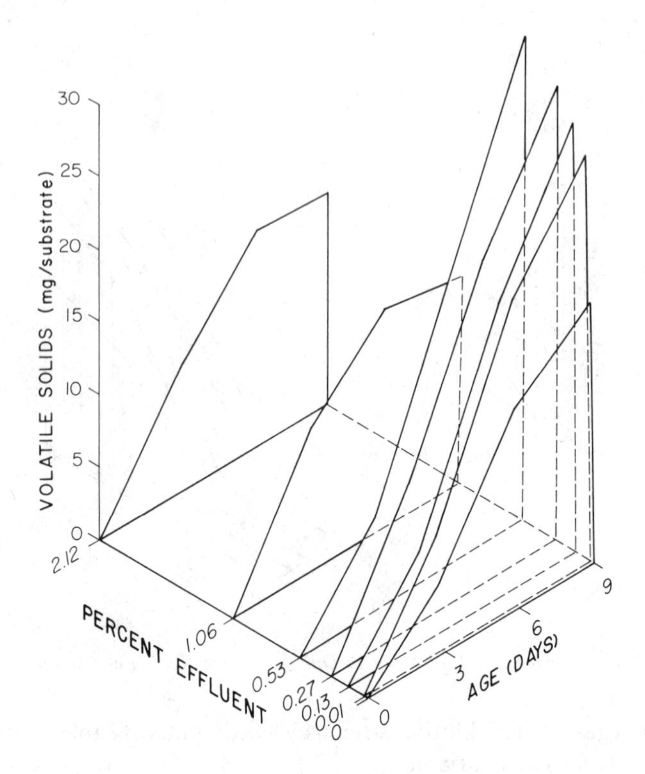

FIG. 3—*Aufwuchs volatile solids versus Omega-9 water dilution and sample time.*

The analysis of variance for percent volatile solids contrasts that performed on the other two biomass variables in that, although the Omega-9 water dilution was still very significant ($P = 0.001$), the sampling time was not an important factor but the stream identity was ($P = 0.004$). No two-way interactions were perceived by the analysis. Of the total variance in the data, 59.8 percent resulted from the effect of these three factors in concert. The initial percent volatile solids value appropriate for the beginning of the experimental run could not intuitively be assigned the quantity zero as had been done with the other parameters, so a linear regression was carried out on the control stream data versus time to obtain a Day 0 intercept. This value of 58.7 percent volatile solids was used in making the plates in Fig. 4 for each Omega-9 water dilution. The percent volatile solids of the aufwuchs was increased at all the dilutions and sampling times tested. Generally, the volatile fraction grew with increasing effluent concentration.

The sampling time, the Omega-9 water dilution, and the interaction of these two factors were shown to be very significant ($P = 0.001$) in the

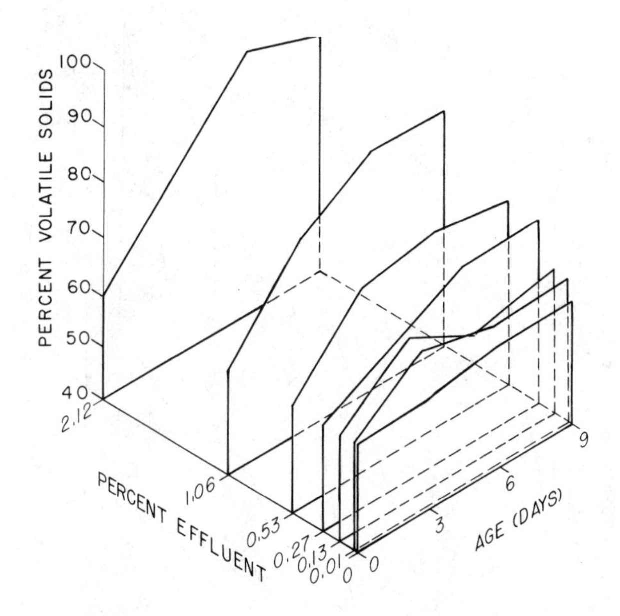

FIG. 4—*Aufwuchs percent volatile solids versus Omega-9 water dilution and sample time.*

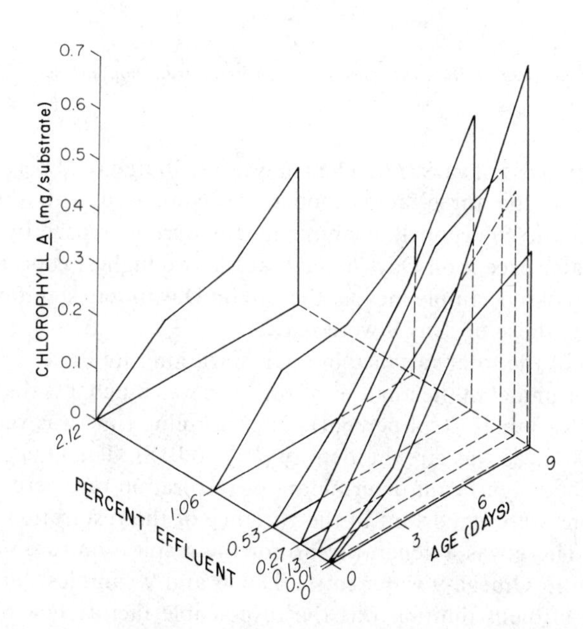

FIG. 5—*Aufwuchs chlorophyll* a *versus Omega-9 water dilution and sample time.*

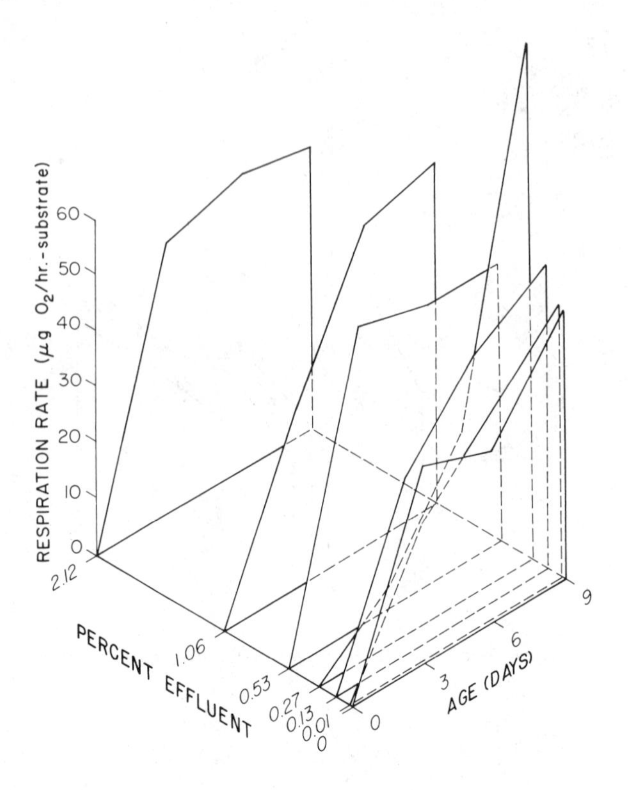

FIG. 6—*Aufwuchs respiration rate versus Omega-9 water dilution and sample time.*

chlorophyll *a* analysis of variance. The analysis attributed 69.3 percent of the total variance to the three-factor model. As compared with the control streams' mean, the chlorophyll *a* sample means were increased by exposure to Omega-9 water except on Days 6 and 9 at the two highest concentrations. Of the four depressed sample means, the two on Day 6 were significant (P = 0.004), whereas those on Day 9 were not.

The portion of the total variance in the respiration rate data explained by the three-factor model in the analysis of variance was much less than that for the other four variables (34.4 percent). The sampling time was very significant (P = 0.001), as was stream identity (P = 0.018). The Omega-9 water dilution was not a significant determinant of respiration rate here. The two-way interactions were important in the variance of the respiration rate data (P = 0.003). There was a general trend for the respiration rate to increase upon exposure to Omega-9 water in the Day 6 and 9 samples, but it is not clear that the effluent dilution was the responsible factor. Few significant divergences from the control streams' mean were found by the *t*-test.

TABLE 2—*Control stream parameter results.*

Parameter	Sampling Day	Number of Measurements	Mean	Minimum	Maximum	Coefficient of Variation, %
Total solids, mg/substrate	3	27	8.7	2.0	17.7	44.8
	6	27	23.2	12.0	32.8	23.5
	9	27	30.4	12.5	43.8	28.3
Volatile solids, mg/substrate	3	27	4.9	1.3	8.6	35.5
	6	27	13.9	8.8	20.3	21.5
	9	27	18.4	8.7	26.8	27.2
Percent volatile solids	3	27	59.2	45.1	77.7	14.1
	6	27	61.2	43.1	75.3	14.5
	9	27	61.0	46.1	71.3	10.4
Chlorophyll *a*, mg/substrate	3	27	0.095	0.027	0.178	46.4
	6	39	0.300	0.176	0.452	24.0
	9	27	0.381	0.127	0.660	35.6
Respiration rate, μg O_2/h/substrate	3	14	36.7	9.3	57.4	35.3
	6	22	33.3	7.9	72.5	48.3
	9	22	50.5	29.7	87.5	34.4

TABLE 3—t *Test significance of Omega-9 water effects.* [a]

Parameter	Dilution, %	Day 3	Day 6	Day 9
Total solids	0.013	0	+++	+++
	0.13	0	+++	+++
	0.27	+++	+++	+++
	0.53	0	0	+++
	1.06	+	0	---
	2.12	0	0	-
Volatile solids	0.013	+++	+++	+++
	0.13	0	+++	+++
	0.27	+++	+++	+++
	0.53	0	+++	+++
	1.06	+++	0	0
	2.12	+++	0	0
Percent volatile solids	0.013	0	0	0
	0.13	+	0	0
	0.27	0	+	+++
	0.53	+++	+++	+++
	1.06	+++	+++	+++
	2.12	+++	+++	+++
Chlorophyll *a*	0.013	0	+	+++
	0.13	+++	0	0
	0.27	+++	+++	0
	0.53	0	0	+
	1.06	+	---	0
	2.12	0	---	0
Respiration rate	0.013	0	0	0
	0.13	0	0	0
	0.27	---	0	+++
	0.53	+	0	0
	1.06	0	+++	0
	2.12	0	+	0

[a] Symbols represent *t*-test significance of the difference between the control streams' mean and the mean for the indicated dilution: (+++) Omega-9 water increases value ($P \leq 0.01$); (+) Omega-9 water increases value ($P \leq 0.05$); (0) no significant difference between means; (−) Omega-9 water decreases value ($P \leq 0.05$); (−−−) Omega-9 water decreases value ($P \leq 0.01$).

Discussion

The model stream bioassays presented address two distinct problems. First, how much confidence can be placed in measurements of aufwuchs growth and metabolism using the apparatus and methodology described here? This issue can be reduced into two more basic questions: how reproducible are the results, and how accurately do the experimental communities reflect natural communities? The latter consideration is beyond the scope of this project. Of necessity, the assumption is made that results obtained from this model stream study are a first approximation of impacts of oil-shale-related discharges to rivers and creeks. It cannot be overly stressed that field confirmation of these results must be obtained before they are utilized.

The reproducibility of measurements in the model streams is subject to quantitative and qualitative differences in the aufwuchs community between sampling events and to error in the analytic techniques. Table 2 lists the coefficient of variation for each variable over all the control streams' samples to quantify the variability in sample measurements. The most striking aspect of the aufwuchs growth and metabolism variables is that the variability in the biomass measurements is much less than that of either the chlorophyll *a* or the respiration rate measurements. It is assumed that the degree of variability in the treatment streams' data is similar to that of the control streams.

The smallest coefficient of variation was observed for percent volatile solids data (Table 2). If the total variation is the sum of variation due to spatial and temporal heterogeneity within and between the model streams and the variation due to error in the analytic technique, then this variable may be used to give a rough approximation of the relative contributions of the sources of variation. Heterogeneity within and between the streams seemed to result mostly from differential colonization and growth rates and from sloughing of the aufwuchs mat in the turbulent flow regime. Furthermore, it can be assumed that the stream heterogeneity occurs in terms of both species composition and biomass density. Stream density heterogeneity was most pronounced on Day 9 when the sloughing phenomenon was observed most frequently. By assuming that the entire percent volatile solids variability was the result of stream heterogeneity and that the analytic error was negligible for this variable, a minimum stream heterogeneity of 10 to 15 percent variability is suggested. The low analytic error assumption for percent volatile solids is warranted because the same filter disk was used for the weighing after each temperature; thus, no error from transferring the aufwuchs from the substrate to the disk was involved. In addition, the weighing error estimated from the filter disk blanks carried through the procedure was quite low, less than 1 percent of the sample weight. Probably most of the 10 to 15 percent coefficient of variation in the percent volatile solids data resulted from differences in species composition in the aufwuchs. Microscopic examination of aufwuchs samples showed that increases in the percent volatile solids were strongly correlated with decreases in the contribution of diatoms to the community population. The nonvolatile silica frustules of the diatoms are undoubtedly instrumental in this relationship. Since the sloughing phenomenon would not necessarily alter the volatile fraction of the remaining aufwuchs, the species composition heterogeneity predominating in the percent volatile solids variability is probably distinct from the density heterogeneity controlling in the other two biomass variables.

The total solids and volatile solids variability reflects analytic error as well as stream heterogeneity. These variables respond to sloughing and differential growth rates more readily than differences in population makeup.

Analytic error is probably a relatively minor contributor to these coefficients of variation, compared with the variability resulting from stream biomass heterogeneity. At any rate, if the analytic error for these variables is very small, then the spatial/temporal heterogeneity in the stream biomass can be no greater than about 25%.

The coefficients of variation for the chlorophyll *a* and respiration rate variables are substantially larger than those for the biomass variables. This excess variation is probably the result of analytic error. Two reasons for this assertion are evident. First, the chlorophyll *a* content and respiration rate probably parallel the available quantity of biomass when no effluent stress is present. Second, subjective observations on the methodologies indicate that extraneous variability was arising through analytic error. In the case of chlorophyll *a*, 30-s immersion in boiling methanol did not always give complete extraction of the green pigments, especially when the aufwuchs growth was dense. Circulation of the methanol through the aufwuchs matrix was probably impeded by the tortuous hydraulic pathways and the presence of interstitial stream water. In addition, the substrate acted as a heat sink, so that the cells lying deepest in the matrix probably were not heated enough for efficient pigment extraction in the available time. The assays for chlorophyll *a* probably could be improved by removing the aufwuchs from the substrates onto filter paper before extraction, longer immersion in boiling methanol, or sonication to assist in release of the pigments. The inherent drawback of the respiration rate measurement technique is that little or no water movement occurs in the BOD bottle during the 2-h incubation period. Aufwuchs are adapted to a lotic environment in which the current prevents the formation of large concentration gradients of dissolved gases, nutrients, and waste products. The quiescent conditions inside the incubation bottles probably disrupted the normal respiration rate. Mitigation of this problem is not immediately obvious within the constraints of a limited budget.

The second problem addressed by data generated in this research is, How sensitive are the measures of the model stream aufwuchs to dilutions of Omega-9 water? The variables total solids, volatile solids, and chlorophyll *a* consistently show that the total community growth is stimulated by concentrations as high as 0.5 percent Omega-9 water and depressed at concentrations of 1 percent and greater. Furthermore, the biomass variables suggest that this effect becomes more pronounced as the community ages. It is encouraging to note that despite the large coefficient of variation in the chlorophyll *a* data, the *t* tests using this variable substantiate the stimulation/inhibition phenomenon. It seems that a toxic response by the aufwuchs to high effluent concentrations is replaced by growth enhancement with sufficient effluent dilution.

The percent volatile solids results strongly suggest that the contribution of diatoms to the total aufwuchs biomass is reduced in the presence of Omega-9 water. This effect becomes greater with increasing effluent loading. The aufwuchs organisms surviving beyond the tolerance limit of the diatoms may be other primary producers, such as green or blue-green algae or heterotrophic fungi or bacteria. The abundant organic matter in Omega-9 water may well promote heterotrophic organisms.

The results of the respiration rate measurement are not adequate for drawing any firm conclusions. There is a general tendency for enhanced respiration under effluent loading, as would be expected with stimulated heterotrophic growth, but the high variability in the data precludes any confidence in this assertion.

The methodology could be most readily validated by operating the model streams in the field, where ambient insolation, temperature, and a once-through water supply with a continuous source of seed organisms is provided. Such confirmation is incumbent before judicious use can be made of model stream data.

Summary and Conclusions

The laboratory scale model stream apparatus and methodology can yield reproducible results for the biomass variables—total solids, volatile solids, and percent volatile solids. Growth and composition of the aufwuchs varies in the range of 10 to 25 percent under the no-load condition. The variance in the chlorophyll *a* and respiration rate data is enlarged through analytic error. Simple improvements to increase pigment extraction efficiency could lower the variation in the data. The respiration rate methodology is in need of more extensive refinement.

Very significant changes in biomass and chlorophyll *a* values are obtained upon exposure of aufwuchs to Omega-9 water during colonization and development. In the model streams, low concentrations of Omega-9 water stimulated aufwuchs production at 0.5 percent or less and suppressed production at 1 percent or greater effluent loading. Omega-9 water lowered the proportional contribution of diatoms to the aufwuchs biomass, as evidenced by microscopic examination, and higher percent volatile solids measurements. The Omega-9 water effects, particularly the aufwuchs growth inhibition, became more pronounced with time.

Proper field calibration of model stream aufwuchs responses to industrial discharges is necessary before the methodology can be used as a predictive tool. Once refinement has been accomplished, however, an economic way will be available to standardize the toxic or biostimulatory potential of chemicals in lotic waters.

Acknowledgment

This research was supported largely by grants from the Department of Energy Laramie Energy Technology Center and the U.S. Department of the Interior.

References

[1] Murray, D. K. and Haun, J. D. in *Guidebook to the Energy Resources of the Piceance Creek Basin, Colorado,* D. K. Murray, Ed., Rocky Mountain Association of Geologists, Denver, Colo., 1974, pp. 29–40.

[2] *United States Energy Prospects: An Engineering Viewpoint*, Task Force on Energy, National Academy of Engineering, Washington, D.C., 1974.

[3] Kilburn, P. D., *Environmental Conservation*, Vol. 3, No. 2, Summer 1976, pp. 101–115.

[4] Ericsson, N. R. and Morgan, P., *The Bell Journal of Economics*, Vol. 9, No. 2, Autumn 1978, pp. 457–487.

[5] Miller, W. E., Greene, J. C., and Shiroyama, T., *The* Selenastrum capricornutum *Printz Algal Assay Bottle Test—Experimental Design, Application, and Data Interpretation Protocol,* U.S. Environmental Protection Agency, Corvallis, Ore., July 1978.

[6] Nichols, H. W. in *Handbook of Phycological Methods, Culture Methods and Growth Measurements*, J. R. Stein, Ed., Cambridge University Press, London, 1973, pp. 7–24.

[7] Talling, J. F. in *A Manual on Methods for Measuring Primary Production in Aquatic Environments*, IBP Handbook No. 12, R. A. Vollenweider, Ed., Blackwell, Oxford, England, 1969, pp. 22–25.

[8] Farrier, D. S., Poulson, R. E., Skinner, Q. D., Adams, J. C., and Bower, J. P., *Proceedings,* Second Pacific Chemical Engineering Congress, Vol. 2, 1977, pp. 1031–1035.

[9] Fox, J. P., Farrier, D. S., and Poulson, R. E., *Chemical Characterization and Analytical Considerations for an* in Situ *Oil Shale Process Water,* Laramie Energy Technology Center, Laramie, Wyo., Nov. 1978.

R. B. Bogardus[1]

Ecological Factors in the Selection of Representative Species for Thermal Effluent Demonstrations

REFERENCE: Bogardus, R. B., **"Ecological Factors in the Selection of Representative Species for Thermal Effluent Demonstrations,"** *Ecological Assessments of Effluent Impacts on Communities of Indigenous Aquatic Organisms, ASTM STP 730*, J. M. Bates and C. I. Weber, Eds., American Society for Testing and Materials, 1981, pp. 49-67.

ABSTRACT: A number of various federal and state guideline documents that have been issued for use in demonstrations under Section 316(a) of the Federal Water Pollution Control Act contain criteria to be applied in selecting organisms that are representative of the aquatic community at the proposed power plant site. This paper reviews three Section 316(a) case histories in which these selection criteria are applied and evaluates the usefulness and practicality of the methodology. The conclusion is reached that the basic approach of using representative species is practical but that the reliability and accuracy of the selection process is only as good as the available data on the seasonal distribution and life histories of the aquatic community in the receiving waters. A baseline field monitoring program designed to gather adequate data for the Section 316(a) representative species selection and evaluation process is proposed.

KEY WORDS: representative important species (RIS), thermal effluent demonstrations, once-through cooling, thermal effects, evaluation criteria, thermal plume, baseline field sampling program, aquatic biological community, nuisance species, commercial species, recreational species, community integrity, Lake Superior, Block Island Sound, Wabash River, ecology, effluents, aquatic organisms

Under the 1972 Amendments to the Federal Water Pollution Control Act (the Act), operators of steam electric power generating units must comply with requirements for applicable technology based on effluent limitations promulgated by the administrator of the U.S. Environmental Protection Agency (EPA). These limitations are published in the *Code of Federal Regulations,* Vol. 40, Part 423. In addition, compliance with effluent limitations calculated to achieve water quality standards is required under Section 301 (b)(1)(C) of the Act.

[1]Vice president, WAPORA, Inc., Washington, D.C. 20015.

With respect to the discharge of heat, an exemption from any of those limitations is available if the operator can make a successful demonstration under Section 316(a) of the Act. Whenever the operator can successfully demonstrate that the effluent limitation proposed for the control of the thermal component of any discharge is more stringent than necessary to assure the protection and propagation of a balanced, indigenous community of shellfish, fish, and wildlife in and on the receiving waters, the EPA may impose an alternative effluent limitation.

A number of guideline documents for use in determining the effects of existing or proposed thermal effluents on the aquatic biota have been issued by various federal and state regulatory agencies [*1-3*].[2] Each of these documents contains a discussion of general criteria to be applied in the selection of organisms that can be used as representative of the aquatic community of the power plant site. These "representative important species" (RIS) include those that are:

nuisance species,
commercially or recreationally valuable,
threatened or endangered,
primary producers—particularly those communities supporting relatively long-lived fixed-location species of multiple use (form and stabilize habitats, produce organic matter, provide cover), and
necessary for the well-being of the community complex (the dominant food base for important higher-level trophic components; scavengers and decomposers critical to the breakdown and utilization of organic matter).

This paper presents three case histories that illustrate the methodology utilized in applying representative important species selection criteria to thermal effluents located in different major water bodies.

Case History No. 1—Charlestown, R.I., Nuclear Plant

Power Plant Location

The New England Power Company's proposed site is located in Charlestown, Rhode Island, adjacent to Block Island Sound but separated from it by a series of saltwater beach ponds and a narrow barrier beach.

Cooling System Characteristics

Two nuclear reactors each having a gross electrical output of about 1200 MW and each using a single once-through cooling system are proposed for this site. The cooling water will be drawn from Block Island Sound through

[2]The italic numbers in brackets refer to the list of references appended to this paper.

an offshore ocean intake, and warm water will be returned to the sound through an offshore ocean discharge [*4*]. The total water use will be about 3217 m^3/min, and the maximum ΔT across the condensers will be 20.56 deg C (37 deg F). The velocity cap intake structure and the multiport diffuser discharge will be located offshore at the 9 to 12-m depth contour. The exact siting of these structures will be refined during the regulatory agency permitting process.

The dimensions and location of the plume are predicted to be highly variable, depending on tidal flows, but the plume will be basically a surface phenomenon. The volume of the 3.3-deg C (6-deg F) isotherm under slack tide conditions is predicted to be 37 005 m^3 (30 acre ft), while the maximum 1.7-deg C (3-deg F) isotherm will be about 925 117 m^3 (750 acre ft) [*4*].

Physical Setting

Block Island Sound encompasses approximately 1024 km^2 (400 $miles^2$) of continental shelf and is bounded on the north by the coast of Rhode Island, on the south by Block Island and the Atlantic Ocean, on the east by Rhode Island Sound, and on the west by Long Island Sound. The average water depth of the sound is 36.6 m (120 ft), with a maximum depth of 91 m (300 ft) near Fisher's Island. Shoreward, the depth lessens, until, in the area of Block Island Sound within 1.6 km (1 mile) of the proposed site, the water is less than 15 m (50 ft) deep [*4*].

The waters of Block Island Sound are derived from a mixture of continental shelf water from offshore and brackish water from coastal river runoff. The currents in Block Island Sound are predominantly tidal in nature, with two full tidal cycles completed per day. The current speeds reach the highest velocities, approximating 0.3 m/s, during each flood and ebb, and then diminish as the direction shifts [*4*].

The surface water temperature in Block Island Sound cycles annually from a high of about 20°C (68°F) in August to a low of about 2.8°C (37°F) in February or March. During late fall, winter, and early spring, Block Island Sound water is essentially homiothermic (isothermic). However, in summer, due to the differential heating of surface and bottom waters, the bottom waters are generally 4 to 5 deg Celsius cooler than the surface waters [*4*].

The substrate adjacent to the site can be divided into five distinct substrate types. The two most prominent bottom types—wave-rippled sand, and gravel with wave-rippled patches of sand—are most commonly located seaward and shoreward of the 12-m depth contour, respectively. Boulder and gravel substrate patches are widely distributed landward of the 9-m depth line. A bottom displaying numerous trawl marks is present in depths greater than 18 m. Diver observations at the 6-m depth revealed an extremely rough bottom, with boulders 0.5 m in diameter separated by small gravel and sand accumulations [*4*].

Biological Setting

Phytoplankton in Block Island Sound cycle annually, with the highest densities usually occurring during spring and a poorly defined increase during fall. High phytoplankton densities were correlated with low nutrient concentrations, as is typical in temperate marine waters. Diatoms and dinoflagellates comprise the majority of the species identified [*4*].

Zooplankton densities in the sound attain a maximum in terms of both abundance and dry weight during summer. Crustacea are the dominant holoplankters throughout the year, while larval polychaetes, bivalve mollusks, and gastropods are present during the warmer months [*4*].

The benthos of the sediment and rock substrates in the waters adjacent to the site are diverse. Quarterly sampling identified 269 taxa. Of those identified, 149 were polychaete worms, 87 were crustaceans, and 33 were gastropods. Other invertebrates collected included representatives of the groups Protozoa, Cnidaria, Bryozoa, Nemertea, Nematoda, Platyhelminthes, Echinodermata, Chordata, Porifera, and Sipunculide. Included in the benthos were the commercially fished lobster, mahogany quahog, and surf clam [*4*].

From April 1974 through March 1976, a total of 49 species of finfish were collected in the sound waters adjacent to the site by beach seining, gill netting, and trawling. Common in the catch were silver hake, butterfish, windowpane, scup, little skate, winter flounder, ocean pout, and longhorn sculpin. Although not collected efficiently in the gear listed above, bluefish and striped bass are important recreational fishes in Block Island Sound [*4*].

The most abundant species of ichthyoplankton (fish eggs and larvae) present was the Atlantic mackerel, which accounted for an average of 41 percent of the eggs and 27 percent of the larvae. The cunner-tautog-yellowtail flounder group was the second most abundant group in the egg collections. This group, together with the mackerel, accounted for an average of 92 percent of the eggs and 69 percent of the larvae. The anchovy, winter flounder, and sand lance accounted for 14, 4, and 4 percent, respectively, of the averaged total larval collections. These last three species were not important in the egg collections [*4*].

Representative Important Species—Table 1 is the representative important species (RIS) list for this site. The decision process leading to the selection process included consideration of surveys of each species collected in the site area and was refined through the application of ecological, political, and practical criteria. The following paragraphs outline the major factors used in arriving at the designated list.

Threatened or Endangered Species—No threatened or endangered species were identified as occurring in Block Island Sound. If any life stage of such a species were to occur within the zone of effect of either the thermal effluent or intake structure, then it would automatically appear on the designated list.

TABLE 1—*Representative important species list for Block Island Sound.*

Common Name	Scientific Name
Plankton	
Eelgrass	*Zostera marina*
Atlantic menhaden	*Brevoortia tyrannus*
Bay anchovy	*Anchoa mitchilli*
Silver hake	*Merluccius bilinearis*
Scup	*Steonotomus chrysops*
Cunner	*Tautogolabrus adspersus*
Atlantic mackerel	*Scomber scombrus*
Butterfish	*Peprilus triacanthus*
Winter flounder	*Pseudopleuronectes americanus*
American sand lance	*Ammodytes americanus*
Striped bass	*Morone saxatilis*
Bluefish	*Pomatomus saltatrix*
Sand shrimp	*Crangon septimspinosus*
Blue mussel	*Mytilus edulis*
Hard clam	*Mercenaria mercenaria*
Long-finned squid	*Loligo pealei*
American lobster	*Homarus americanus*

Nuisance Species—The only potentially significant nuisance organism identified as occurring in Block Island Sound or the adjacent salt water ponds was *Gonyaulax*, the causative agent of red tide. Stone and Webster [*5*] reported no increased incidence of this organism as a result of the operation of the Pilgrim Nuclear Power Station in Plymouth, Mass., and Prakash [*6*] found that the temperature was not as important as salinity in controlling the summer abundance of *Gonyaulax tamarensis*. Moreover, the volume of water affected by a thermal increase in excess of 0.83 deg C (1.5 deg F) is very small compared with the amount of Block Island Sound water passing the discharge area; thus, an artifically induced bloom of red tide organisms should not occur. Therefore, although a potential nuisance organism was known to occur in the zone of thermal influence, preliminary investigation revealed the potential for thermally induced increases to be so low that the organism was not placed on the designated list and was not given further consideration in the thermal effects section of the Environmental Impact Statement (EIS).

Commercial or Recreational Species—Species of fish and invertebrates found in Block Island Sound and associated salt ponds that are important components of the recreational and commercial fishery are listed in Table 2 [*4*].

Species Important for the Maintenance of Community Integrity—Three of the species listed in Table 2—the tautog, Atlantic mackerel, and winter flounder—have comparatively high larval densities in Block Island Sound. Lobster larvae are also abundant. While larvae of the other listed species may

TABLE 2—*Important components of the recreational and commercial fishery of Block Island Sound and associated salt ponds.*

Species	Component Type[a]	Species	Component Type[a]
Atlantic menhaden	C, P	Red hake	C
Atlantic herring	C	Little skate	C
Scup	C	Lobster	C
Atlantic mackerel	P	Squid	C
Butterfish	C	Mahogany quahog	C
Summer flounder (fluke)	C	Surf clam	C
Windowpane	C	Oyster	C
Winter flounder	C, P, R	Bay scallop	C
Atlantic cod	C	Hard clam	C
Silver hake (whiting)	C	Soft clam	C
Striped bass	R	Bluefish	R
Tautog	R, P		

[a]C = component of the commercial catch, R = important to sport fishery, P = high planktonic larval or egg densities.

be present, their low densities suggest that neither Block Island Sound nor the connected saltwater ponds is a major spawning or nursery area [*4*].

In characterizing the biological community of Block Island Sound and its associated saltwater ponds, consideration was given to the dominant forage species and to habitat formers. The later include plants and animals that have a relatively sessile life state, have an aggregated distribution, or perform an essential function within the community (that is, are a source of food, stabilize sediments, or are important in nutrient cycling). Species within the sound or its salt ponds that are dominant forage species or habitat formers include: (1) the cunner, which is the second most abundant species of fish larvae (behind mackerel); (2) the bay anchovy, whose larvae were the third most abundant; (3) the sand lance, which is abundant in the shallower waters of the sound; (4) sand shrimp; (5) the blue mussel, which is dominant in the meroplankton of Block Island Sound; and (6) eelgrass, which is abundant in the salt ponds [*4*].

The taxa eventually selected as representative important species were usually those having the highest potential for power-plant-induced impact, as well as those with important food-chain, habitat-forming, or recreational roles. Political considerations, however, were a major factor in the inclusion of the striped bass. Since no striped bass eggs or larvae were collected near the site, and since the size and swim speeds of the individuals present in the area should preclude entrainment and impingement problems, the striped bass would not need to be included as an RIS were it not for the tremendous interest in the species by the local recreational fishing community.

Case History No. 2—Breed Plant, Fairbanks, Ind.

Power Plant Location

The Breed Plant of the Indiana and Michigan Electric Co is located on the west bank of the Wabash River at River Mile 183.5,[3] near Fairbanks in Sullivan County, Indiana, approximately 32 km downstream from Terre Haute, Ind.

Cooling System Characteristics

The Breed Plant operates with one coal-fired, steam electric unit, which provides 400 MW in the summer and 420 MW in the winter. The plant is a base load station that has been in operation since 1960 [7].

The plant employs continuous once-through cooling, drawing its water from the Wabash River through conventional shoreline-sited rotating vertical intake screens and subsequently returning the heated effluent through a short (68-m) discharge canal to the river. Four circulating water pumps deliver 1215 m^3/min of cooling water to the condensers during the summer (July and August). Only three pumps are utilized during the remainder of the operations year, and 911 m^3/min are pumped to the condensers during this period [7].

The time of travel from the intake point to the point of discharge is variable with the number of pumps in operation, but the value is generally only a few minutes. The maximum temperature rise during a 10-year operating period (1965 through 1974) was 8.4 deg C (15.1 deg F) during the summer months and 10.8 deg C (19.4 deg F) during the winter [7].

Under conditions of average normal low flow 42.5 to 113.3 m^3/s (1500 to 4000 ft^3/s), the thermal plume tends to fan out slightly past mid-river as it leaves the discharge canal but then returns toward the plant bank as it moves downstream. As the plume approaches the bend above the downstream riffle, it moves toward the far bank and mixes with ambient river water as it crosses the riffle. Because of the shallowness of the Wabash River during these flow conditions, the plume does extend to the substrate over much of its width [7].

Physical Setting

The Wabash River Basin encompasses approximately 53 269 km^2 (33 100 $mile^2$), of which 513 km^2 (319 $mile^2$) are located in eastern Illinois, and 38 975 km^2 (24 218 $mile^2$) are in western Indiana. From its origin near

[3]Note that 1 mile = 1.609 km.

Celina, Ohio, the river flows about 764 km to its junction with the Ohio River [7].

The Wabash River adjacent to the Breed Plant is in an intermediate stage of development. The aging of the stream has been accelerated during the past century by the conversion of woodlands in the basin to row crop agricultural and urban uses. The stability of the banks of the river similarly has been weakened by the clearing of shoreline vegetation [7].

Many miles of the Wabash River shoreline essentially are devoid of tree cover. The absence of bank vegetation has resulted in, and is continuing to result in, the widening and shallowing of the river. The bottom substrate characteristics, cross-sectional areas, and depths fluctuate from year to year [7].

In the areas studied downstream of the Breed Plant, at river stage 0.40 m (1.3 ft), [41.77 m^3/s (1475 ft^3/s)], the water depths range from about 1.22 m (4 ft) at the bank on the plant side to about 0.30 m (1 ft) on the opposite side. The water depths occasionally reach 3.05 m (10 ft). Relatively broad, shallow, sandy areas begin on the west bank and extend across nearly 75 percent of the stream. The bottom gradually slopes downward to its greatest depth near the east shoreline. A bank-to-bank riffle area is located 2.8 km downstream from the Breed Plant and is the most important single habitat for fishes in perhaps a 80.47 km (50-mile) segment of the river. The riffle appears to be stable, having existed in its present location and approximate formation at least since early in this century. A deep pool has formed immediately downstream [7].

Based on U.S. Geological Survey and Breed Plant flow gaging station records for the years 1965 to 1974, the minimum and maximum flows at the Breed Plant were about 28.32 and 2775.05 m^3/s (1000 and 98 000 ft^3/s), respectively. The flows increase rapidly after rainfalls of only a short duration, primarily because of the agricultural land use in the basin and the urban runoff contributions to upstream tributaries [7].

Water quality data for the Wabash River near the Breed Plant indicate that the stream has many of the characteristics of water bodies that drain agricultural watersheds. In addition, the Wabash River is used by several upstream wastewater treatment plants as a primary discharge point. The Wabash adjacent to the Breed Plant could be generally classified as a hard, turbid, slightly alkaline, moderately to highly conductive, nutrient-enriched stream with occasional early morning low (about 4 mg/litre) concentrations of dissolved oxygen [7].

The substrate from the intake point downstream 2.8 km to the head of the riffle is approximately 80 percent sand or coarse sand; 5 to 10 percent mud along the shoreline; less than 5 percent granule, gravel, or pebble; and about 5 percent detrital material. The large downstream riffle contains a variety of substrate ranging from fine muds and detrital accumulations in the pool below the riffle to large rock and occasional boulders in the swift flowing chute in the center of the river adjacent to the riffle [7].

Biological Setting

Populations of phytoplankton generally peak in August at approximately 50 million cells per litre. This relatively high concentration indicates the nutrient-rich condition of the Wabash River in the vicinity of the site. The dominant algal genus during the high-density period was the diatom *Cyclotella,* which was the dominant phytoplankter during most of the annual cycle. Potentially nuisance-causing blue-green algae were not a problem at this site and occupied a high percentage of the phytoplankton community only during the late winter, when the total algal population was low [7].

Zooplankton densities ranged from a low of $301/m^3$ in September 1974 to a high of $4\ 095/m^3$ in November of the same year. These numbers are very low when compared with counts of around $50\ 000/m^3$ in the Ohio River [8] and much higher counts in many ponds and lakes. Rotifers were the dominant group of zooplankters during most of the year, with the genera *Brachionus* and *Keratella* common in the collections [7].

The macroinvertebrate community is diverse and substrate-specific. Sand substrates are dominated by midges of the family Chironomidae, mud substrates by aquatic worms (mostly family Tubificidae), and gravel substrates by caddis flies of the genera *Potamyia* and *Hydropsyche*. The substrate intergrades contained mixed populations of these major organism groups. The numbers of caddis flies ranged as high as $50\ 000/m^2$ in the gravel substrate at the riffle [7].

During the period 1971 to 1978, over 80 species of fish were collected in the river adjacent to the Breed Plant by seines, hoop nets, trawls, gill nets, and electrofishing. Common fish species included the channel catfish, shovelnose sturgeon, longnose and shortnose gar, gizzard shad, carp, speckled chub, silver chub, emerald shiner, river shiner, spotfin shiner, mimic shiner, bullhead minnow, river carpsucker, buffalo (smallmouth and bigmouth), mountain madtom, spotted bass, mud darter, and slenderhead darter [7].

Representative Important Species—Table 3 is a list of those animals considered representative at the Breed Plant site. The Section 316(a) demonstration for the proposed continuation of once-through cooling at the Breed Plant was based on an evaluation of the issue of prior appreciable harm from an existing thermal source and was not primarily a predictive demonstration. The RIS list was, therefore, compiled by the author without regulatory agency input in order to evaluate thermal conditions that had not been sampled in the field. The following paragraphs outline the decision process upon which selection of these species was based.

Threatened or Endangered Species—No species that are currently on the

TABLE 3—*Animals considered representative at the Breed Plant site.*

Common Name	Scientific Name
Macroinvertebrates	
Gizzard shad	*Dorosoma cepedianum*
Carp	*Cyprinus carpio*
Emerald Shiner	*Notropis atherinoides*
Ribbon shiner	*Notropis fumeus*
Spotfin shiner	*Notropis spilopterus*
Speckled chub	*Hybopsis aestivalis*
Bullhead minnow	*Pimephales vigilax*
Northern hogsucker	*Hypentelium nigricans*
Blue sucker	*Cycleptus elongatus*
Channel catfish	*Ictalurus punctatus*
Mountain madtom	*Noturus eleutherus*
Bluegill	*Lepomis macrochirus*
Spotted bass	*Micropterus punctulatus*
Logperch	*Percina caprodes*
Rainbow darter	*Etheostoma caeruleum*
Greenside darter	*Etheostoma blennioides*
Eastern sand darter	*Ammocrypta pellucida*

federal list of "endangered and threatened wildlife and plants" compiled by the U.S. Fish and Wildlife Service were collected adjacent to the site. However, specimens of three species designated by Indiana as endangered (Indiana Department of Natural Resources) were collected. These species are the eastern sand darter (*Ammocrypta pellucida*), the ribbon shiner (*Notropis fumeus*), and the blue sucker (*Cycleptus elongatus*). The prime habitat adjacent to the site for the sand darter would primarily be over sandy substrates on the western (Illinois side) of the river. Only a single specimen of this species was collected during the sampling effort. It is reported to be only locally abundant throughout its range, and its classification appears justified for the main stem of the Wabash River. The ribbon shiner is abundant throughout parts of its range, and it would appear to be classified incorrectly for the Wabash River. It is most likely that the low abundance of this species is due to the fact that the Wabash River is on the edge of its range. The blue sucker was certainly collected only infrequently adjacent to the site and, based on its past recorded abundance, is properly classified for the Wabash [7].

Nuisance Species—The carp and the gizzard shad are often considered undesirable when they occur in large numbers. They were included in the RIS list because of their common occurrence in the catch, but their numbers have not been high enough over the sampling period (1971 through 1978) for them to be construed as nuisance species. The nutrient levels in the Wabash River would be high enough to support nuisance populations of phytoplankton or aquatic macrophytes were it not for the high turbidity during most of the primary growing season. Because the physical processes in the river limit the full utilization of nutrient resources and, hence, development of the nuisance potential of the primary producers in the river, primary producers were not included on the list.

Commercial or Recreational Species—This category was only marginally applicable to this section of the Wabash River because of its low commercial or recreational use. Commercial interest is limited to a few local part-time fishermen who fish trotlines and occasionally D nets and who sell their catch to local small-town grocery stores. The river in this stretch is only occasionally fished recreationally by a few local townspeople, and most of this catch is used for household consumption. Therefore, the commercial or recreational considerations were not considered as important criteria for inclusion on the RIS list.

Species Important for the Maintenance of Community Integrity—The majority of the species were included because they were abundant in the river adjacent to the site. These species include several omnivores/scavengers (gizzard shad, carp, young-of-the-year and juvenile channel catfish), several forage species (emerald shiner, spotfin shiner, speckled chub, and bullhead minnow), and several top predators (bluegill, spotted bass, and adult channel catfish). In addition, several species (northern hogsucker, mountain madtom, logperch, rainbow darter, and greenside darter) were included as representative of species that are generally restricted to riffle habitats in the Wabash. Other riffle species, such as the mud darter and slenderhead darter, were much more commonly collected than the darters included in the RIS list, but very little life history or thermal data were available, and practicality precluded their inclusion.

Case History No. 3—Presque Isle Power Station, Marquette, Mich.

Power Plant Location

The Presque Isle Power Station is located in Marquette, Michigan, on Presque Isle Harbor, Lake Superior, just north of the mouth of the Dead River.

Cooling System Characteristics

At the time of the submittal of the Section 316(a) demonstration in 1976, the Presque Isle Power Station consisted of six coal-fired steam units. The thermal effluent from these units was the subject of an "absence of prior appreciable harm" demonstration. Units 7 through 9 were proposed as new once-through coal-fired units, and they were the subject of a predictive Section 316(a) demonstration.

Units 1 Through 6—Units 1 through 6 have a combined net generating capacity of 340 MW. The cooling water is drawn from Presque Isle Harbor

275 m offshore at a depth of approximately 2.4 m through an intake fitted with a velocity cap. Units 1 through 6 use approximately 611.34 m^3/min (161 500 gal/min) for condenser cooling [9].

The cooling water from Units 1 through 4 is discharged via a surface discharge canal, which terminates at the harbor shoreline about 150 m north of the mouth of the Dead River. Units 5 and 6 discharge via a submerged multiport diffuser located about 485 m offshore at a depth of 5.5 m. This discharge is located about 240 m south of the intake structure [9].

Typical spring plumes from the surface discharge of Units 1 through 4 covered 24.28 to 60.70 hectares (60 to 150 acres) (including plume from Dead River); summer plumes covered 12.14 to 24.28 hectares (30 to 60 acres), and fall plumes covered about 8.09 hectares (20 acres). These plumes were restricted to the upper 0.61 m (2 ft) of water column in most areas. Winter surface plumes usually covered less than 8.09 hectares (20 acres), but bottom plumes were also present as a result of temperature-related density changes. Bottom temperatures in the water sinking plumes were usually between 2 and 4°C (35.6 and 39.2°F). The thermal plume from Units 5 and 6 was very small and usually undetectable at the surface. The only noticeable temperature increases were in the multiport jets themselves [9].

Proposed Units 7 Through 9—The proposed units will be constructed on the same site as the existing units and will use the existing intake structure. These units will add 240 MW of net generating capacity to the facility and will increase the circulating water flow by 843 m^3/min. The cooling water will be returned to the harbor via a new submerged multiport diffuser, which will be located in about 9 m of water approximately 275 m south of the existing diffuser for Units 5 and 6 [9].

The thermal plume from Units 7 through 9 will be much smaller than the state-allowed 28.33-hectare (70-acre) mixing zone, and surface temperatures are projected to be around 0.56 deg C (1 deg F) above ambient temperatures [9].

Physical Setting

Presque Isle Harbor, also called Upper Marquette Harbor, is a natural indentation of the south Lake Superior shoreline. The harbor is approximately 2.7 km long north to south and 1.1 km wide west to east. The mainland of Michigan's upper peninsula forms its western and northern boundaries, while an 805-m-long breakwater separates the harbor from Lake Superior to the northwest. Picnic Rocks, a series of granite outcroppings, are generally considered to be the harbor's southern extremity [9].

An ore dock is located approximately 410 m to the west-southwest of the breakwater and extends about 500 m out into a dredged basin. Over the years, the harbor has been the recipient of numerous dredging operations.

The Dead River is the major natural drainage into Presque Isle Harbor, and it enters Presque Isle Harbor about 150 m south of the thermal discharge canal for Units 1 through 4. The flows in the river are entirely regulated by upstream hydroelectric dams, and, thus, there are times when flows are actually reversed, and harbor water flows into the river. The average river flow is approximately 5.66 m^3/s (200 ft^3/s) from September through March and 1.98 m^3/s (70 ft^3/s) from April through August. The river temperature is normally 2.8 deg C (5 deg F) above the ambient Lake Superior water temperature from September through March and 11.1 deg C (20 deg F) above ambient from April through August [9].

Background temperatures within the harbor vary greatly because of its shallowness, the influence of lake seiches, and input from the Dead River. Recorded data indicate that nearshore natural temperatures may fluctuate from a low of −1.67°C (29°F) in the winter to a high of 20.6°C (69°F) in the summer. Temperatures also may change as much as 5.6 deg C (10 deg F) in 24 h.

Drogue studies conducted by WAPORA, Inc., in 1975 through 1976 showed that overall water movements were extremely variable and often nondescript; however, this study did establish that a general clockwise gyration often existed in the harbor. During high winds (15 to 20 knots), surface and 3.05-m (10-ft) deep drogues moved at about 0.12 m/s (0.4 ft/s). Under calm conditions, drogue speed ranged from 0.03 to 0.06 m/s (0.1 to 0.2 ft/s) [9].

Substrate mapping in the summer of 1975 revealed that most of the harbor bottom was fine sand. There were, however, some small areas of coarse sand, granules, gravel, and cobble. In addition, a few outcrop areas containing granite boulders were located in the nearshore middle portion of the harbor. Winter storms in 1975 changed the shoreline and bottom areas drastically from those described earlier in the summer. For example, the outcrop boulder area just described was completely buried in sand after the storms [9].

Biological Setting

In overview, Presque Isle Harbor is a relatively productive area in Lake Superior, and it is important as a seasonal feeding, spawning, and nursery area for many of the valuable cold-water sport fishes that form the base of the recreational fishery in the lake.

A 1975 through 1976 study showed that densities of total phytoplankton in the harbor were variable from station to station and season to season. Densities at control stations (not affected by power plant operations) ranged from a low of 40 000 per litre in December to a high of 190 000 per litre in February. Densities in April and July were closer to 80 000 per litre. *Chroomonas acuta* was often the most common taxon [9].

Zooplankton populations in the harbor were most dense in the summer (about 40 000 organisms/m^3) and least dense in the spring (about 1000 organisms/m^3). *Keratella bostoniensis, K. longispina, Conochilus unicornis,* and *Bosmina longirostris* were the common species. The zooplankton populations were generally similar throughout the harbor during each season, except when recruitment from the Dead River altered the compositions in the northern portion. Changes in composition or reduction in population density resulting from the existing power plant operations were not evident in the samples collected [*9*].

Samples collected in 1975 through 1976 showed that the harbor macroinvertebrate community had a clumped distribution. The estimated mean densities ranged from 11 to 8385 organisms/m^2 (1 to 779 organisms/ft^2). In general, the densities were higher in the northern portion of the harbor, where there was some degree of organic enrichment. Although large numbers of tolerant organisms were present in this area, many intolerant organisms, such as *Pontoporeia affinis, Ephemerella aestiva, Ephemera simulans, Phylocentropis placidus,* and species of *Polycentropus* were also present. The overall harbor community was considered well-balanced and typical of nearshore lake areas [*9*].

Although 36 Lake Superior fish species were captured in Presque Isle Harbor, most were not year-round residents. Most were present for only two or three seasons of the year (coho salmon, chinook salmon) and some for only one (pink salmon). Some species that were present for more than one season were actually present in different life stages (lake trout). The following species were abundant in the harbor catch during at least one season in 1975 through 1976: coho salmon, rainbow smelt, lake trout, round whitefish, chinook salmon, brown trout, rainbow trout (steelhead), lake whitefish, white sucker, burbot, longnose sucker, slimy sculpin, mottled sculpin, yellow perch, and ninespine stickleback. Whitefish (at least three species), sculpins (four species), lake trout, yellow perch, burbot, and ninespine stickleback successfully spawned in the harbor in 1975 through 1976. Rainbow smelt and white suckers successfully spawned in the Dead River in 1975 through 1976 [*9*].

Representative Important Species—Table 4 is a list of those plants and animals considered representative in Presque Isle Harbor. The RIS list for Presque Isle Harbor was refined from a generic State of Michigan RIS list for Lake Superior based upon an evaluation of site-specific sampling data. Some fish species included on the state list were omitted because they were not collected in recent sampling efforts. These include lake sturgeon (*Acipenser fulvescens*), shortnose cisco (*Coregonus reighardi*), shortjaw cisco (*C. zenithicus*), blackfin cisco (*C. nigripinnis*), kiyi cisco (*C. kiyi*), bloater (*C. hoyi*), sauger (*Stizostedion canadense*), and Atlantic salmon (*Salmo salar*). Other state list species were omitted because of low relative abundance or infrequent occurrence of these species in the harbor. These include spottail

TABLE 4—*Plants and animals considered representative in Presque Isle Harbor.*

Common Name	Scientific Name
Rooted aquatic plants	*Chara* sp. and *Nitella* sp.
Plankton	
Macroinvertebrates	
White sucker	*Catostomus commersoni*
Longnose sucker	*Catostomus catostomus*
Lake herring	*Coregonus artedii*
Lake whitefish	*Coregonus clupeaformis*
Round whitefish	*Prosopium cylindraceum*
Slimy sculpin	*Cottus cognatus*
Mottled sculpin	*Cottus bairdi*
Spoonhead sculpin	*Cottus ricei*
Burbot	*Lota lota*
Ninespine stickleback	*Pungitius pungitius*
Rainbow smelt	*Osmerus mordax*
Yellow perch	*Perca flavescens*
Lake trout	*Salvelinus namaycush*
Rainbow trout	*Salmo gairdneri*
Brown trout	*Salmo trutta*
Coho salmon	*Oncorhynchus kisutch*
Chinook salmon	*Oncorhynchus tshawytscha*

shiner (*Notropis hudsonius*), emerald shiner (*N. atherinoides*), walleye (*Stizostedion vitreum*), johnny darter (*Etheostoma nigrum*), and brook trout (*Salvelinus fontinalis*) [9].

Threatened or Endangered Species—No species currently on the federal list of "endangered and threatened wildlife and plants" complied by the U.S. Fish and Wildlife Service is listed for Lake Superior. However, two species that the Michigan Department of Natural Resources considers "threatened" in the lake were collected in the harbor and were included on the RIS list. These are the lake herring and the spoonhead sculpin. Lake herring adults were collected in the harbor only in July and February. Lake herring larvae were collected in the spring, so it is assumed that they spawned in the harbor sometime during the winter. Spoonhead sculpins were collected in the harbor sporatically in low numbers throughout the year. It is assumed that they spawned in the harbor, since larvae were abundant in nearshore areas in the spring and early summer [9].

Nuisance Species—Rooted aquatic plants of *Chara* and *Nitella* species have the potential to develop nuisance populations should they occur in densities high enough to interfere with recreational fishing, contribute sufficient biomass during the annual die-off to upset the oxygen balance in the harbor, or upset the food chain balance by altering food availability and fish species competition. The distribution of these plants in the harbor appears to be limited by the organic content of the harbor sediments. The thermal effluent from Units 1 through 4 and the Dead River also contribute to a longer growing season. However, the present extent of rooted aquatic plant growth in the harbor is limited to a relatively small area in the north end, and the extent of growth is not projected to expand because of the physical limitations of the depth and substrate character. Therefore, because of the spawning habitat provided by the plants for some fish species and the abundance of macroinvertebrates living on or associated with the plants, the presence of the rooted macrophytes was judged to be beneficial on an overall basis [9]. No other organism groups or species were included on the list because of their nuisance potential.

Commercial or Recreational Species—Species of fish that were included on the list primarily because of their commercial (C) or recreational (R) value are listed in Table 5. The trout and salmon fishery in the harbor, particularly in the thermal plume areas, provides excellent sport fishing opportunities during much of the ice-free seasons and contributes to the local economy. The commercial fishery for the suckers and whitefish is not important in Presque Isle Harbor, but it is important lakewide.

Species Important for the Maintenance of Community Integrity—The plankton and the macroinvertebrate communities were included because of their importance as a food base for the large number of larval and juvenile fishes that utilize the harbor as an early-stage nursery and because of their importance as a food base for forage fishes. The primary thermal evaluation criterion would be the maintenance of normal seasonal population abundance patterns in synchronization with the occurrence of larval and juvenile fish populations in the harbor.

The slimy sculpin, mottled sculpin, burbot, and ninespine stickleback were included on the RIS list primarily because of their abundance and year-round residence in the harbor and resultant food-base importance. Thermal evaluation criteria would center on the maintenance of reproducing populations at levels sufficient to sustain predator pressures.

TABLE 5—*Commercial and recreational fish.*

Species	Type
White sucker	commercial and recreational
Longnose sucker	commercial and recreational
Lake whitefish	commercial
Round whitefish	commercial
Rainbow smelt	recreational
Yellow perch	recreational
Lake trout	recreational
Rainbow trout	recreational
Brown trout	recreational
Coho salmon	recreational
Chinook salmon	recreational

Discussion

The case histories that have been sketched in this paper illustrate that the concept of selecting species that are representative of aquatic communities for the purpose of applying thermal tolerance criteria is practical and workable. The degree to which the selected species are indeed representative is, however, directly a function of the completeness and accuracy of the available life history and local distribution information. Each of the described case history selection processes suffered to varying degrees from a lack of site-specific seasonal distribution data and from a lack of properly derived thermal tolerance data for different life history stages. A few specific examples will serve to illustrate the point.

The proposed intake and discharge locations for the nuclear units at Charlestown, R.I., are located in a rock-and-boulder-strewn open coastal area that is very difficult to sample with conventional fish sampling gear. Therefore, although the species selected may well represent the general fish community at the site, no data are available that specifically describe the seasonal and diurnal abundance distribution of various life stages of these species immediately in the zone of thermal influence. The proper application of thermal criteria requires that the biologist have a good understanding of the seasonal occurrence of fish in the thermal influence zone. The scientist faced with a lack of such specific information is forced to make judgments based on an understanding of similar systems or, often, from inferences in the literature regarding life history data from another water body or even from closely related species. This approach is often the best available when faced with the prospect of rendering judgments within regulatory agency or plant construction time schedules, and the decisions in some cases, such as this particular example, may not be difficult because of the relative size of the water body and of the plume. However, there is little doubt that good information on the fishes within the proposed zone of thermal influence would change perceptions of the relative importance of various species. The soundness and accuracy of the scientist's judgments on thermal effects would certainly be strengthened with the availability of such data.

Another example of the need for thorough site-specific field sampling information is drawn from the Breed Plant case history. Several years of intensive sampling in the Wabash River with a variety of gear by several investigators failed to document the importance or, in some cases, even the presence of shovelnose sturgeon. Had the RIS list been prepared prior to the results of the 1977 WAPORA, Inc., study, this species would not have appeared on the list. However, when gill nets were placed in the highest current velocity chute areas in the center of the river, the catch data indicated that, shovelnose sturgeon were one of the dominant species in the fish community. In this case, simply fishing commonly used collecting gear in an unconventional and difficult-to-sample habitat completely altered the perception of species dominance that had been developed over several sampling years.

Scientists investigating large water bodies need to be continually aware of the shortcomings of the available literature information and of the limits of their understanding of the life histories of the animals present in such difficult-to-sample systems. Regulatory agency and utility company scientists involved in the early planning stages of power projects that may involve thermal effluent demonstrations under Section 316(a) should carefully consider the current state of knowledge concerning the biology of the receiving water body in designing preoperational site-monitoring studies. Sampling regimes should be designed to be thorough enough to allow an evaluation of effluent location and design options with some assurance that the organisms collected are indeed representative of the habitats sampled. Consideration should also be given to multiyear sampling programs because of the known variability of aquatic populations over time. The need for an understanding of this variability is particularly important when spawning success is considered. For example, some of the large river fishes, such as the shovelnose sturgeon, bigmouth buffalo, and paddlefish, apparently do not spawn every year, because of their relatively long life spans, delayed maturity, and particular flow and substrata requirements. In situations where these and similar species may be important, multiyear sampling programs may be a necessity rather than an option.

Another weakness in the RIS predictive evaluation process is the lack of thermal tolerance data for certain life stages of the selected species or, in the case of several species listed, an almost total lack of such data. The increasing emphasis on the use of cooling towers for new power plants by regulatory agencies has contributed to a general slowdown in research on thermal tolerances due to the reluctance of utilities to become involved in the uncertainties of the Section 316(a) process. It is certainly true, however, that there are power plant sites available where once-through cooling may be the least costly environmental option. In such cases, where the RIS data contain poor or incomplete information on thermal tolerances, the inclusion of laboratory bioassay studies should be an integral part of the total Section 316(a) procedure. The availability of properly derived thermal tolerance data would seem to be essential to the formation of sound judgments.

Conclusions

The representative important species approach to evaluating proposed thermal sources is practical and can provide realistic judgments, provided that adequate data are available on which to base the selection of such species. Weaknesses in the process occur when inadequate life history, seasonal distribution, or thermal tolerance data are available when the selection of the RIS is made. Properly designed field and laboratory studies are essential to provide an adequate decision data base. Planning for Section 316(a) demonstrations for new sources should begin early in the site selection and design phases because of the potentially extended nature of the research and because of the need for early regulatory agency input and decisions.

Acknowledgments

The author wishes to acknowledge the substantial technical contribution of G. R. Finni, J. T. Hatch, and T. N. Seng to the primary documents on which this paper is based. The Charlestown, R.I., Nuclear Plant studies were done under contract with the U.S. Environmental Protection Agency. Financial support for the studies at Breed Plant have been provided by American Electric Power Service Corp. and by Indiana and Michigan Electric Co. Studies at the Presque Isle Station were funded by the Upper Peninsula Generating Co. and Cleveland Cliffs Iron Co.

References

[1] "Interagency 316(a) Technical Guidance Manual and Guide for Thermal Effects Sections of Nuclear Facilities Environmental Impact Statements," Office of Water Enforcement, Permits Division, Industrial Permits Branch, U.S. Environmental Protection Agency, Washington, D.C., May 1977.

[2] "Guide for Demonstrations Under Section 316 of the Federal Water Pollution Control Act Amendments of 1972 (P.L. 92-500)," Division of Water Quality, Minnesota Pollution Control Agency, Roseville, Minn., July 1974.

[3] "Proposed Guidelines for Demonstrations Under Section 316 of Public Law 92-500," Water Resources, Bureau of Water Management, Michigan Department of Natural Resources, Lansing, Mich., May 1974.

[4] "Water Quality Appendix, Draft Environmental Impact Statement, Proposed Charlestown, Rhode Island, Nuclear Power Plant," Wapora, Inc., Washington, D.C., May 1978.

[5] "316 Demonstration Pilgrim Nuclear Power Station Units 1 and 2, Boston Edison Company," Stone and Webster Engineering Corp., Boston, 1975.

[6] Prakash, A., *Journal of the Fisheries Research Board of Canada*, Vol. 24, 1977, pp. 1589-1606.

[7] "Additional 316(a) Considerations, Breed Plant," technical report submitted to the American Electric Power Service Corp., Wapora, Inc., Washington, D.C., March 1978.

[8] "Continuing Ecological Studies of the Ohio River," technical report, Wapora, Inc., Washington, D.C., January 1974.

[9] "316(a) Demonstration for Proposed Units 7, 8, and 9, Presque Isle Power Station, Marquette, Michigan," technical report submitted to the Upper Peninsula Generating Co., Wapora, Inc., Washington, D.C., October 1976.

L. E. Sage[1] *and M. M. Olson*[1]

Power-Plant-Related Estuarine Zooplankton Studies

REFERENCE: Sage, L. E. and Olson, M. M., **"Power-Plant-Related Estuarine Zooplankton Studies,"** *Ecological Assessments of Effluent Impacts on Communities of Indigenous Aquatic Organisms, ASTM STP 730,* J. M. Bates and C. I. Weber, Eds., American Society for Testing and Materials, 1981, pp. 68–88.

ABSTRACT: Increased emphasis is being placed on formulating a particular study design that will generate the greatest amount of relevant information for a particular problem. This is a case history of a study design as it evolved from one level of information to another, yielding both quantitative and biologically meaningful information.

In-plant studies examining the effects of entrainment on zooplankton and field studies examining zooplankton abundance, composition, and distribution in the Chesapeake Bay in the vicinity of Calvert Cliffs Nuclear Power Plant (Baltimore Gas and Electric Co.) have been conducted from 1974 to the present. The evolution of these studies, with particular emphasis on design and statistical treatment, is discussed. Entrainment study designs evolved from discrete sampling episodes at 4-h intervals over 24 h to a time-series sampling design in which sampling took place every 30 min over 24 and 48-h periods. The near-field study design and sampling methods have included replicated net tows, using 0.5-m nets, and replicated and nonreplicated pumped samples, using a high-speed centrifugal pump. The relative collecting characteristics of sampling methods are described; study designs and applicable data analysis techniques are evaluated with respect to their ability to detect plant effects within defined limits of personnel and budgetary constraints.

KEY WORDS: zooplankton, estuarine, entrainment studies, environmental impact assessment, sampling methodologies, statistical treatment, ecology, effluents, aquatic organisms

Environmental assessment studies have undergone a considerable evolution in philosophy of purpose and direction. Initially, these studies were descriptive and qualitative, relying on the trained eye and the objectivity of the biologist. With greater federal regulations imposed on industry, greater personal detachment from the study was required of the biologist, and the era of rigid quantification of results appeared. Neither of these

[1]Director and biologist, respectively, Benedict Estuarine Research Laboratory, Academy of Natural Sciences of Philadelphia, Benedict, Md. 20612.

strategies was satisfactory, for separately, they served to address only isolated segments of the problem posed. If these strategies are combined, they complement each other. This constitutes the more recent phase of environmental assessment, emphasizing a more balanced approach of quantification and biological understanding.

A critical aspect of impact assessment is the generation of data that are meaningful in addressing an environmental question from both quantitative and biologic perspectives. The data should be able to withstand rigorous statistical treatment, but, more important, should provide an understanding of the dynamics of the aquatic assemblages under study. Designs that concentrate on one aspect are unbalanced and frequently lead to misdirected or erroneous conclusions in assessing the extent and consequences of impacts.

In our experience of assessing the impact of a large electrical generating plant on an estuarine zooplankton community, the study design evolved from one preoccupied with quantification to one more balanced, displaying a greater awareness of the biology of the system under study. This assessment program, initiated in 1974, was continually revised in design, analysis, and methodologies, ultimately resulting in more informative studies. Iterative process in study development is time-consuming and thus resource-consuming but occasionally is necessary to achieve an adequate understanding of the biological system under study when preparing an impact assessment.

The objective of our study was to evaluate the environmental impact of operation of a large nuclear power plant on the zooplankton assemblage of adjacent Chesapeake Bay waters. This assessment focused on (1) the zooplankton in the receiving waters and (2) the zooplankton pumped through the plant (entrainment). Both aspects of the study consisted of a qualitative evaluation describing the organisms present, when they were present, how they were distributed, how they were affected by plant operations, and a quantitative phase enumerating the densities of each species and age class, their relative distribution (vertically, temporally, and spatially), and the extent of the impact.

Specifically, the objectives of the combined program include the following:

(*a*) to characterize the species and age class composition, abundance, and spatial and temporal distribution of near-field zooplankton, including larval stages of shellfish throughout an annual cycle;

(*b*) to examine the population dynamics of the summer zooplankton community, specifically the weekly fluctuations in abundance and structure and to investigate the relationship of these to physicochemical and biological factors;

(*c*) to provide quantitative data on entrained zooplankton by species and by age class where possible;

(*d*) to provide for each species an estimate of entrainment survival; and (*e*) to provide information on periodic phenomena over 24-h periods (or longer) resulting from diel, diurnal, and tidal variation and on the possible correlation between periodic phenomena and entrainment effects.

There are many considerations that must be weighed in designing a zooplankton assessment program, particularly in estuarine waters. An estuary is dynamic, especially in a hydrographic sense, and this feature has a profound effect on any "planktonic" assemblage. Therefore the spatial, vertical, and temporal distribution is dependent on the circulation patterns and the salinity and thermal regimes, which affect vertical density characteristics. Along a horizontal axis, the zooplankton are aggregated in patches where their densities can exceed background (interpatch) densities by more than fivefold, a phenomenon that can introduce a large source of variation to any quantitative data. In addition to spatial variability, extensive variability corresponds to tidal, diurnal, and diel periodicities. Vertical stratification of zooplankton results from behavioral characteristics of some of the organisms as well as from hydrographic mechanisms. Finally, fluctuations in community density as well as in community structure exist on a seasonal basis. These distribution and periodic density patterns make *reliable* assessment complex and place increased importance on the design phase of the zooplankton program.

Due to the distribution characteristics of zooplankton, especially in an estuarine environment, different sampling techniques have distinct advantages and disadvantages. Essential requirements to be considered for methods of sampling zooplankton communities are replicability, sampling velocity, volume sampled per unit of time, organism avoidance, and, in special cases, pump-induced mortality. All methodologies should be tested and their characteristics documented prior to adoption into assessment or other research programs.

The zooplankton collections were coordinated with water chemistry studies and phytoplankton and zooplankton entrainment studies. Zooplankton near-field surveys, which were conducted at night, were scheduled concurrently with macroplankton surveys and on the night before or following phytoplankton productivity and finfish field studies. This holistic approach of integrating the various elements will yield the greatest amount of information for management, regulatory, and research purposes.

Study Site

The plant is located along the middle reach of the Chesapeake Bay approximately 165 km north of the mouth of the bay (Fig. 1). This site is located immediately north of the mouth of the Patuxent River, which is the first of a series of major tributaries entering the lower bay area. The area of the bay north of the Patuxent River is dominated by one river system, the

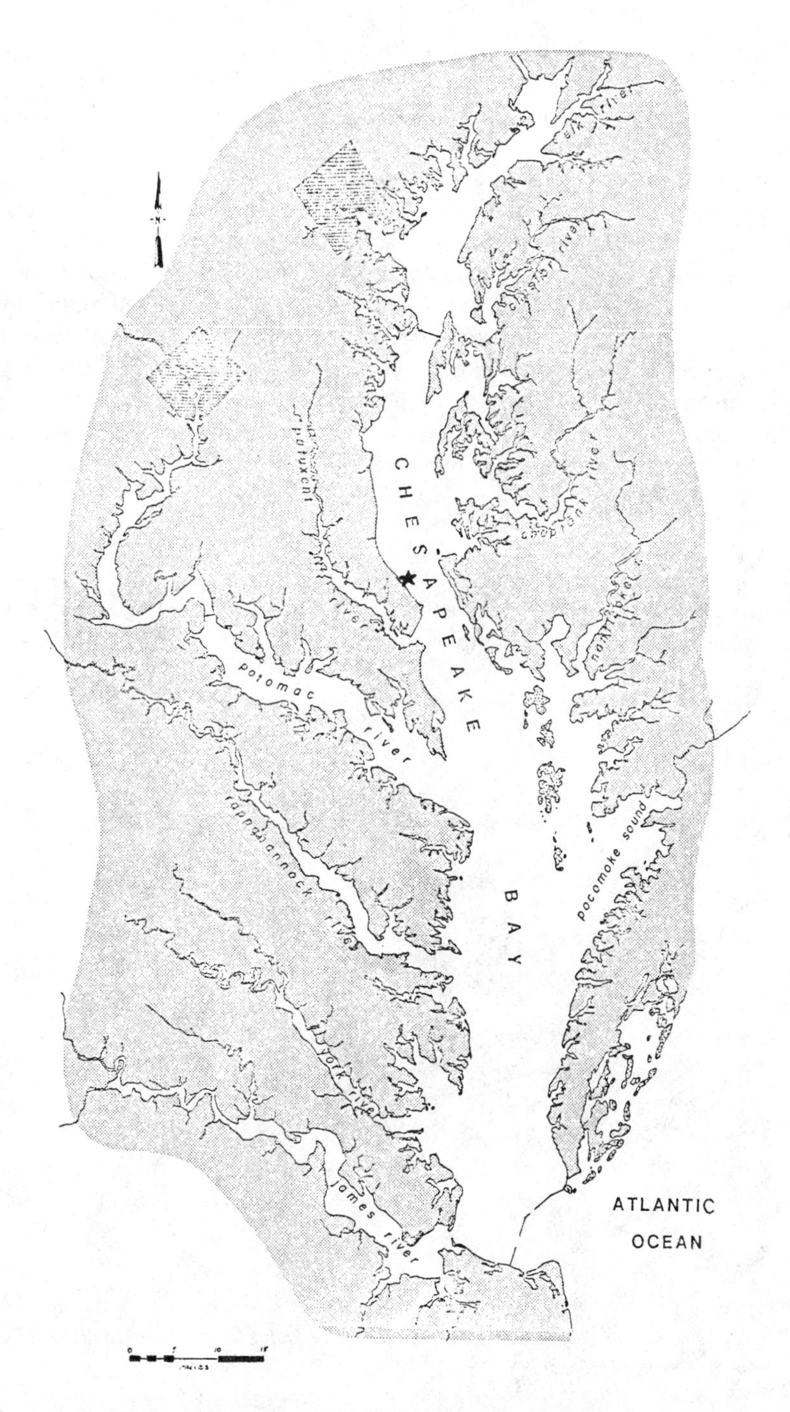

FIG. 1—*Map of Chesapeake Bay indicating the study location (★).*

Susquehanna, which is the largest tributary to the entire bay system. The Chesapeake Bay and its tributaries support areas of intense urbanization and industrialization, which use the waters of the Chesapeake Bay for waste release.

Figure 2 is a schematic diagram of the facility. The intake area includes a large forebay bounded by a curtain wall structure that extends to a depth of 9 m in a water column of 15 m. This plant uses 156 m^3/s (5500 ft^3/s) of bay water, elevating the entrained water to a ΔT of 5.5°C. The entrained water experiences a residency time of 4 min before it is discharged to the bay through a submerged high-velocity discharge. At present, this plant does not apply biocides to the main cooling water system, but rather controls fouling in the condensers via the Amertap mechanical cleaning system.

The methods and some of the results of these studies are only partially described here. The results are presented and discussed in detail in other publications [*1*-*5*].[2]

Entrainment Studies

The earliest design of the entrainment studies consisted of seven, and subsequently five, replicate samples taken at each station and depth at 4-h

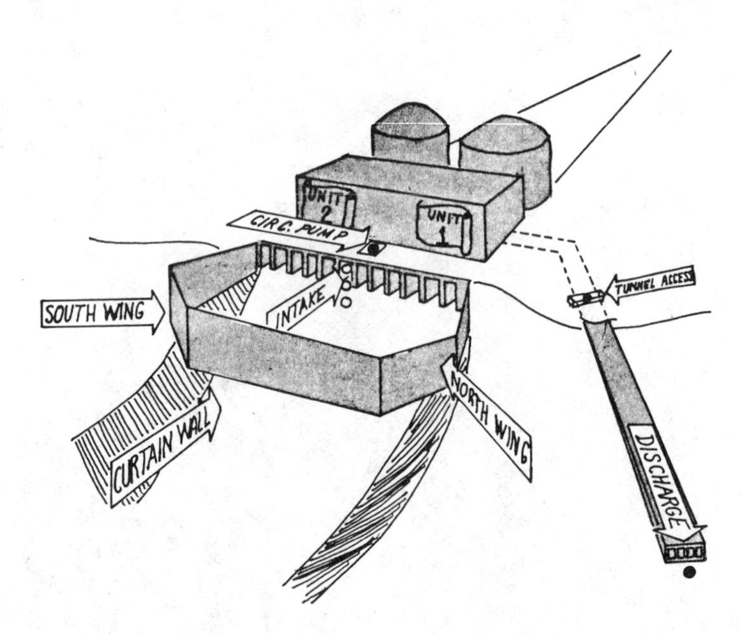

FIG. 2—*Schematic of the generating facility with the sampling locations indicated (●). The discrete depths sampled at the intake are also shown (○).*

[2]The italic numbers in brackets refer to the list of references appended to this paper.

intervals, six times during a 24-h period. Means of the replicates were used as the units for analytical purposes. However, the results were not totally satisfactory because of the large variation between sampling periods, a phenomenon also noted at other sites [*6, 7*]. However, in this study, the agreement within replicates was good (Fig. 3). Since only a few points were produced through 24 h from the many replicated samples, recognization of a "real" biological phenomenon could be confused with sampling error or sample variation. The data set was void of multiple points through time, which serves to verify population fluctuations. For example, dense populations of zooplankton occur in relatively large spatial patches in the Chesapeake Bay. Large fluxes in the density of entrained organisms due to such patchiness are difficult to distinguish from sampling error. In addition, the early studies suggest that zooplankton survival itself can fluctuate significantly over time (that is, between sampling periods in 24 h and month to month depending on species composition densities) (Fig. 4).

An entrainment study design should take into account both long- and short-term variability. Because long-term variability exists, it is necessary to make specific abundance and survival estimates at as many different times of the year as is economically feasible. Furthermore, it is most desirable to employ a design that can identify and describe quantitatively any systematic variation in the data. When this variation is removed by appropriate sta-

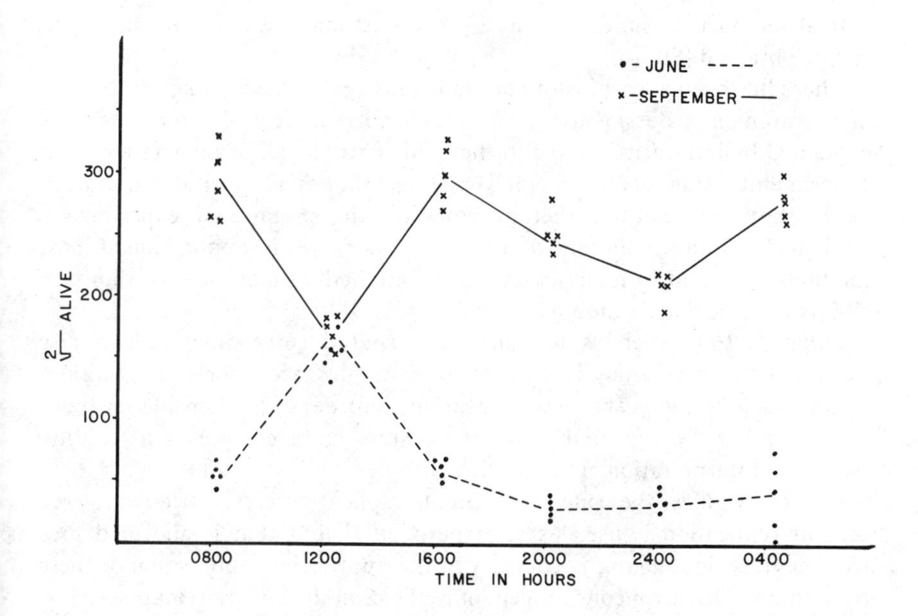

FIG. 3—*Total live zooplankton recorded at the intake during June and September 1976. The data have been smoothed by square root transformation of the total number alive.*

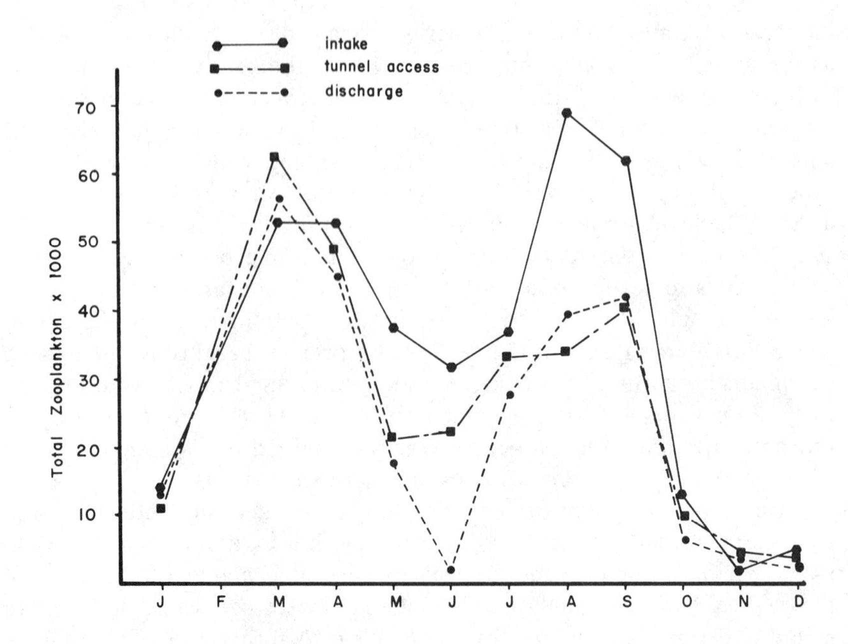

FIG. 4—*Total zooplankton (alive and dead) collected at the three stations during the 1975 monthly entrainment studies.*

tistical methods, a valid and more precise estimate of entrainment effects can be obtained [*8*].

There has been a long history of inadequately designed studies in the area of entrainment assessment [*8*]. Although a great deal of effort has been expended by innovative people in the field, virtually all of the studies apply experimental statistical principles and techniques to a situation that is observational in nature, thereby violating the premise of experimental statistical techniques that the design be random [*9*]. The application of these methods was, and in many cases still is, applied although there is no possibility of achieving a random design.

Other factors, such as temporal and spatial correlations, which also invalidate the usual analysis of variance techniques, are ignored. In addition, factors such as the year, season, location, and depth are considered treatments that make the analysis of the results more complex and confuse biological interpretation of the results.

In these studies, the concept of sample replication is not strictly adhered to, but replication is used as an experimental unit that is allocated to a treatment, which, again, is contrary to the appropriate application of these techniques. This erroneous concept of replication then leads to inappropriate statistical analysis of the data. These multiple measurements should be considered observational units, and an average value of these measurements should be the actual statistical unit used in statistical analyses. In our 1975 and 1976 entrainment studies, therefore, the replicates were averaged, which resulted in only six observations per 24-h period. Unhappily, precise estimates of survival are not possible from so few observations. An even greater disadvantage of using a design incorporating discrete data sets is the difficulty in interpreting the biological significance of such data.

In a team effort with statisticians and biologists, a systematic sampling design was developed that alleviated most of the problems just described to arrive at a reasonable estimate of survival of entrained organisms. Further, this design permitted the generation of data that was more useful for biological interpretation, such as the detection of periodic zooplankton behavior. Therefore, in 1977, we implemented a time series entrainment study designed to replace the discrete data sets generated in 1975 and 1976.

Statistical Treatment

We know the most characteristic property of a complex system such as an aquatic ecosystem is that functional relationships between the various component systems are nonlinear. A crucial characteristic of nonlinear systems is a disposition toward periodic behavior or a cyclic organization in space, time, or both. Information on the frequency of all the cycles of a system enhances the biological interpretation (Fig. 5).

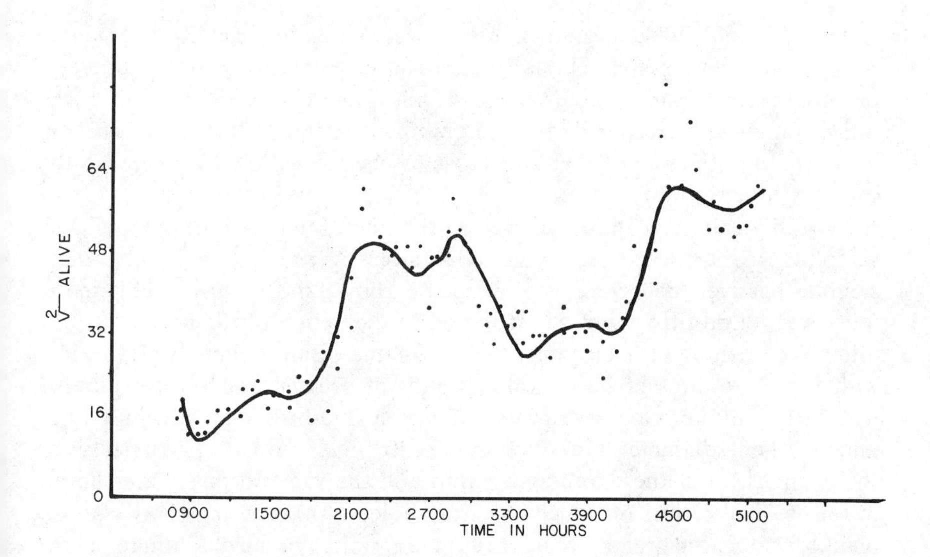

FIG. 5—*Abundance of live polychaete larvae collected 6-7 June 1977 at the intake, illustrating the cyclic behavior of the individual samples and the resultant curve based on the smoothed data.*

A basic analytical tool for the time series design is power spectral analysis, a form of analysis of variance (ANOVA) that partitions the variance of the time series into the contributions at various frequencies. Spectral analysis is the statistical method by which the frequency and relative magnitude of the cycles are identified and defined [*10*]. Periodograms may be computed for each variable of interest, such as species density, live/dead ratio, and total density. A periodogram depicts the simultaneous least-squares estimate of a finite number of sine and cosine functions of different frequencies to a data set and can be used to describe the periodic components of the data, indicating the intensity at the frequencies corresponding to the various periodicities. Next, sinusoidal functions corresponding to each of the important periodicities found by spectral analysis are fitted to the data.

The residuals, which are defined as the observed data minus the values predicted by the functions, can then be tested for random error. If the residuals are random error, then a valid estimate of precision can be calculated. For our purposes, this variance estimate can be used to obtain statistically valid confidence intervals for estimates of entrainment effects.

In order to enhance the visual impact of low-frequency periodicities in the data, a smoothing technique was applied to the square root transformation of the raw data. Smoothing is an integral part of the time series analysis in order to define patterns.

Entrainment Design

The 1977 and 1978 entrainment design was altered to a collection of single, unreplicated samples at the intake and discharge points every 30 min throughout a 48-h time span. The discharge samples were collected 4 min after the intake samples, corresponding to the approximate transit time of the cooling water through the plant. The samples were collected with a 30.3-litre (8-gal)/min diaphragm pump.

Mortality caused by passage through the diaphragm-type pump used for sampling was analyzed. The pumped samples were compared with water samples passively collected in submerged carboys, and no discernable mortality was found that could be attributed to the action of the pump. In addition, the fishing efficiency of this low-volume pump compared favorably with that of pumps of larger volume and, of greater importance, pumps with larger intake velocities (a 5-cm diameter, 300-litre/min vacuum pump and a 10-cm diameter, 760-litre/min centrifugal pump). The results of the comparison of the diaphragm pump and the vacuum pump are shown in Fig. 6. The results of the comparison indicate that the mortality of zooplankton appeared greater in the vacuum pump than in the diaphragm pump (Fig. 6*a*). The lower-velocity diaphragm pump did not increase avoidance by the particular microzooplankton community that we were studying (Fig.

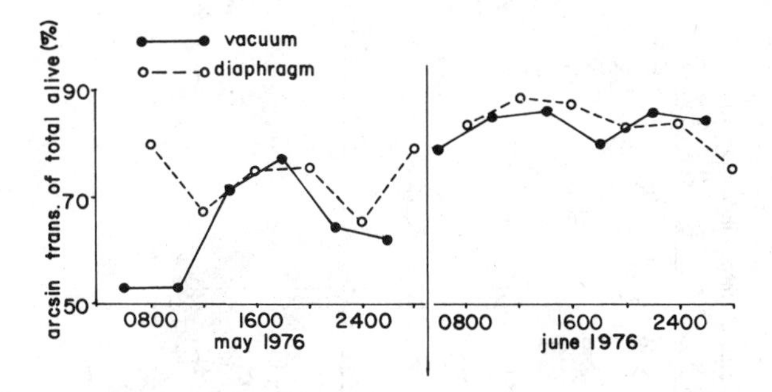

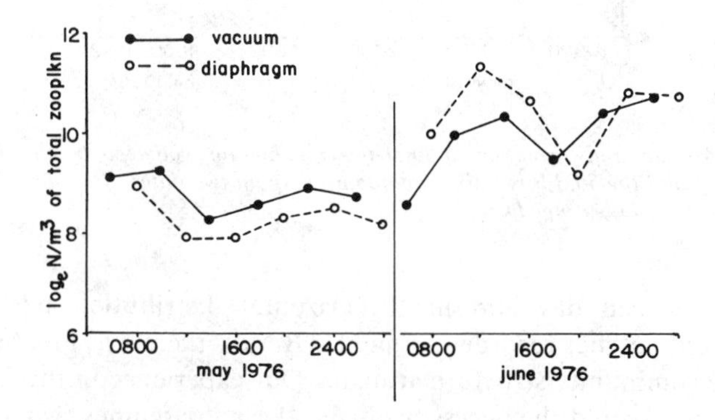

FIG. 6—*(a) Comparison of zooplankton mortality (percent alive) in samples collected with the 30.3 litre/min diaphragm pump and the 300 litre/min vacuum pump. (b) Comparison of zooplankton density (*$\log_e$ *N/m*^{3}*) in samples collected with the 30.3 litre/min diaphragm pump and the 300 litre/min vacuum pump.*

6*b*). Figure 7 indicates the variability between replicate samples of the vacuum pump and the slower diaphragm pump.

It is also important to minimize the effects of short-term spatial and temporal patchiness, which was found in the previous studies. Temporal patchiness is characteristic of zooplankton populations. Because micro-zooplankton are known to aggregate in natural waters, as noted in coastal waters by Sameoto [*11*] and Haury [*12*], the design of any study must accommodate this characteristic. Cassie [*13*] examined the results of plankton studies of several areas by different authors and determined that the coefficient of variation frequently is between 22 and 44 percent, although much higher values were observed. Most studies conclude that the typical patch size is small, less than 1000 m. In addition, Haury indicated a substantial

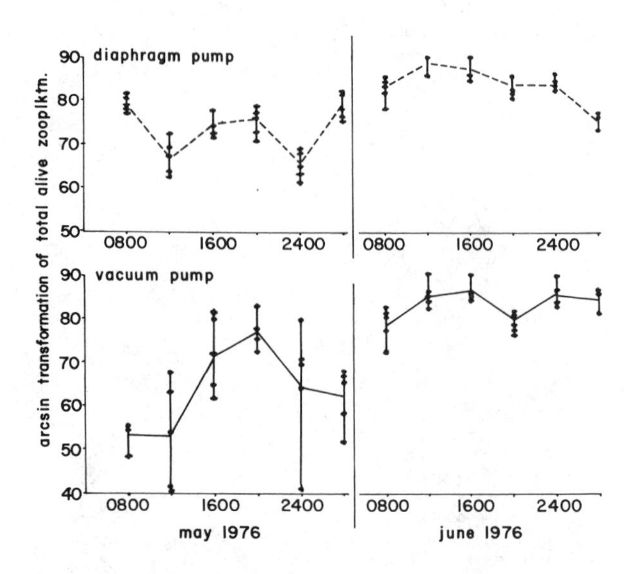

FIG. 7—*Arc sin transformation of total live zooplankton collected by the 300 litre/min vacuum pump and the 30.3 litre/min diaphragm pump on two dates. The individual replicate values are plotted on the range bars.*

variation between day and night horizontal distribution patterns, with smaller-sized patches recorded at night. Overall, there was greater variation in relative community structure at night. Our experience in the Chesapeake Bay has underscored the necessity of considering patchiness when designing a study. We have encountered markedly heterogeneous zooplankton assemblages with densities within patches exceeding the background densities fivefold. These examples serve to emphasize the need for new approaches to the ecological study of natural systems. Figure 7 illustrates the result of sampling over a short interval (~3 s, vacuum pump), in which individual patches are sampled, as opposed to sampling over a longer interval (40 s, diaphragm pump), in which the patches are integrated in each sample.

To minimize the effects of this phenomenon, the elapsed time for collecting the 20-litre sample was extended from 40 s (the maximum volume of the diaphragm pump) to 6 min, which has the effect of integrating small temporal patches and produces a more representative sample. To accomplish this, the flow from the pump was split using a polyvinyl chloride (PVC) custom splitter, which divides the flow into a 60/40 ratio, with the lesser volume used to fill a 20-litre carboy in increments of one-third of the total volume every 2 min, with the total volume being pumped at the end of 6 min. The 20-litre samples were then differentially stained, using a neutral red vital dye in a concentration of 1:150 000, to determine the extent of morality within the plankton community. The samples were allowed to incubate for 1 h, during which the temperature was noted periodically. Following the incubation period, the samples were filtered through a 72-μm mesh net. In the laboratory, the organisms were counted and identified as to species, with the life stage and the number of organisms per cubic metre tabulated. Vitality was determined by the coloration of the organisms. Organisms that took up the dye were counted as alive at the time of collection; those that did not take up the dye were considered dead.

Over all the years of study at this power plant, the primary entrainment effect has been a reduction in numbers of organisms between the intake and discharge stations (cropping) rather than differences in relative numbers of dead organisms between the intake and discharge points. The mean number of zooplankton at each station during each month of the 1975 study is plotted in Fig. 4. The graph illustrates the increased cropping of these organisms during the months of May through September, with the cold-water months showing little effect of entrainment. In Fig. 8, comparison of the live/dead ratios at the intake with those at the discharge point demonstrates a slight reduction in the percentage of zooplankton alive at the discharge when compared with the mortality occurring in the natural community (as measured at the intake). This live/dead reduction through the plant is usually less than 10 percent. The cropping effect is approximately 30 percent for the

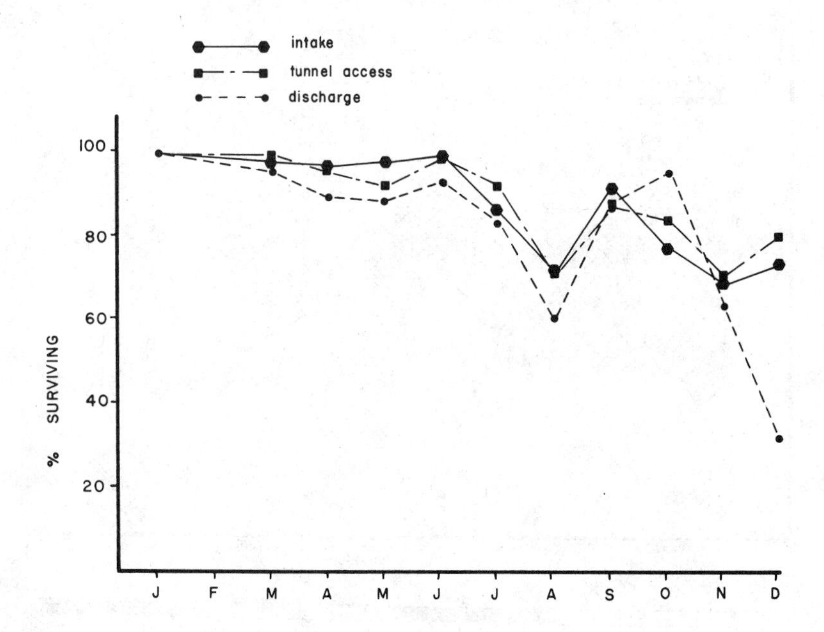

FIG. 8—*The alive/dead ratio of total zooplankton expressed as percent alive at the three stations in 1975.*

year, although, as mentioned previously, a precise estimate of survival was impossible with the original discrete data set design.

The reduction in the number of organisms was noted again in 1976, and the pattern of temporal variation in the cropping effect paralleled that established in 1975. Analysis of the data by species, and in specific cases by age class, indicated that the cropping effect was greatest on the copepod nauplii. Similar trends, though not as pronounced, were also observed for the copepodite stage and adult *Acartia tonsa*, a calanoid copepod that is the dominant summer zooplankton species in the Chesapeake Bay. A review of the 1975 to 1976 study results left many questions unanswered and hastened the formulation of a new, more responsive design.

The results of the 1977 entrainment studies using time-series analysis indicated a cyclic pattern of cropping and mortality ranging between 24 and 52 percent, with a higher incidence of cropping during the nighttime hours when the greatest densities of zooplankton were entering the plant (Figs. 9, 10, and 11). Moreover, as was noted in the discrete data set design, the greatest effect of this reduction was noted in the juvenile stages of *Acartia tonsa* (Fig. 10). A comparison of the live/dead ratio at the intake and discharge stations indicated only a very slight increase in the numbers of dead organisms at the discharge.

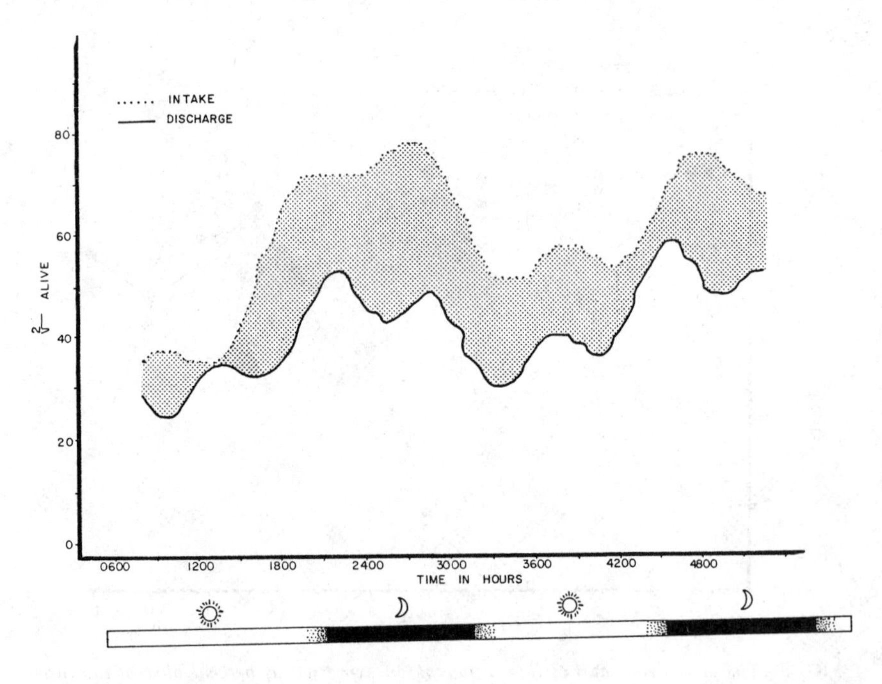

FIG. 9—*Smoothed data curve of alive dye-sensitive organisms (dye-specifics) at the intake and discharge stations (June 1977).*

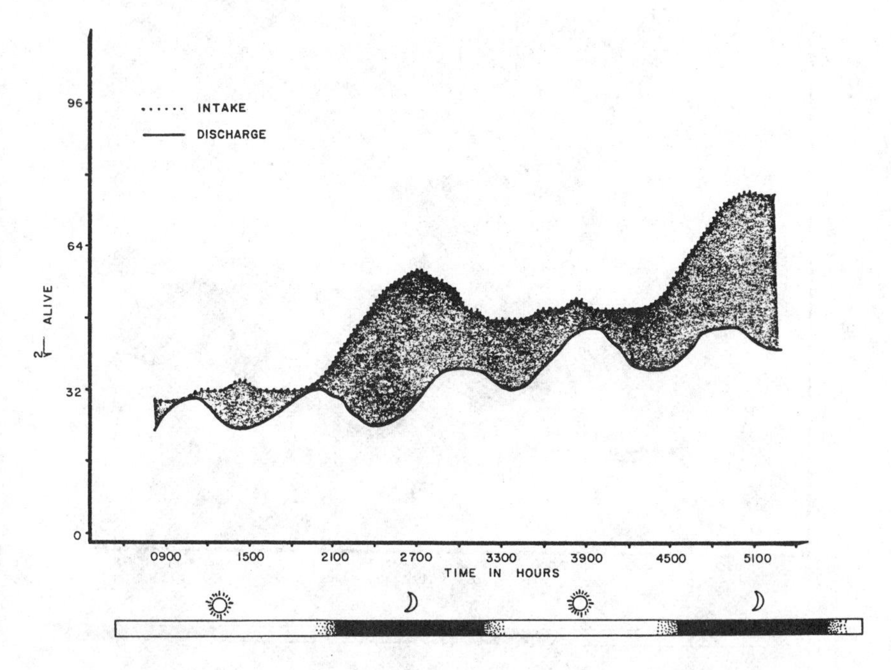

FIG. 10—*Smoothed data curve of copepod nauplii at the intake and discharge stations (June 1977).*

Spectral analysis on data collected over the 48-h span indicated a diel periodicity in the densities of most entrained organisms, with the density maximum occurring during the nighttime hours. In contrast, the analysis revealed that cirriped (barnacle) nauplii consistently reach peak densities during the daylight hours (Fig. 12). Periodicities of 6 and 12 h, corresponding to tidal fluctuations, were also noted (Table 1). Identification of these periodic components gave meaning to the variation in entrained zooplankton densities throughout the sampling period, a condition that was difficult to interpret with the previous entrainment designs.

We could also demonstrate that survival had periodic components. Entrainment effects were greater during the night, with the greater survival observed during the daylight hours. Tidal periodicity was evident for most species, indicating a variable cropping effect depending on tidal stage. Correlating cropping losses with a specific tidal stage was not attempted, since it would have required a sampling program of at least 96 h.

Although we can obtain information from examination of the densities of total zooplankton in these studies, patterns are obscured by the inclusion of many species and life stages, all of which exhibit the behavior characteristics of their own group. Since our ultimate goal in these investigations is to gain a better understanding of the effects of entrainment and to be able to

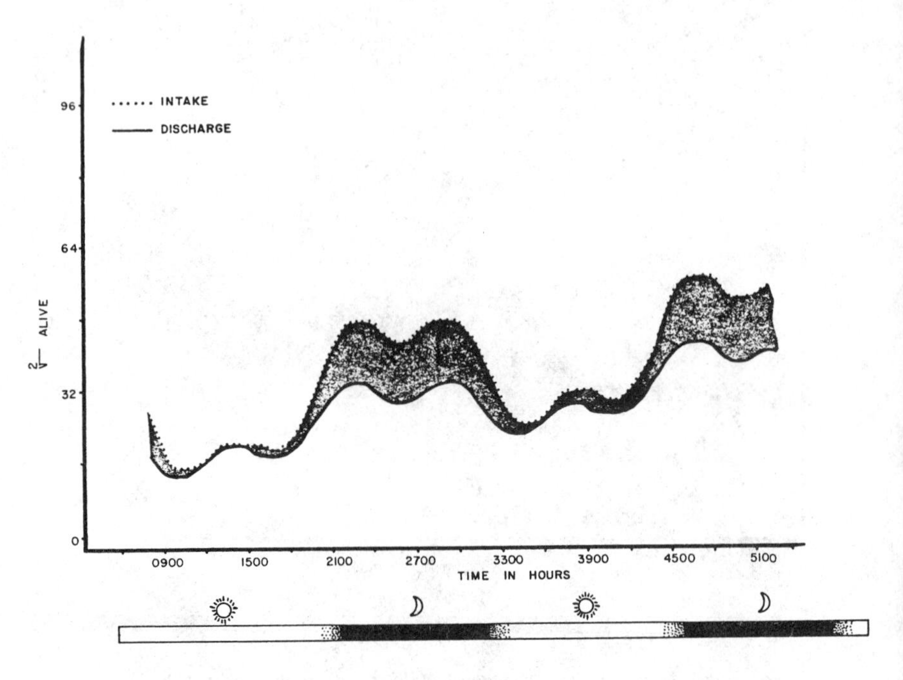

FIG. 11—*Smoothed data curve of polychaete larvae at the intake and discharge stations (June 1977).*

relate this to the total impact of power plant operations on the structure of the near-field biota and its associated food chain, we cannot overemphasize the importance of identifying precisely which organisms and life stages are affected. For this reason, periodograms and plots were constructed of each species of interest, especially other dye-sensitive species, as well as of the life stages of the dominant calanoid copepods.

It is evident that there are many advantages to using the time-series design, not the least of which is that it meets the criteria of the appropriate statistical tests, a glaring deficiency in any discrete data set design. From a biologist's viewpoint, moreover, it is particularly useful to have increased data that have greater biological significance to aid in the interpretation of the biologic phenomena.

Near-Field Studies

The near-field studies have also evolved over the years since the program was initiated in 1974. The immediate objectives were to characterize the zooplankton community in the area, the seasonal changes in the community, the vertical distribution, and distribution throughout the plant site and nearby stations. Most important was the determination of whether effects

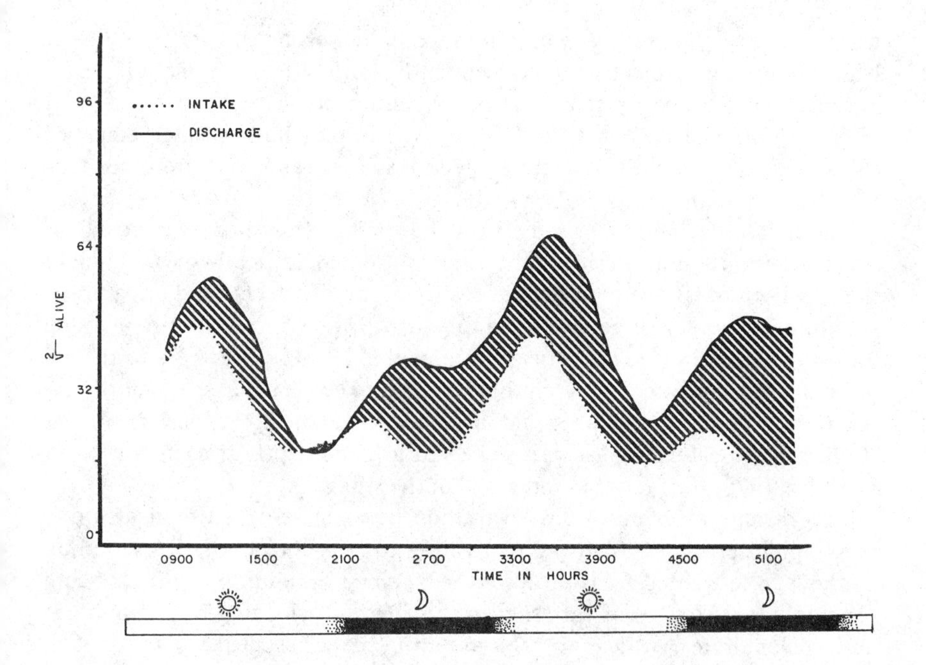

FIG. 12—*Smoothed data curve of cirriped nauplii at the intake and discharge stations (June 1977).*

TABLE 1—*Periodicities of zooplankton abundance determined by spectral analysis.*

Organism	Period (Cycle) Length, L		
	June	August	September
Polychaete larvae	48, 24, 8	24. 10	24
Copepod nauplii	24. 12	48, 24, 12, 6	24, 16, 12
Cirriped nauplii	24. 12	48, 24, 16	24, 16
Acartia copepodite	48, 24, 16	24, 16, 12	48, 24
Acartia tonsa	...	24, 16, 12	48, 24

from entrainment or from the large-volume heated effluent of the power plant could be detected in near-field populations.

The initial study design included a station in the vicinity of the plant and two reference stations north of the plant. Stations were not established south of the plant because of construction activity 6.5 km (4 miles) below the plant. Farther south, the inflow of the Patuxent River, which is biologically distinct from the study area, provides further complicating influences to the sampling sites.

All the routine near-field sampling efforts were carried out at night, when vertically migrating zooplankton are higher in the water column. Night sampling is inconvenient, frequently difficult, and sometimes hazardous, but its importance cannot be overemphasized. Analysis of day and night samples have shown that the night samples have greater densities (Figs. 9 through 12) and a greater species richness and thus are more representative samples.

The sampling methods used in the first years consisted of a string of three 0.5-m, 150-μm mesh plankton nets oriented vertically on a chain so that the bottom (10 m), mid-depth area (5 m), and surface could be sampled simultaneously. These nets were equipped with flow meters positioned one fourth of the distance across the mouth of the net. The nets were bridleless in that they consisted of a gimbal arrangement, leaving the mouth unobstructed to minimize net avoidance, characteristically a concern in net studies. Triplicate samples were collected at each depth at each station in 1974 and 1975, the first year and a half of the study.

The chronic problem of large variation between so-called replicates due to sampling error provided the motivation to look for alternative sampling methods. We wanted to avoid towed nets, since, in addition to their being some inherent sources of variation in this method, clogging by ctenophores and coelenterates, which are abundant in the Chesapeake Bay through the summer and fall, was severe and completely unacceptable. Nets of mesh finer than 150 μm, used for sampling small life stages and species, were utterly impractical under these conditions.

We had other reservations about the nets. A primary objective of these studies was to look for differences, probably very subtle differences, between locations and to interpret these differences in terms of possible power plant impact. The filtering efficiency of nets varies with the density of organisms, suspended matter, and so forth—characteristics that can be expected to differ among depths and stations. Data gathered for the purpose of measuring differences between locations, but collected by methods influenced by these very differences, are difficult to sort out and interpret. Another source of variation is the decrease of net filtration efficiency as a function of the length of the sample tow [*15,16*], and, conversely, nets towed for equal lengths of time often show large differences in the total water volume filtered.

Inconsistencies in individual flow meters, variation between flow meters, and the difficulty of keeping calibration factors up to date caused us to have doubts about the accuracy of our filtered water volume measurements. After considering the large number of variables inherent in the net collecting method, in an admittedly imprecise science, we felt it imperative to minimize known variances wherever possible.

The system that best met our requirements was a high-volume pumping system. The sampling system finally devised and currently in use is relatively portable and practical for a sampling vessel of small size and has many features that make it suitable for sampling other aquatic environments as well.

The actual collecting device consists of a large clear plastic tube approximately 1.5 m in length, open at both ends, with a 5-cm intake orifice located at the periphery of the tube midway on its long axis. The tube is open at both ends to permit continuous flow and to minimize turbulence, which would increase organism avoidance. The tube is stabilized as it is towed in the water by fins at its trailing edge and a depressor suspended from the bottom of the tube. A centrifugal trash pump draws water in the tube through a 5-cm-diameter plastic hose of appropriate length, through the pump itself, and pumps it subsequently through a solenoid valve and a water meter with a large readout. The solenoid valve is connected to a timer and functions to control the pumping rate or volume if such control is desired. In our studies, the sample volume or total filtered water volume was typically 1000 litres (1 m^3). After going through the meter, the water is flushed through a 73-μm mesh plankton net, and the net is then washed in a net washer (Fig. 13).

Before adopting this system, we ran comparative tests between the gim-

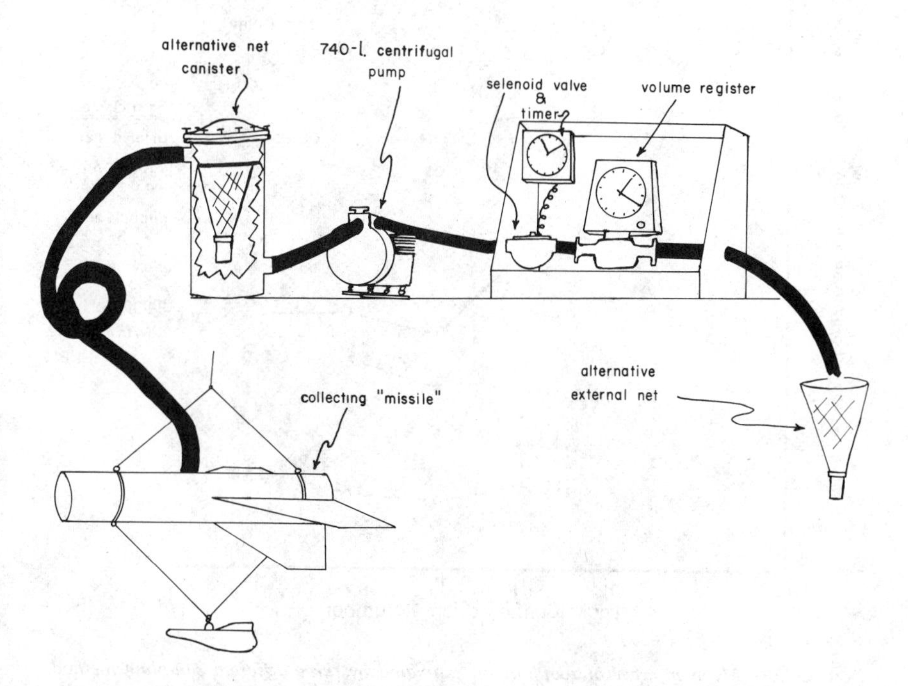

FIG. 13—*Near-field zooplankton sampling gear, the high-speed centrifugal pump system as an alternative to towed nets.*

balled net, a bridled net, and the pump system (Fig. 14). The various devices were oriented horizontally in a frame and were towed in the various position combinations to eliminate any position effect that might appear, and the results were analyzed. The bridled net was less effective in its catching ability than the gimballed net. The pump system appeared to collect fewer organisms, but the relative abundance of species and life stages was the same, indicating to us that avoidance was not a serious problem. In the final analysis, we concluded that the difference in total numbers was an effect of the high accuracy of the metering system in the pump system in comparison with the highly inaccurate system of the flow meters. Because the true density of zooplankton was not known, the method giving the "correct" answer was impossible to determine. Samples collected with this high-speed pump compare very well, however, with samples collected with the low-speed diaphragm pump, the pump used in the entrainment studies described earlier.

The pump system was initiated in 1976. As a check on the new system and for continuity with the previous year's data, both pumps and nets were used

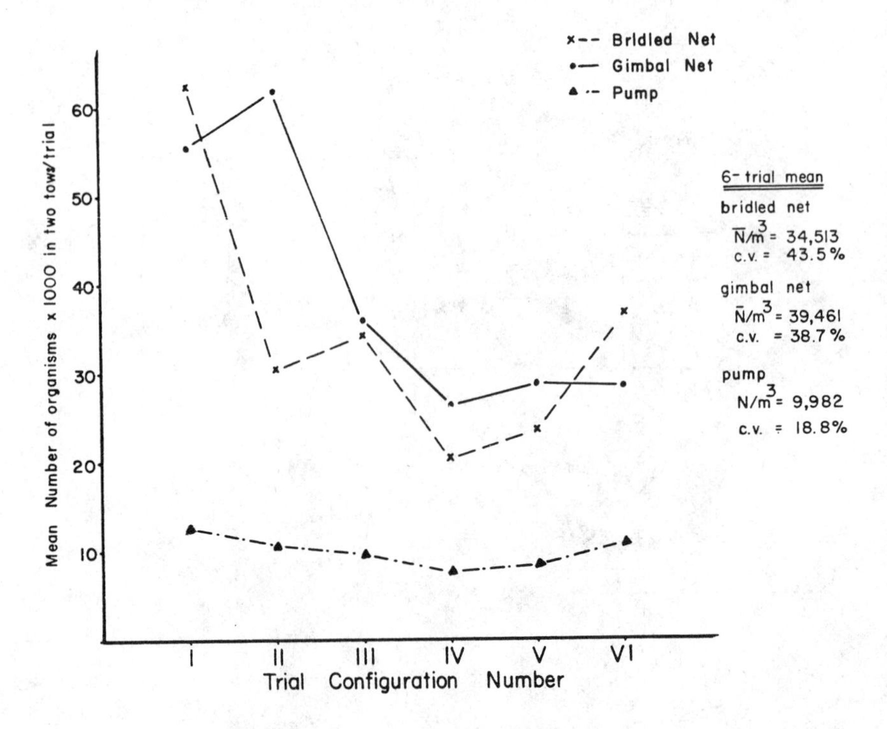

FIG. 14—*Mean number of zooplankton estimated by three different sampling methods—bridled net, gimbal net, and pump. The mean for each method represents the mean of two tows per configuration.*

for that year. The species composition and relative abundance estimates determined by both methods were generally similar. When clogging was not a problem, the differences in density estimates between collecting methods were of a consistent magnitude but otherwise unresolvable.

The basic design of three replicate samples at three depths at three stations was changed in 1978. The change came as a result of frustrating attempts to analyze the data statistically or to interpret the results intuitively. As more specific questions were asked of the near-field data, the number of sampling locations became inadequate to characterize the impact area, and the overall design violated many of the assumptions of rigorous statistical methods.

The design that was adopted consists of six stations on a longitudinal transect through the study area. Unreplicated samples were collected at three depths at each of the stations. To minimize the effects of patchy zooplankton distribution, a "station" was defined as a rather large area. The sampling apparatus was towed over this distance at a boat speed of 5 km/h (3 knots) during the sample collection. To minimize tidal effects, the tow pattern formed a large ellipse. It took approximately 2 to <2.5 min to pump the 757-litre sample volume at each depth, and only one depth could be sampled at one time. The time required to complete one station was approximately 15 to 20 min. Associated hydrographic data, such as temperature, salinity, and dissolved oxygen, were easily collected by direct measurement or from a bucket filled by the pump. To permit possible correlations with the phytoplankton community, continuous fluorometric measurements were made at each depth.

Because of the large variations in the density and relative abundance of zooplankton, which were demonstrated in the entrainment time-series data, the near-field densities obtained from the single-sample-per-month monitoring studies have not been emphasized. Instead, station comparisons within each time frame have been examined, and the consistency of the relative differences has been emphasized.

This design is more appropriate to nonparametric statistical methods and has provided more interpretable information than the earlier studies, within the same budget constraints. Another advantage of the near-field pump system is that it is compatible with the entrainment sampling method, thus adding to a more integrated total sampling program.

Acknowledgments

The authors would like to acknowledge the Baltimore Gas and Electric Co. for supporting these studies. We are also grateful to A. L. Gauzens, M. E. Thompson, N. Roberts, and Capt. H. Shorter for their assistance. Special recognition must be accorded to Thomas Capizzi for the statistical preparation of the data.

References

[1] Sage, L. E. in "Non-Radiological Environmental Monitoring Report," Calvert Cliffs Nuclear Power Plant, Baltimore Gas and Electric Co., Baltimore, 1976, Section 13.2.
[2] Sage, L. E. and Olson, M. M. in "Non-Radiological Environmental Monitoring Report," Calvert Cliffs Nuclear Power Plant, Baltimore Gas and Electric Co., Baltimore, 1977, Section 13.2.
[3] Sage, L. E. and Bacheler, A. G. in "Non-Radiological Environmental Monitoring Report," Calvert Cliffs Nuclear Power Plant, Baltimore Gas and Electric Co., Baltimore, 1978, Section 12.2.
[4] Olson, M. M. and Sage, L. E., Academy of Natural Sciences Report No. 78-18, Philadelphia, Pa., 1978.
[5] Olson, M. M. and Sage, L. E., Academy of Natural Sciences Report No. 78-32D, Philadelphia, Pa., 1978.
[6] Carpenter, E. J., Peck, B. B., and Anderson, S. J., *Marine Biology*, Vol. 24, 1974, pp. 49–55.
[7] Heinle, D. R., *Environmental Pollution*, Vol. 11, 1976, pp. 39–57.
[8] Small, R. D., Capizzi, T. P., and Sage, L. E., "Statistical Design and Analysis of Power Plant Entrainment Studies," Statistical Symposium, U.S. Department of Energy, Las Vegas, Nev., 1978.
[9] Tukey, J. W., *Exploratory Data Analysis*, Addison-Wesley, Reading, Mass., 1977.
[10] Platt, T. and Denman, K. L., *Annual Review of Ecology and Systematics*, Vol. 6, 1975, pp. 189–210.
[11] Sameoto, D. D., *Journal of the Fisheries Research Board of Canada*, Vol. 32, 1975, pp. 347–366.
[12] Haury, L. R., *Marine Biology*, Vol. 37, 1976, pp. 137–157.
[13] Cassie, R. M., *Oceanography and Marine Biology Annual Review*, Vol. 1, 1963, pp. 223–252.
[14] Crippen, R. W. and Perrier, J. L., *Stain Technology*, Vol. 49, 1974, pp. 97–104.
[15] UNESCO, *Monographs on Oceanographic Methodology*, UNESCO Press, Paris, 1968.
[16] Sage, L. E. in "Methods for the Assessment and Prediction of Mineral Mining Impacts on Aquatic Communities," FNS/OBS-78/30, U.S. Department of the Interior, Washington, D.C., 1978.

S. M. Gertz[1] *and I. H. Suffet*[2]

Method for Evaluating the Impact of Radioactivity on Phytoplankton

REFERENCE: Gertz, S. M. and Suffet, I. H., **"Method for Evaluating the Impact of Radioactivity on Phytoplankton,"** *Ecological Assessments of Effluent Impacts on Communities of Indigenous Aquatic Organisms, ASTM STP 730,* J. M. Bates and C. I. Weber, Eds., American Society for Testing and Materials, 1981, pp. 89–100.

ABSTRACT: Concentration factors are often used to assess the impact that radionuclides may have on aquatic organisms. The use of literature values or even measured concentration factors may not be appropriate for a given situation, since these factors are dependent on many environmental parameters. However, if the environmental stresses that affect a concentration factor are known and quantified, a better and potentially more efficient estimate of impact assessment can be made. The authors report here the use and results of a laboratory technique that could be used to evaluate the environmental impact of radioactive effluents on phytoplankton.

The target alga is *Chlamydomonas reinhardii*, a common green flagellate that is amenable to growth in the laboratory. The model radionuclide is cesium-137, an important constituent of the radioactive liquid effluents from nuclear power plants. The organisms are grown in a chemostat, which consists of four growth chambers. The chemostat allows algae to be maintained in their log growth phase while various environmental parameters are manipulated. It was found that the cesium-137 concentration factors are controlled by the phosphate concentration, the sodium concentration, and the algal biomass. Temperature, algal reproduction rate, potassium concentration, and chloride concentration have no effect on the observed concentration factors. The cesium concentration factor ranges from 115 to 586 on a dry weight basis under the various environmental conditions studied.

KEY WORDS: ecology, effluents, aquatic organisms, algae, phytoplankton, radioactivity, cesium, flow culture, impact assessment, concentration factor, chemostat, water quality

Before any commercial nuclear power plant may begin operating, the Nuclear Regulatory Commission requires the sponsoring utility to provide an evaluation of the radiological impact of the plant on the indigenous phytoplankton community. These evaluations are usually accomplished by con-

[1]Project manager, Roy F. Weston, Inc., West Chester, Pa. 19380.
[2]Professor of environmental science, Environmental Studies Institute, Drexel University, Philadelphia, Pa. 19104.

sidering the radionuclides to be released in the plant's liquid effluents, the dilution effects of the receiving water, and algal radionuclide concentration factors, in order to calculate the radiation dose to the phytoplankton. Of those terms used to calculate the radiation dose, the concentration factor has the greatest error.

The concentration factors recommended by the Nuclear Regulatory Commission are based on observed and theoretical average values and, as such, are necessarily limited in their direct application to impact assessment. For example, the listed concentration factor for cesium-137 in algae is 500 [*1*],[3] yet reported concentration factors range from 25 to 25 000 [*2-4*]. Thus, any impact assessment can be significantly under- or overestimated.

The wide range of reported concentration factors for algae are due to species differences and various environmental stresses. The most important environmental stress that affects the ultimate cesium-137 concentration factor in algae is the chemical environment. Algal concentration factors for cesium-137 are constant over a wide range of cesium-137 activities if the chemical environment remains constant [*4-10*]. If the chemical environment changes, there is a concomitant change in the concentration factor. Potassium, sodium, phosphate, and algal biomass all affect the concentration factor [*4,8,9*]. Species differences are also found for cesium accumulation [*4,9-12*]. Algal concentration factors for strontium also change with changes in the calcium and magnesium concentrations [*13-16*]. Therefore, it is apparent that site-specific concentration factors are superior for impact assessment, and, in fact, they are recommended for this purpose [*17*].

However useful, site-specific data are difficult to obtain. One cannot contaminate a water body with radioactive materials to determine site-specific concentration factors, as this is impractical, if not illegal, nor can one use stable element studies at the required concentrations. The concentrations of the stable element must necessarily be greater than the respective radionuclide's concentration because of analytical requirements, and vastly different concentration factors may result because of different uptake mechanisms [*17*]. Thus, laboratory studies of algal radionuclide uptake in simulated environments remain the method of choice for determining site-specific concentration factors.

Algae can be grown in the laboratory in batch and flow cultures. In batch culture systems, algae are grown in flasks containing nutrient media that is not changed during the test period. These experiments are necessarily of limited duration, and, during this time, the algae pass through all phases of their growth curve. Initially, the culture medium will have a high level of nutrients, and the reduction of nutrient material may become the growth-limiting factor as the experiment continues [*18*]. It is therefore impossible to maintain a steady-state culture simulating the aquatic environment under

[3]The italic numbers in brackets refer to the list of references appended to this paper.

these conditions. The environmental applicability of the data derived from batch culture experiments is questionable [*7,10*].

The chemostat, a type of flow culture, overcomes the difficulties inherent in batch culture techniques. Fresh nutrient material is introduced into the growth vessel at a constant rate, and a portion of the total medium and population is removed. This permits the maintenance of constant nutrient levels and biomass over long time periods. The chemostat uses two principal mechanisms for stabilizing population and nutrient levels: (1) the flow rate is a constant value, less than that for the maximum cellular growth rate under the experimental conditions, and (2) the growth medium contains an excess of all nutrients except one, which is present in low enough concentrations to limit the growth rate.

As a consequence of flow-regulated growth rates, the algae are maintained in their exponential phase of growth, their most active metabolic state. This is analogous to many cases of existing environmental conditions. Therefore, chemostatic flow culture studies of steady-state uptake are thought to be the best laboratory method for environmental simulation.

The advantage of the chemostat over other algal culturing methods is that, within the metabolic capacities of the organism, the variables of population density, growth rate, and nutrient or pollutant levels are adjustable and can be controlled independently. The concentrations of all substrates in the chemostat are controlled by their respective concentrations in the nutrient reservoir. The concentration of the growth-limiting factor in the chemostat determines the algal biomass at equilibrium.

This report shows development of a laboratory model that uses flow culture methodology to facilitate the study of water quality effects on cesium-137 uptake by a common freshwater alga, *Chlamydomonas reinhardii.* The water quality parameters chosen for this study are sodium and potassium, because of their chemical similarity to cesium; chloride, because it can alter all membrane permeability; and phosphate, because it effects active transport [*14,19,20,21*]. The methodology employed is simple and capable of infinite variation. This method can be used for other radionuclides and other pollutants, such as pesticides and heavy metals. This method, therefore, is a technique that could be developed for evaluating site-specific environmental impact by laboratory studies.

Materials and Methods

Culture System

The algal culture system used for this study is a chemostat designed with four separate, parallel, continuous-culture units [*8*]. The flow culture units, 500 ml Erlenmeyer flasks, are immersed in a constant-temperature water bath to maintain a temperature of 15 ± 0.1°C. In this study a flow culture experiment required 4 to 6 weeks to reach equilibrium.

Model Organism

Chlamydomonas reinhardii Strain 89, obtained from Indiana State University (Bloomington, Ind.), is the organism chosen to test the approach. This unicellular green alga is selected because it is common in polluted and eutrified water systems. It has a very high growth rate and is amenable to growth under laboratory conditions. *Chlamydomonas* is an important food chain organism, and this lends significance to its choice as a test alga.

Algal Growth Medium

The nutrient medium in which the organisms are grown was developed and used successfully at the Marine Biological Laboratories in Wood's Hole, Mass. (unpublished data). Its composition is given in Table 1.

Lighting

Lighting is provided by two 20-W cool white fluorescent and two 40-W incandescent bulbs. This lighting combination gives a luminous intensity of 31 lux (330 footcandles) at the water's surface in the constant-temperature bath. The bulbs are set to provide a 16-h day and 8-h night.

Radioactivity Determinations

Cesium-137 determinations are made on a Baird-Atomic 132 multiscaler coupled with an 810A well scintillation detector. This provides a counting efficiency of 33.5 ± 0.18 percent (1 sigma standard deviation) based on counting statistics only. The cesium-137 was obtained from Amersham Searle Corp. as the chloride. The cesium-137 standard was certified as having less than a 1 percent standard deviation.

Algal Radioactivity—The algae are harvested by centrifuging a glass counting tube with a sample at 800 g for 10 min. The centrifuged cells are washed three times with distilled water and then dried to a constant weight at 75°C. The samples are then weighed, and the activity is determined.

Nutrient Media Radioactivity—The nutrient medium is counted directly in 3-ml portions.

Chemical Determinations

Sodium, potassium, and total phosphate are determined by following the respective procedures in *Standard Methods for the Examination of Water*

TABLE 1—*Composition of the algal growth nutrient medium.*[a]

Component	Concentration, mg/litre	Component	Concentration, mg/litre
Macronutrients		Copper sulfate	0.01
Calcium chloride	36.76	Zinc sulfate	0.022
Magnesium sulfate	36.97	Cobalt chloride	0.01
Sodium bicarbonate	12.60	Manganese chloride	0.18
Potassium biphosphate	8.71	Sodium permanganate	0.006
Sodium nitrate	85.01	Vitamins	
Sodium silicate	28.42	Thiamine hydrochloride	0.01
Boric acid	24.00	Biotin	0.0005
Micronutrients		B_{12}	0.0005
Sodium ethylenediamine		Buffer	
tetraacetate (Na_2EDTA)	4.36	Tris	500
Ferric chloride	3.15		

[a]All the chemicals are ACS certified except the vitamins.

and Wastewater [22]. The algae are ultrasonically disrupted prior to determination of the total phosphate concentration in the media.

Results

The results of a series of experiments to determine the cesium-137 uptake by algae under different water quality conditions are shown in Table 2. The quantity P_a, introduced in the tables, is the ratio of the total molar phosphate concentration in the reactor vessel (total $[PO_4^{3-}]$) to the kilograms (dry weight) of algae per litre. The total molar phosphate concentration is the sum of the moles of phosphate in both the water and algal phases. This representation allows a meaningful interpretation of the effect of phosphate on cesium accumulation by algae.

The first experimental series, Runs 1 through 4, examines the effect of different cesium-137 levels on the concentration factor. It is shown that cesium-137 concentration factors are constant with respect to external concentrations of cesium-137 at the constant water quality conditions studied and cesium-137 levels tested. The concentration factor averaged 369. This result is expected, as the cesium concentrations were below a saturation value previously reported [*9*].

A comparison of the first three series of experiments, Runs 1 through 12, shows the effect of chloride ion, potassium ion, and total phosphate concentration on algal concentration factors for cesium. The concentrations of all three chemical species in Runs 5 through 12 are either equal to or greater than their respective concentrations in Runs 1 through 4.

The effect of chloride ion on the algal concentration factor for cesium is shown in a comparison of Runs 11 and 5. It is noted here that similar concentration factors resulted with a fourfold difference in chloride molarity. In Runs 9 through 11, large differences in the concentration factors are found with identical chloride and potassium ion concentrations. Comparison of Run 10 with Runs 1 through 4 shows a sixfold increase in potassium ion concentration with no change in the concentration factor. Based on these observations, it appears unlikely that potassium or chloride ion, acting separately or together, could have caused the resultant concentration factors. Therefore, it is reasonable to assume that only the P_a caused the noted concentration factors.

The effect of P_a on cesium accumulation is investigated in Runs 9 through 12. The findings indicate that increases in P_a caused increases in the algal concentration factors for cesium-137.

Sodium concentration, in Runs 13 through 16, is investigated to ascertain its effect on cesium accumulation. When the sodium concentration to be tested is less than 1.3×10^{-3} *M*, the anion deficiencies are corrected by adding the appropriate acids. Sodium is added as the hydroxide when the tested concentration is greater than 1.3×10^{-3} *M*. It has been determined

TABLE 2—*Cesium-137 concentration factors for* Chlamydomonas reinhardii *under various environmental conditions.*

Run No.	Temperature, deg C	Flow Rate, ml/h	Algal Dry Weight, g/litre	Na	PO_4	Cl	K	P_a	Cesium-137 DPM/ml of culture media	CF[b]
1	15	2.51	0.196	1.32E − 3	5.40E − 5	5.00E − 4	1.08E − 4	0.276	102	368
2	15	2.63	0.182	1.32E − 3	5.40E − 5	5.00E − 4	1.08E − 4	0.297	288	375
3	15	2.63	0.183	1.32E − 3	5.40E − 5	5.00E − 4	1.08E − 4	0.295	510	357
4	15	2.70	0.213	1.32E − 3	5.40E − 5	5.00E − 4	1.08E − 4	0.253	655	376
5	15	2.47	0.144	1.30E − 3	1.50E − 4	2.00E − 3	4.95E − 4	1.04	547	526
6	15	2.63	0.178	1.30E − 3	2.98E − 4	5.00E − 4	4.95E − 4	1.67	584	538
7	15	2.61	0.152	1.30E − 3	3.64E − 4	5.00E − 4	7.28E − 4	2.39	605	543
8	15	2.50	0.134	1.30E − 3	3.64E − 4	7.50E − 4	9.69E − 4	2.72	606	534
9	15	2.50	0.182	1.33E − 3	1.67E − 5	5.00E − 4	6.41E − 4	0.0918	509	154
10	15	2.62	0.190	1.33E − 3	8.20E − 5	5.00E − 4	6.41E − 4	0.432	514	381
11	15	2.65	0.196	1.33E − 3	1.61E − 4	5.00E − 4	6.41E − 4	0.821	510	511
12	15	2.50	0.164	1.33E − 3	2.61E − 4	5.00E − 4	6.41E − 4	1.59	525	558
13	15	2.50	0.054	3.90E − 4	4.36E − 5	5.00E − 4	8.72E − 5	0.807	433	540
14	15	2.63	0.118	6.96E − 4	4.36E − 5	5.00E − 4	8.72E − 5	0.369	451	400
15	15	2.63	0.254	2.56E − 3	4.36E − 5	5.00E − 4	8.72E − 5	0.172	514	163
16	15	2.52	0.203	5.10E − 3	4.36E − 5	5.00E − 4	8.72E − 5	0.215	551	115
17	15	2.66	0.083	3.90E − 4	3.20E − 4	5.00E − 4	6.41E − 4	3.86	551	586
18	15	2.58	0.098	6.96E − 4	3.20E − 4	5.00E − 4	6.41E − 4	3.27	508	560
19	15	2.52	0.232	2.56E − 3	6.40E − 4	5.00E − 4	6.41E − 4	2.76	509	473
20	15	2.61	0.218	5.10E − 3	6.40E − 4	5.00E − 4	6.41E − 4	2.94	515	369

[a]DPM = disintegrations per minute.
[b]CF = concentration factor.

in this study that increases in sodium concentration decrease the concentration factors for cesium. This effect was also investigated at high total phosphate levels, and the results are similar (Runs 17 through 20).

Discussion of Results

Effect of Phosphate

Figure 1, based on Runs 1 through 12, shows that increases in the P_a ratio causes higher levels of cesium-137 to be accumulated by the algae. There is an initial rapid rise in the concentration factors for cesium-137 as the phosphate-to-organism ratio increases. Above a P_a level of 1.26, a maximum concentration factor is reached, and the P_a ratio causes no further increases in cesium accumulation. These concentration factors are determined at a constant sodium molarity of 1.32×10^{-3} *M*.

The resultant curve of concentration factor versus P_a can be fitted mathematically by a variety of equations, which may be parabolic, hyperbolic, elliptical, logarithmic, or first-, second-, or n_{th}-order relationships. The data were fitted to each general equation, and the resultant fit was analyzed both visually and by calculating the sum of squares. The best fit, the lowest sum of the squares of deviations, is obtained with a first-order, second-degree function.

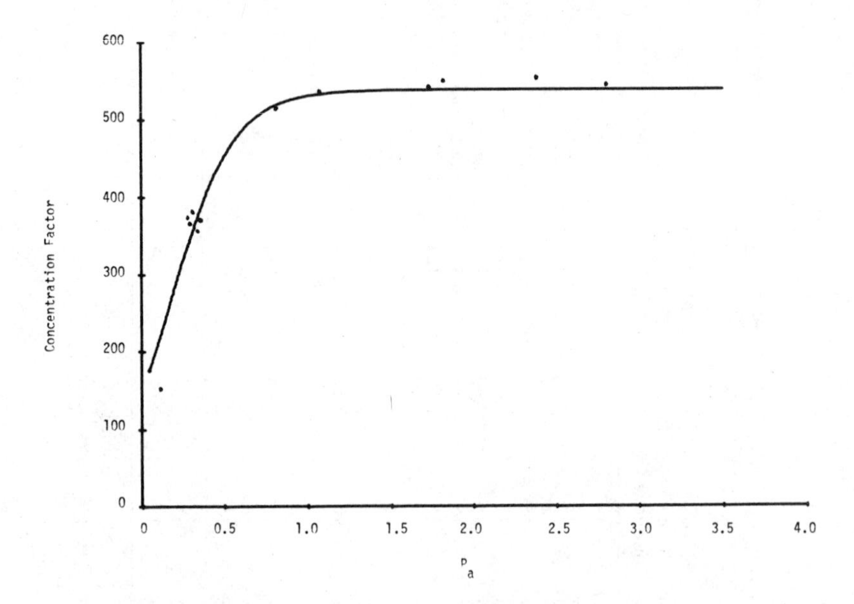

FIG. 1—*Cesium-137 concentration factors in* C. reinhardii *as a function of* P_a *at* $[Na^+] = 1.32E-3$.

The second-degree function that best describes the data is similar in form to the logistic growth equation [*23*]

$$dC/dP_a = rC(1 - C/K_s) \tag{1}$$

whose solution is

$$C = \frac{K_s}{1 + [(K_s - C_0)/C_0] \exp[-r(P_a - P_{a0})]} \tag{2}$$

where C is an equilibrium concentration factor, K_s is the maximum attainable concentration factor under the given conditions, r is a constant related to the increase of C with increasing P_a, and C_0 and P_{a0} are a C and P_a of a lesser value than any data point. These values are arbitrarily chosen as a reference point that best describes the data.

Analysis of the experimental data using this format gives rise to the expression

$$C = \frac{K_s}{1 + [(K_s - 150)/150] \exp -5.29(P_a - 0.0065)} \tag{3}$$

in which C is the concentration factor for different P_a values, and K_s equals 574.

Effect of Sodium

Sodium, as evidenced in Runs 13 through 16, had a depressant effect on the concentration factors for cesium. This effect is evaluated in Runs 17 through 20, in which P_a is >2.0, well into the asymptotic region of the fitted concentration factor curve in Fig. 1, and the depressant effect is quantified. Evaluation of the noted concentration factors yields an inverse linear relationship between sodium concentration and cesium concentration factors, which is shown in Fig. 2. The regression equation for this line is

$$C = 598 - (4.56E + 4)[Na^+] \quad \text{at} \quad P_a > 2.0 \tag{4}$$

where C is the noted concentration factor and $[Na^+]$ is the aqueous sodium molarity.

As a check on the validity of this relationship at low P_a levels, the results from Runs 13 through 16 are used. The experimental values for these runs are substituted back into Eqs 3 and 4, where C in Eq 4 replaces K_s of Eq 3, and the concentration factors are calculated. Good agreement, within 10 percent, is found between the observed and calculated cesium concentration factors. Thus, we are assured that Eq 4 is correct at low P_a levels.

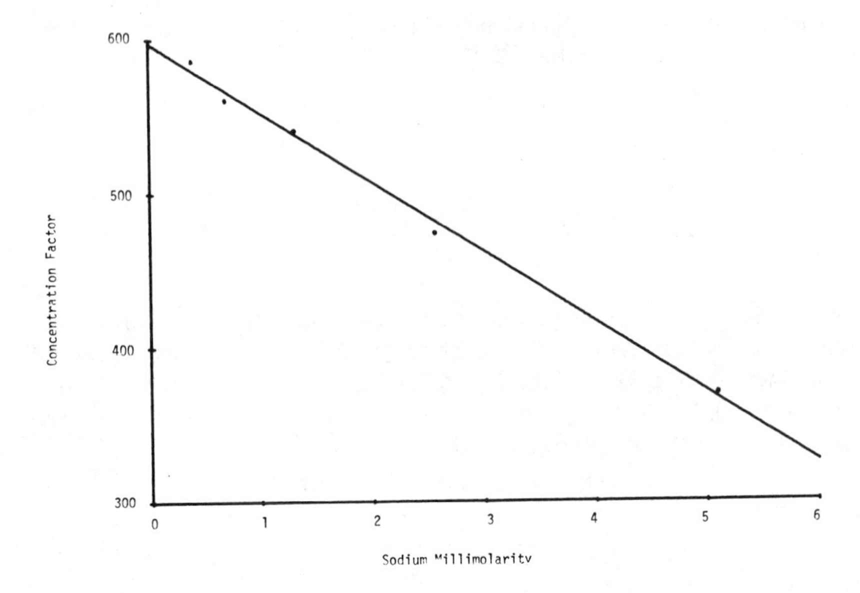

FIG. 2—*Cesium-137 concentration factors in* C. reinhardii *as a function of the aqueous sodium molarity.*

Effects of Other Parameters

Potassium appears to have no effect on the concentration factors for cesium over the range of $(8.72E - 4)$ to $(2.0E - 3)$ M. Previous work has shown that temperatures up to 25°C and different reproductive rates have no effect on the concentration factors [*8*].

Resultant Equation Representing Cesium Uptake

The following equation represents concentration factors of cesium-137 by *C. reinhardii* as a function of water quality

$$\frac{dC}{dP_a} = rC\left[1 - \frac{C}{f[\mathrm{Na}^+]}\right] \tag{5}$$

The solution to Eq 5 is

$$C = \frac{598 - (4.56E + 4[\mathrm{Na}^+])}{1 + \{[448 - (4.56E + 4[\mathrm{Na}^+])]/150\}\exp[-5.29(P_a - 0.0065)]} \tag{6}$$

Thus, the saturation value, K_s, is a function of sodium ion concentration and is estimated at 598. The use of this equation requires that the algal biomass and aqueous sodium and phosphate molarities be known before the concentration factor can be calculated. This calculated concentration factor is then multiplied by the aqueous cesium-137 concentration to obtain the levels of cesium-137 to be found in the algae.

Discussion of Method

The algal growth system used for this study is a chemostat. The chemostat is a continuous-flow apparatus that allows the maintenance of cell populations and substrate concentrations at constant levels over long time periods. The chemostat uses two principle mechanisms for stabilizing population and nutrient levels: (1) the flow rate is a constant value below that for the maximum cellular growth rate under the experimental conditions, and (2) the growth medium contains an excess of all nutrients except one, which is present in low enough concentration to limit the growth rate. This nutrient is called the "growth-limiting factor." The growth-limiting factor need not be identified, for in any growth medium, one nutrient will always assume this role.

The advantage of the chemostat over other algal culturing methods is that, within the metabolic capacities of the organism, the population density, growth rate, and nutrient or pollutant levels can be selected arbitrarily. These variables can be controlled independently of each other. For example, if non-growth-limiting temperatures and lighting conditions are employed, photosynthetic algal growth rates are controlled by the rate of flow through the system. The doubling time of the culture, a measure of growth rate, is determined by dividing the flow rate into the growth vessels capacity. The concentrations of all substrates in the chemostat are controlled by their respective concentrations in the nutrient reservoir. The concentration of the growth-limiting factor in the chemostat determines the algal biomass at equilibrium.

As a consequence of flow-regulated growth rates, the algae are maintained in their exponential phase of growth, their most active metabolic state. This is analogous to existing environmental conditions. Therefore, chemostatic flow cultures are thought to be the best laboratory method for environmental simulation of biological fate [*7,10,18*].

The laboratory studies that can be done using a chemostatic approach are virtually unlimited. This particular research investigated the effect of environmental conditions on the uptake of a single pollutant by a single species. This method of approach could be used for other radionuclides, such as tritium, cobalt-60, or zinc-65, and their effects on algae can be determined. This system can also be used to determine the effects of other pollutants, such as pesticides or heavy metals. Pollutants to be investigated can also be studied together to determine the existence of synergistic or antagonistic effects. For example, two different pesticides, such as lindane and parathion, can be examined both separately and together to ascertain any possible changes in their individual effects. Thus, this approach can be developed to evaluate environmental impact at the first trophic level for a variety of pollutants and environmental conditions.

References

[1] "Proposed Rule Making Action: Numerical Guides for Design Objectives and Limiting Conditions for Operation to Meet the Criterion 'as Low as Practicable' for Radioactive Material in Light-Water-Cooled Nuclear Power Reactor Effluents," WASH 1258, Vol. 2, Appendix F, Directorate of Regulatory Standards, United States Atomic Energy Commission, Washington, D.C., July 1973, p. F-51.

[2] Eisenbud, M., *Environmental Radioactivity*, 2nd ed., Academic Press, New York, 1973, p. 153.

[3] Reichle, D. E., Dunaway, P. B., and Nelson, D. J., *Nuclear Safety*, Vol. 11, No. 1, Jan.-Feb. 1970, pp. 43-55.

[4] Williams, L. G. and Swanson, H. D., *Science*, Vol. 127, No. 3288, Jan. 1958, pp. 187-188.

[5] Harvey, R. S., in *Proceedings*, Second National Symposium on Radioecology, D. J. Nelson and R. C. Evans, Eds., U.S. Atomic Energy Commission, 1969, pp. 266-272.

[6] King, S. F., *Ecology*, Vol. 45, No. 4, Summer 1964, pp. 852-859.

[7] Watts, J. R. and Harvey, R. S., *Limnology and Oceanography*, Vol. 8, No. 1, Jan. 1963, pp. 141-145.

[8] Gertz, S. M., "The Fate of a Radionuclide in the Aquatic Environment—A Mathematical Equation Representing Radionuclide Uptake by Algae," Ph. D. thesis, Drexel University, Philadelphia, June 1973.

[9] Williams, L. G., *Limnology and Oceanography*, Vol. 5, No. 3, May 1960, pp. 372-375.

[10] Harvey, R. S. and Patrick, R., *Biotechnology and Bioengineering*, Vol. 9, No. 4, April 1967, pp. 449-456.

[11] Krumholz, L. A. and Foster, R. F. in *The Effects of Atomic Radiation on Oceanography and Fisheries*, R. Revelle, Ed., National Academy of Sciences, Washington, D.C., 1957, p. 551.

[12] Auerbach, S. E. and Crossley, D. A. in *Proceedings*, Second United Nations International Conference on the Peaceful Uses of Atomic Energy, Vol. 18, United Nations, New York, 1958, pp. 494-499.

[13] Austin, J. H., Klett, C. A., and Kaufman, W., *International Journal of Oceanography and Limnology*, Vol. 1, No. 1, Jan. 1967, pp. 1-28.

[14] Ophel, I. L. and Fraser, J. W., *Nature*, Vol. 51, No. 2, 1970, pp. 324-327.

[15] Pickering, D. C. and Lucas, J. W., *Nature*, Vol. 193, No. 4820, March 1962, pp. 1046-1047.

[16] Agnedal, P. O. in *Radioecological Concentration Processes*, B. Aberg and F. P. Hungate, Eds., Pergamon Press, New York, 1967, pp. 879-896.

[17] Gertz, S. M. and Suffett, I. H. in *Fate of Pollutants in the Air and Water Environment*, Part II, I. H. Suffett, Ed., Wiley, New York, 1978, pp. 223-237.

[18] Tempest, D. W. in *Methods in Microbiology*, Vol. 2, J. R. Norris and D. W. Ribbons, Eds., Academic Press, New York, 1970, pp. 335-391.

[19] Lilley, R. M. and Hope, A. B., *Biochimica et Biophysica Acta*, Vol. 226, 1971, pp. 161-171.

[20] Krauss, R. W., *Annual Review of Plant Physiology*, Vol. 9, 1958, p. 351.

[21] Scott, G. T. and Hayward, H. R., *Journal of General Physiology*, Vol. 36, No. 4, April 1953, pp. 427-436.

[22] *Standard Methods for the Examination of Water and Wastewater*, 13th ed., American Public Health Association, Washington, D.C., 1971.

[23] Pielou, E. C., *Mathematical Ecology*, Wiley, New York, 1977, pp. 20-27.

C. I. Weber[1] *and B. H. McFarland*[1]

Effects of Copper on the Periphyton of a Small Calcareous Stream

REFERENCE: Weber, C. I. and McFarland, B. H., **"Effects of Copper on the Periphyton of a Small Calcareous Stream,"** *Ecological Assessments of Effluent Impacts on Communities of Indigenous Aquatic Organisms, ASTM STP 730,* J. M. Bates and C. I. Weber, Eds., American Society for Testing and Materials, 1981, pp. 101-131.

ABSTRACT: The effect of the continuous addition of copper to a small calcareous stream at a concentration of approximately 120 μg/litre was studied for a year. Of the parameters evaluated in this study, the species composition of the periphyton was found to be the most sensitive and informative measure of the effects of copper. Two of the dominant species of algae in the stream were eliminated from the periphyton: the diatom *Cocconeis placentula* var. *euglypta,* and the filamentous green alga, *Cladophera glomerata. Cocconeis placentula,* which commonly contributed 85 to 98 percent of the summer diatoms in the treated reach of the stream prior to the addition of copper, was replaced by three species of diatoms, *Nitzschia palea, Navicula nigrii,* and *N. seminulum* var. *hustedtii.* Other species of algae that were more abundant in the copper-treated reach than in the control reach of the stream were the filamentous blue-green alga, *Schizothrix calcicola,* and the desmids, *Cosmarium granatum* and *C. subprotumidum.*

KEY WORDS: periphyton; algae; diatoms; water pollution; copper; metals; stream; species diversity; indicator organisms, ecology, effluents, aquatic organisms

This report describes an experimental study of the effects of a controlled discharge of copper on aquatic organisms under natural stream conditions, and deals principally with changes in the species composition of periphytic diatoms during the period 16 Feb. to 21 Dec. 1970. The study was part of a research project on low-level, chronic copper toxicity conducted by the Newtown Fish Toxicology Laboratory, U.S. Environmental Protection Agency, Newtown Ohio [*1*],[2] and was carried out to evaluate methods of measuring the effects of toxic effluents on natural periphyton communities. The stream used in this study (Shayler Run) is a small tributary of the East

[1]Chief Aquatic Biology Section, and aquatic biologist, respectively, Biological Methods Branch, Environmental Monitoring and Support Laboratory, U.S. Environmental Protection Agency, Cincinnati, Ohio 45268.

[2]The italic numbers in brackets refer to the list of references appended to this paper.

Fork of the Little Miami River in Clermont County, Ohio, 16 km east of Cincinnati, which has a normal flow during the summer months of approximately 50 litres/s. It originates in large part from a small domestic waste treatment plant, and flows approximately 1.6 km through a series of limestone riffles and pools before reaching the study site. The stream gradient in the study reach was approximately 27 m/km. The biomass and community structure of the periphyton in the control reach of the stream indicate an oligo-saprobic to beta-mesosaprobic condition [*2–4*]. A brief summary of the chemical and physical water quality parameters is given in Table 1.

Fish barriers (weirs) were constructed at the upper and lower ends of the study reach. The stream-gauging equipment and copper discharge were housed at the upper weir. Beginning 16 Feb. 1970, a solution of copper sulfate was metered into the stream by a flow-controlled apparatus so as to maintain a constant nominal concentration of approximately 120 μg copper/litre at the point of discharge at all rates of stream flow of less than 250 litres/s. Copper was added to the stream 75 percent of the time during the study period. The background concentration of copper in the stream was 6 μg/litre.

The parameters examined in this study included cell density, dry weight (DW), ash-free weight (AFW), AFW/DW, chlorophyll content, AFW/chlorophyll *a* (autotrophic index), pheopigment content, species composition and community diversity. Only the cell density, species composition, and community diversity data are discussed in this report.

Methods

Two riffle and two pool stations were maintained in a 0.7-km reach of the stream—one each above and below the copper discharge (Fig. 1). Current velocities at the riffle and pool stations during normal flow were approximately 2 m/s and 0.2 m/s, respectively. Periphyton was collected on plain glass microscope slides 25 by 75 mm held just below the surface of the pools in floating samplers (Fig. 2) [*5,6*], or in similar plexiglass racks staked to the stream bed in the riffles at a depth of approximately 10 cm. Eight slides were exposed in each sampler for four-week intervals between 16 Feb. and 21 Dec. 1970. Four (alternate) slides from each collection were placed in a single 100-ml bottle and preserved in 5% formalin for cell counts and identifications [*5*]. The remaining slides were taken for chlorophyll and biomass determinations.

Slides taken for counts and identifications were scraped, the scrapings were dispersed in the preservative, and aliquots were taken for Sedgwick-Rafter cell counts and for Hyrax mounts and diatom-species proportional counts [*5*]. The species diversity of diatoms was calculated according to Wilhm [*7*]. Qualitative and quantitative samples were also taken from natural substrates in the control and copper-treated riffles.

TABLE 1—*Summary of chemical and physical data, from Geckler* [1]. *The peak instantaneous discharge was 67.961 m^3/s, on 2 April 1970.*

Month	Dissolved Oxygen, mg/litre	pH[a]	Hardness, mg/litre as $CaCO_3$	Alkalinity, mg/litre as $CaCO_3$	NO_3-N, mg/litre	Acidity, mg/litre as $CaCO_3$	Temperature, deg. C	Total Phosphorus	Flow, m^3/s
February	12.4	7.8	244	151	1.5	2.7	2	1.0	0.498
March	11.3	8.0	226	142	1.2	2.8	5	0.6	0.799
April	10.2	8.3	226	155	0.5	1.6	12	1.5	1.985
May	8.8	8.1	275	188	1.6	0.4	20	1.9	0.239
June	7.0	8.1	306	212	4.1	0.6	23	4.0	0.051
July	6.9	8.1	278	191	2.9	0.4	24	4.4	0.057
August	6.7	8.1	253	184	3.4	0.1	23	4.0	0.108
September	5.9	8.0	310	206	7.5	0.1	21	5.4	0.032
October	7.6	8.0	292	202	5.5	0.8	13	5.1	0.077
November	9.2	8.1	284	202	2.0	1.0	8	2.5	0.131
December	10.4	8.0	286	200	2.5	1.5	5	2.3	0.498

[a]The pH ranged from 7.7 to 8.5 during the study period.

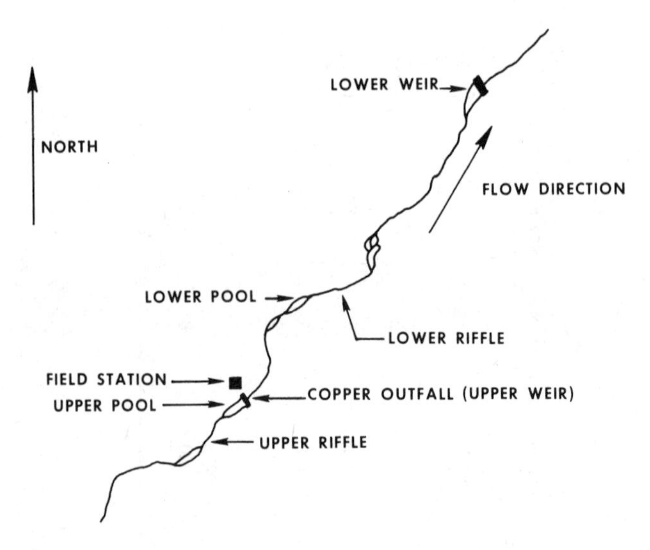

FIG. 1—*Station location. Scale: 0.5 cm = 80 metres.*

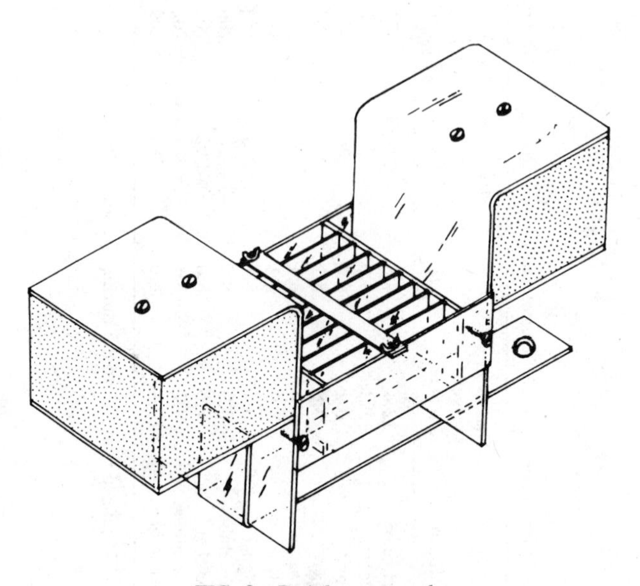

FIG. 2—*Periphyton sampler.*

Photomicrographs (PMs) were taken with a Zeiss photomicroscope equipped with Nomarski interference contrast optics. Scanning electron micrographs (SEMs) were taken with an ISI Mini-SEM, Model MSM-2, or Cambridge Stereo-Scan. Transmission electron micrographs (TEMs) were taken with an RCA Model 3-G.

Results and Discussion

When copper was first added to the stream, 16 Feb. 1970, the riffles were covered with a very heavy, dark-brown layer of diatoms, consisting principally of *Gomphonema olivaceum, G. parvulum,* and *Surirella ovata,* which were the typical winter forms. After 6 days of copper discharge, the brown color of the treated reach began to fade. Within 2 weeks, the diatoms sloughed off the riffles in the treated reach, and the color of the stream bed changed from brown to green as the underlayer of crustose blue-green algae was exposed. During this period, the appearance of the control reach remained unchanged. The cell motility of diatoms in rock scrapings taken from the natural substrate declined noticeably after 4 days of copper treatment, and ceased within 2 weeks. However, the pheopigment content of the rock scrapings changed only slightly during this period, and the autotrophic index [*8*] was not affected. Some minor shifts were observed in the diatom species composition, the most notable of which was a relative decline in the abundance of *Gomphonema olivaceum* in rock scrapings taken from the treated reach. The copper continued to suppress the growth of periphyton on both artificial and natural substrates in the treated reach through the period, 1 April to 5 May 1970.

With the occurrence of normal seasonal changes in the taxonomic composition of the periphyton community in May and June, differences in the species compositions in the control and treated reaches became more pronounced, although the standing crop of organisms colonizing the glass slides in the treated reach increased to approximately that of the control reach. In the control reach, *Gomphonema* and *Surirella* were replaced by *Cocconeis placentula* var. *euglypta, Nitzschia amphibia, N. frustulum, N. palea, Navicula luzonensis, N. nigrii,* and *N. seminulum* var. *hustedtii* (Fig. 3). *Cladophora glomerata* grew profusely on the natural substrate, but was not observed on glass slides because of the short exposure period (Tables 2 to 6). However, other algae occurred in approximately the same relative abundances on both types of substrate.

Of the 43 species of diatoms that occurred on the glass slides, 19 contributed 10 percent or more of the diatom count in at least one sample (Figs. 4 to 8). Seven species showed some sensitivity to copper (Table 7). The relative abundances of *Cocconeis placentula, Gomphonema parvulum, G. olivaceum, Nitzschia amphibia, N. frustulum, Amphora veneta,* and

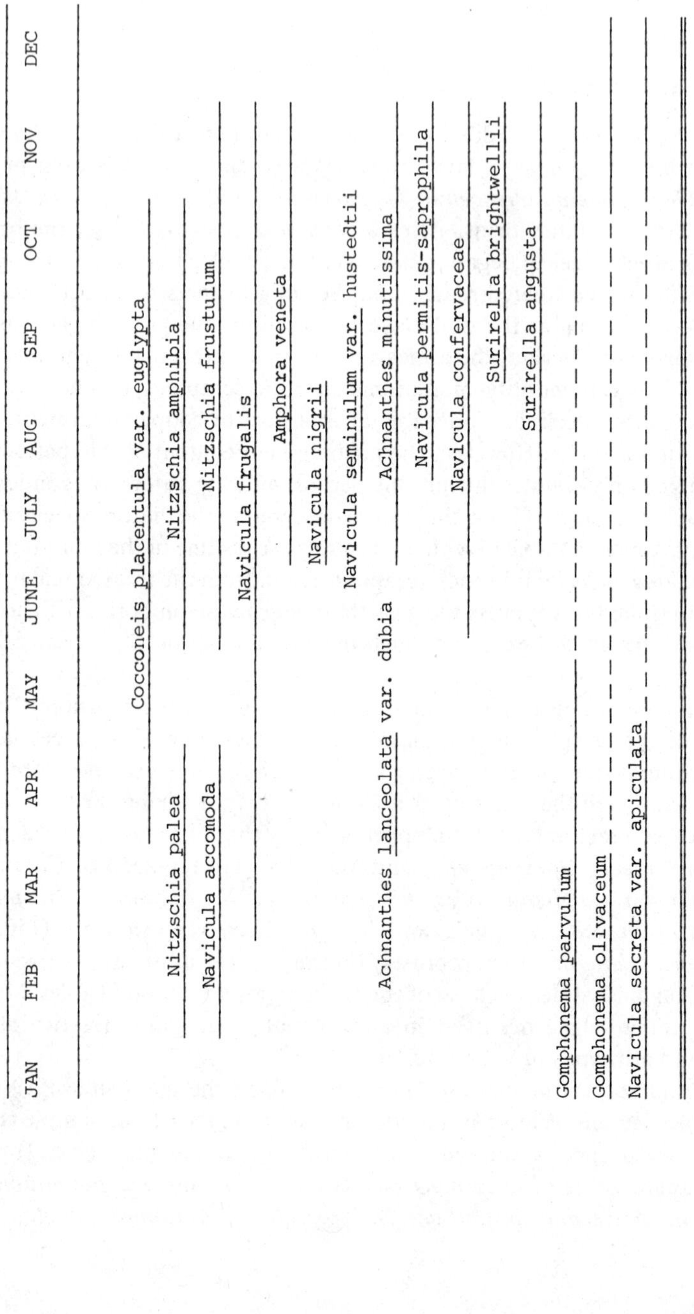

JAN FEB MAR APR MAY JUNE JULY AUG SEP OCT NOV DEC

Cocconeis placentula var. euglypta

Nitzschia palea

Nitzschia amphibia

Navicula accomoda

Nitzschia frustulum

Navicula frugalis

Amphora veneta

Navicula nigrii

Navicula seminulum var. hustedtii

Achnanthes lanceolata var. dubia

Achnanthes minutissima

Navicula permitis-saprophila

Navicula confervaceae

Surirella brightwellii

Surirella angusta

Gomphonema parvulum

Gomphonema olivaceum

Navicula secreta var. apiculata

FIG. 3—*Seasonal distribution of the dominant species of diatoms.*

Achnanthes lanceolata var. *dubia* were greatly reduced in the treated reach, when compared with their occurrences in the control reach (Fig. 8). Species of diatoms that occurred in greater abundance in the treated reach than in the control reach ("tolerant species") included *Nitzschia palea, Navicula seminulum* var. *hustedtii, N. nigrii, Surirella ovata, Achnanthes minutissima, N. permitis,* and *S. angusta,* in that order (Fig. 8). The seasonal distributions of the indifferent forms (those that occurred in both reaches in approximately equal relative abundances) are shown in Fig. 8, and included *Navicula frugalis, N. accomoda, N. confervacea, N. secreta* var. *apiculata,* and *Surirella brightwellii* (Table 7).

The most striking effects of copper on the non-diatoms were seen in the occurrence of *Cladophora glomerata,* which was completely eliminated from the treated reach, and the filamentous blue-green alga, *Schizothrix calcicola,* the coccoid green alga *Chlorococcum* sp., and the desmids *Cosmarium granatum* and *C. subprotumidum* (Fig. 7), which were more abundant in the treated reach than in the control reach (Table 8).

The diatoms listed above that have been reported as tolerant of copper include *Nitzschia palea* [*3,9,10*] and *Gomphonema parvulum* [*9*] (Table 9). Another diatom, *Navicula seminuloides,* was reported by Blum [*10*] to be "found nowhere in the unpolluted portion of the stream (Saline River), but is present on virtually every rock in the river below the (copper-laden) industrial outfall . . . for at least two miles downstream." He later [*11*] identified this diatom as *Navicula atomus.* Because of our interest in copper-tolerant organisms, we examined a sample of rock scrapings from Blum's collections from the Saline River (Station 14), located just below the discharge of a plating plant effluent containing 6 to 26 ppm cyanide, 8 to 24 ppm hexavalent chromium, and 4 to 42 ppm copper. By the use of optical and scanning electron microscopy, we found that his sample consisted almost exclusively of *Navicula nigrii* (=*N. minima*), rather than *N. seminuloides* or *N. atomus,* which supports our observations of this taxon in Shayler Run [*12*]. According to Granetti [*13*], the name *N. nigrii* De Notaris 1872 has priority over *N. minima* Grunow [*14*], and we have adopted this nomenclature.

Regarding the non-diatom algae, *Cladophora glomerata,* which was absent from the copper-treated reach of Shayler Run, was also absent from a 21-km reach of the Saline River, below the Saline, Michigan, industrial outfall [*10*]. Also, cultures of this taxon were found to be inhibited by copper concentrations as low as 50 μg/litre [*15*], and were killed by a concentration of 200 μg/litre [*16*].

Chlorococcum, which was common in the copper-treated reach of Shayler Run, was reported by Butcher [*2*] to be among the first to appear below a copper outfall in the River Trent, and was common in the polluted reach of the Saline River, Michigan [*10*]. Its occurrence in the latter stream, however, was not specifically identified with the copper-laden industrial outfall at Saline. No published information is available regarding the copper sen-

TABLE 2—*Periphyton cell densities (cells/mm²) on glass*

Organisms	15 Feb.- 16 March 1970	7 April- 5 May 1970	5 May- 2 June 1970	2 June- 1 July 1970
Filamentous blue-green algae				
Schizothrix calcicola	...	...	...	...
Coccoid green algae				
Scenedesmus quadricauda	...	...	...	...
Scenedesmus spp.	...	...	...	...
Unknown	...	...	...	...
Filamentous green algae				
Stigeoclonium	...	285	143	...
Green flagellates				
Chlamydomonas spp.	...	...	...	...
Trachelomonas spp.	...	...	...	...
Filamentous bacteria and fungi				
Sphaerotilus natans	1 381	...	...	...
Protozoa				
Unknown	...	28	...	...
Live centric diatoms	...	...	...	12
Live pennate diatoms	1 398	5 707	1 026	1 210
% Dead centrics	...	...	...	16.7
% Dead pennates	...	10.0	0.2	0.2
Total live cell count (cells/mm²)	2 779	6 020	1 169	1 222

sitivities of the other common species of non-diatoms that occurred in the treated reach of Shayler Run. *Schizothrix calcicola* has not been previously reported to be copper tolerant, but Drouet [*17*] described it as the "most hardy of all species of Oscillatoriaceae." *Cosmarium granatum* and *C. subprotumidum* also have not been previously reported to be copper tolerant, although *C. granatum* was cited by Brook [*18*] as the most important eutrophic indicator among the five species of this genus occurring in English lakes. This correlates with its occurrence in Shayler Run, which is highly eutrophic.

During the course of this study, problems encountered in the identification of the small species of *Navicula,* such as *N. seminulum* var. *hustedtii, N. nigrii, N. frugalis,* and *N. saprophila,* in Shayler Run led to an extensive review of the literature relating to these taxa. Three of these species belong to what Lange-Bertalot and Bonik [*19*] previously referred to provisionally as the *Navicula permitis* association, which consists of nine species of *Navicula* and one species of *Nitzschia,* including: *Navicula permitis* Hust., *N. minima* Grun. (=*N. nigrii* De Not.), *N. atomus* Grun., *N. saprophila* Lan.-Ber. & Bon., *N. cloacina* Lan.-Ber. & Bon., *N. lacunolaciniata* Lan.-Ber. & Bon., *N. frugalis* Hust., *N. submolesta* Hust., *N. twymaniana* Arch., and *Nitzschia retusa* Lan.-Ber. & Bon. They observed this association in large numbers only in streams heavily polluted with sewage, and found that it occurred with the filamentous sewage bacterium, *Sphaerotilus natans.* They found that *slides exposed in the untreated (control) riffle (Station 1).*

1 July–31 July 1970	31 July–30 Aug. 1970	30 Aug.–29 Sept. 1970	29 Sept.–27 Oct. 1970	27 Oct.–24 Nov. 1970	24 Nov.–21 Dec. 1970
2 420	101	...	...	...	...
...	5	...	...	...	...
...	...	9	...	...	...
...	...	9	...	...	...
...	...	...	...	...	...
5	...	...	...	...	...
...	...	...	...	4	...
...	...	...	...	...	4 662
5	...	...	...	...	...
...	2	4	...	...	...
356	351	1 366	2 752	2 584	517
...	8.3	...	...	...	...
6.5	3.9	2.2	6.5	1.7	0.9
2 786	459	1 388	2 752	2 588	5 179

within this group, *Navicula frugalis* was often extremely abundant in polluted waters, and was highly resistant to industrial wastes.

The structural details of the frustules of species in this association lie at or below the limit of resolution of the light microscope. Two species, *Navicula atomus* and *N. nigrii,* are well known, but four species are new, and the other four species are relatively unknown. Five of the ten species of this group were found in Shayler Run.

Material collected by Lange-Bertalot and Bonik from the Rhine River in Germany and Ille River in France, containing *N. frugalis, N. permitis, N. atomus, N. accomoda, Gomphonema parvulum,* and other taxa, was examined in our laboratory for comparative purposes. Two species, *N. atomus* and *N. permitis,* are nearly identical except for the striae counts. The material from Shayler Run that resembled these two species matched the description of *N. permitis* more closely than that of *N. atomus* (Fig. 5). The Lange-Bertalot and Bonik interpretation of *N. permitis* is shown in Fig. 5.

Although Shayler Run originates largely from a sewage-treatment plant, self-purification is nearly complete before the water enters the upper reach of the study area, and the stream does not carry as heavy a load of undegraded organic matter as the waters in which Lange-Bertalot and Bonik observed the *N. permitis* association. This is probably why in Shayler Run this association occurs with the filamentous blue-green alga, *Schizothrix calcicola,* rather than with *Sphaerotilus natans.*

TABLE 3—*Periphyton cell densities (cells/mm²) on glass*

Organisms	23 Feb.- 16 March 1970	7 April- 5 May 1970	5 May- 2 June 1970	2 June- 1 July 1970
Filamentous blue-green algae				
Microcoleus vaginatus	...	...	430	...
Schizothrix calcicola	...	...	...	1 375
Coccoid green algae				
Ankistrodesmus falcatus	...	...	...	...
Coelastrum	...	...	...	...
Cosmarium granatum	...	...	...	...
Oocystis sp.	...	...	...	...
Scenedesmus spp.	...	...	123	220
Unknown	144	...	506	...
Filamentous green algae				
Oedogenium spp.	...	...	...	...
Stigeoclonium	6	595	...	...
Ulothrix	20	...	...	...
Unknown	...	287	...	415
Green flagellates				
Euglena spp.	...	...	...	...
Unknown	3	...	...	...
Other pigmented flagellates				
Chrysococcus sp.	...	...	...	...
Filamentous bacteria and fungi				
Sphaerotilus natans	1 026	...	...	...
Protozoa				
Amoeba spp.	...	...	...	...
Stalked	...	...	...	...
Unknown	...	29	...	...
Live centric diatoms	...	67	115	615
Live pennate diatoms	7 289	3 850	4 157	5 820
% Dead centrics	...	...	44.4	12.1
% Dead pennates	2.0	14.9	5.7	13.2
Total live cell count (cells/mm²)	8 488	4 828	8 959	8 445

slides exposed in the untreated (control) pool (Station 2).

1 July– 31 July 1970	31 July– 30 Aug. 1970	30 Aug.– 29 Sept. 1970	29 Sept.– 27 Oct. 1970	27 Oct.– 24 Nov. 1970	24 Nov.– 21 Dec. 1970
...	...	...	...	...	...
1 595	5 939	1 327	...	...	...
...	...	...	...	...	...
...	19	21	...	7	...
...	...	...	2	...	...
...	7	...	...	...	...
...	79	69	69	32	...
289	336	196	790	575	...
36	...	57	...	...	...
...	...	...	...	...	...
...	...	...	...	...	...
...	...	...	...	74	167
...	...	...	...	5	2
...	...	...	...	...	...
...	2	...	...	...	...
...	...	...	...	...	536
...	...	...	12	...	...
...	...	...	...	...	2
...	...	...	...	...	...
14	...	5	40	74	...
1 730	5 322	2 311	1 070	1 724	1 557
16.7	...	...	27.3	...	...
4.0	13.0	11.5	12.7	16.6	0.9
3 664	11 704	3 995	1 983	2 491	2 264

TABLE 4—*Periphyton cell densities (cells/mm²) on*

Organisms	15 Feb.- 16 Mar. 1970	5 May- 2 June 1970	2 June- 1 July 1970
Filamentous blue-green algae			
Schizothrix calcicola	...	135 257	4 332
Coccoid green algae			
Ankistrodesmus falcatus	...	...	...
Cosmarium granatum	...	...	...
Cosmarium subprotumidum	...	...	...
Cosmarium sp.	...	8	43
Crucigenia sp.	...	...	29
Scenedesmus bijuga	...	...	...
Scenedesmus spp.	...	...	...
Unknown	...	884	8 850
Filamentous green algae			
Schizomeris leibleinii	...	...	...
Stigeoclonium	8	...	129
Uronema	...	...	...
Green flagellates			
Chlamydomonas spp.	...	...	...
Euglena spp.	...	...	...
Trachelomonas spp.	...	23	...
Unknown	...	...	...
Other pigmented flagellates			
Chrysococcus sp.	...	...	...
Filamentous bacteria and fungi			
Sphaerotilus natans	2 385	...	...
Protozoa			
Amoeba spp.	...	...	...
Stalked	3	...	...
Unknown	...	...	122
Live pennate diatoms	1 008	5 897	15 286
% Dead centrics	...	...	...
% Dead pennates	2.6	3.4	10.6
Total live cell count (cells/mm²)	3 404	142 069	28 791

glass slides exposed in the treated pool (Station 3).

1 July- 31 July 1970	31 July- 30 Aug. 1970	30 Aug.- 29 Sept. 1970	29 Sept.- 27 Oct. 1970	27 Oct.- 24 Nov. 1970	24 Nov.- 21 Dec. 1970
2 172	10 227	2 642	271	...	...
...	...	5	...	...	...
8	11	...	28	12	...
36	53	...	10	...	...
...	...	10	5	...	...
...	...	...	...	...	...
...	75	30	...	...	...
132	...	30	...	...	...
630	147	215	402	932	...
...	4 275	122	...	...	...
...	...	23	...	...	...
...	...	...	1 060	...	...
8	...	...	...	...	...
...	...	...	...	2	...
...	...	...	...	...	...
...	...	...	...	7	...
...	4	...	...	...	1
...	...	...	...	...	...
...	...	...	34	5	...
...	...	...	5	...	...
...	...	...	...	...	...
3 297	5 110	3 178	3 813	1 023	16
...	100.0	...	...	...	...
27.4	30.2	7.5	1.1	12.1	11.9
6 283	20 135	6 250	5 628	1 981	17

TABLE 5—*Periphyton cell densities (cells/mm²) on*

Organisms	16 Feb.- 16 Mar. 1970	7 Apr.- 5 May 1970	5 May- 2 June 1970	2 June- 1 July 1970
Filamentous blue-green algae				
Microcoleus vaginatus	...	...	...	...
Schizothrix calcicola	...	...	16 660	3 722
Coccoid green algae				
Cosmarium granatum	...	...	...	16
Cosmarium subprotumidum	...	...	...	...
Scenedesmus abundans	...	...	...	21
Scenedesmus bijuga	...	...	...	...
Scenedesmus quadricauda	...	...	...	...
Scenedesmus spp.	...	...	...	...
Unknown	...	...	...	...
Filamentous green algae				
Oedogenium spp.	...	...	...	...
Schizomeris leibleinii	...	...	...	...
Other pigmented flagellates				
Unknown	...	...	...	...
Filamentous bacteria and fungi				
Sphaerotilus natans	36	...	...	...
Protozoa				
Amoeba spp.	...	...	...	...
Stalked	...	...	...	...
Live centric diatoms	...	...	...	...
Live pennate diatoms	32	1 353	6 929	8 742
% Dead centrics	...	...	...	...
% Dead pennates	...	9.6	9.7	25.9
Total live cell count (cells/mm²)	68	1 353	23 589	12 501

glass slides exposed in the treated riffle (Station 4).

1 July- 31 July 1970	31 July- 30 Aug. 1970	30 Aug.- 29 Sept. 1970	29 Sept.- 27 Oct. 1970	27 Oct.- 24 Nov. 1970	24 Nov.- 21 Dec. 1970
...	...	...	...	190	...
1 292	454	3 735	...	381	...
8	6	6	5	2	...
51	13	4	...	...	...
...	...	...	...	...	...
...	59	...	...	...	...
...	19	...	...	...	...
34	58	...	10	...	...
...	68	461	420	133	...
...	...	...	141	...	...
...	...	69	...	...	...
...	...	61	...	...	...
...	...	...	...	...	...
...	...	...	...	2	...
...	...	...	10	...	...
11	3	...	...	...	...
3 984	1 176	593	2 716	1 892	...
33.3	83.3	...	...	...	...
5.9	29.4	7.0	3.6	15.4	...
5 369	1 836	4 929	3 302	2 600	0

TABLE 6—*Percent abundances of diatoms on glass slides: Station 1, control riffle; Station 2, control pool; Station 3, treated pool; Station 4, treated riffle.*[a]

Date	15 Feb.-16 Mar.				7 Apr.-5 May				5 May-2 June				2 June-1 July				1 July-31 July			
Station	1	2	3	4	1	2	3[b]	4	1	2	3	4	1	2	3	4	1	2	3	4
Centric diatoms																				
Cylotella																				
menghiniana	...	...	...	...	...	...	...	...	...	...	...	...	x	3	...	...	...	1	...	x
Melosira																				
varians	...	...	...	...	...	...	...	...	...	5	...	...	...	...	...	...	...	...	...	...
Stephanodiscus																				
invisitatus	...	...	...	...	...	...	...	...	...	...	...	...	...	...	...	...	...	...	...	...
Pennate diatoms																				
Achnanthes																				
lanceolata																				
var. *dubia*	x	...	x	...	...	...	...	...	3	x	...	x	...	x	...	x	...	...	x	4
minutissima	...	...	...	...	...	...	...	...	...	...	...	...	...	...	...	2	...	...	6	...
Amphora																				
ovalis																				
var. *pediculus*	...	...	...	...	...	...	...	...	...	...	...	...	...	...	...	...	...	...	...	...
veneta	...	...	...	...	...	...	...	...	...	3	...	...	...	2	...	...	3	1	x	x
Cocconeis																				
pediculus	...	...	...	...	...	x	...	...	2	...	...	...	...	...	...	...	x	...	...	...
placentula																				
var. *euglypta*	...	...	...	x	...	...	...	...	89	...	...	...	95	x	...	x	86	x	x	x
Fragilaria																				
contruens																				
var. *subsalina*																				
Gomphonema																				
parvulum	42	14	51	43	13	14	...	8	x	14	2	1	x	13	x	...	1	13	...	x
olivaceum	49	78	28	43	22	13	...	13	x	6	1	x	...	...	...	x	...	...	...	...
Navicula																				
accomoda	...	2	1	x	4	6	...	3	...	3	...	x	...	...	...	x	...	...	...	x
confervacea	...	...	...	...	...	...	...	...	...	...	...	...	...	10	...	...	...	3	...	...

cryptocephala																				
var. *veneta*	...	...	...	...	3	...	...	x	...	...	...	x	...	...	...	...	...	...	...	...
frugalis	...	...	...	...	7	14	...	14	1	6	...	8	2	7	3	...	x	11	...	7
nigrii	...	...	...	...	...	...	...	...	...	...	...	...	...	...	...	x	x	28	5	6
pupula	...	...	...	...	...	...	...	...	...	...	...	...	...	2	...	...	...	...	...	...
secreta																				
var. *apiculata*	x	x	x	...	16	3	...	2	...	...	x	...	...	...	...	...	...	...	...	...
seminulum																				
var. *hustedtii*	...	...	...	...	2	x	...	2	x	6	...	6	1	5	57	60	x	2	37	32
tripunctata																				
var. *schizonemoides*	...	...	...	...	x	x	...	...	...	x	...	...	...	x	...	...	...	...	...	...
permitis	...	...	...	...	...	...	...	...	...	...	...	...	...	...	...	...	...	...	...	1
spp.	...	...	...	...	...	...	...	...	x	...	...	...	...	...	...	...	...	...	...	...
Nitzschia																				
acicularis	...	...	...	x	...	x	...	3	...	...	...	...	...	...	...	...	...	...	...	...
amphibia	...	x	x	2	...	...	...	...	...	2	...	...	x	13	x	...	9	37	...	7
apiculata	...	...	...	...	...	...	...	...	...	x	...	...	...	...	...	...	...	...	...	...
capitellata	...	...	...	...	...	...	...	...	...	...	...	...	...	...	...	...	...	...	...	...
dissipata	...	...	...	...	3	2	...	x	...	2	...	...	...	...	...	...	...	...	...	...
frustulum	...	...	...	...	x	...	...	...	...	...	3	x	...	37	8	2	...	x	...	x
var. *perminuta*	...	...	x	...	...	...	...	...	...	...	...	...	...	...	...	...	...	...	...	...
gracilis	...	...	...	...	...	...	...	...	...	...	...	...	...	...	...	...	...	...	...	...
hungarica	...	...	...	...	...	...	...	...	...	...	...	...	...	...	...	...	...	...	...	...
linearis	...	...	...	...	...	x	...	...	...	...	...	...	...	...	...	...	...	...	...	...
palea	...	...	...	1	6	17	...	13	2	7	65	70	...	5	31	34	...	x	51	45
spp.	...	...	...	...	...	...	...	x	...	...	...	...	...	...	...	...	...	...	...	...
Rhoicosphenia																				
curvata	...	...	...	...	2	x	...	...	...	...	...	...	x	...	...	...	...	...	...	...
Surirella																				
angusta	...	...	...	...	...	...	...	...	...	...	...	x	...	...	...	...	...	...	...	...
brightwellii	...	...	...	...	...	...	...	...	...	...	...	...	...	...	...	...	...	...	...	...
ovata	6	5	17	16	37	28	...	40	1	4	7	11	x	...	...	...	...	...	...	...
Synedra																				
fasciculata	...	...	...	...	...	...	...	...	...	x	...	...	...	...	...	...	...	...	...	...
ulna	...	...	...	...	x	...	...	...	...	...	...	...	...	...	...	...	...	...	...	...
Unidentified pennates	...	...	...	...	...	...	...	...	...	...	...	...	...	...	...	...	...	...	...	...

TABLE 6—*Continued.*

Date	31 July–30 Aug.				30 Aug.–29 Sept.				29 Sept.–27 Oct.				27 Oct.–24 Nov.				24 Nov.–21 Dec.			
Station	1	2	3	4	1	2	3	4	1	2	3	4	1	2	3	4	1	2	3	4
Centric diatoms																				
Cylotella																				
menghiniana	x	...	...	...	...	x	...	...	...	3	...	...	...	x	x	...	...	x	2	...
Melosira																				
varians	...	...	...	...	x	...	...	...	...	x	...	...	...	x	...	...	...	...	...	...
Stephanodiscus																				
invisitatus	...	...	...	...	...	...	...	...	...	...	...	...	...	...	...	...	...	...	...	1
Pennate diatoms																				
Achnanthes																				
lanceolata																				
var. *dubia*	x	x	...	...	x	...	...	...	7	1	...	...	4	...	...	...	4	1	3	1
minutissima	...	...	x	6	x	...	x	23	x	...	x	1	x	...	1	...	...	x	1	4
Amphora																				
ovalis																				
var. *pediculus*	...	...	...	...	...	...	...	...	x	...	...	...	...	...	...	...	...	...	...	...
veneta	4	x	1	1	2	1	2	3	7	1	...	...	3	2	1	4	...	...	...	...
Cocconeis																				
pediculus	...	...	...	...	...	...	...	...	...	...	...	...	x	...	...	...	...	...	...	...
placentula																				
var. *euglypta*	80	...	...	x	68	x	1	...	7	x	x	x	x	x	...	1	x	x	1	...
Fragilaria																				
contruens																				
var. *subsalina*																				
Gomphonema																				
parvulum	x	5	...	...	x	x	x	...	5	6	...	...	47	20	x	...	49	33	17	...
olivaceum	...	...	...	...	...	...	...	...	...	x	x	...	x	...	...	...	34	22	2	...
Navicula																				
accomoda	...	...	...	...	...	x	...	1	...	...	...	x	x	1	...	...	...	1	...	1
confervacea	...	4	...	...	4	3	2	...	...	8	...	...	...	...	1	7	...	...	2	7
cryptocephala																				
var. *veneta*	...	...	...	...	...	...	...	...	...	...	...	...	...	x	...	...	x	...	...	...

frugalis	4	11	2	3	6	2	...	1	17	15	...	x	12	14	2	2	2	3	7	5
nigrii	2	13	22	33	2	5	30	14	4	5	12	15	1	3	16	4	...	2	14	18
pupula	...	...	...	...	...	x	...	...	...	x	...	...	...	x	...	...	...	...	1	...
secreta																				
var. *apiculata*	...	...	...	...	...	...	...	...	...	...	...	...	...	x	...	...	...	1	5	1
seminulum																				
var. *hustedtii*	...	2	56	29	x	3	27	41	3	2	15	17	3	1	5	9	...	x	3	25
tripunctata																				
var. *schizonemoides*	...	...	...	...	...	...	...	...	...	...	...	...	...	...	x	...	...	...	...	...
permitis	...	...	...	...	...	2	...	x	...	x	...	x	...	x	37	5	...	...	...	5
spp.	x	...	x	x	x	x	...	...	...	...	...	...	...	...	...	...	...	...	...	7
Nitzschia																				
acicularis	...	...	...	...	...	...	...	...	...	...	...	...	...	...	...	...	...	...	...	...
amphibia	3	60	x	4	6	80	14	3	3	7	...	2	1	3	2	x	...	...	4	16
apiculata	...	...	...	...	...	...	...	...	...	...	...	...	...	x	...	x	x	3	...	...
capitellata	x	...	...	...	...	...	...	...	...	x	...	...	...	...	...	1	...	2	...	...
dissipata	...	...	...	...	...	...	...	...	...	...	...	...	2	8	...	...	...	x	1	...
frustulum	1	2	...	x	3	x	...	1	49	33	...	x	9	21	1	2	2	5	5	1
var. *perminuta*	...	...	...	...	x	...	...	...	...	...	...	...	1	...	...	...	...	...	...	1
gracilis	...	...	1	...	...	...	...	...	...	3	...	...	...	...	...	x	...	x	...	...
hungarica	...	...	1	...	...	...	...	...	...	...	...	...	...	x	...	...	...	...	...	...
linearis	...	...	...	...	...	...	...	...	...	...	...	...	...	...	...	...	...	x	...	...
palea	...	x	15	20	1	x	21	10	5	10	70	60	5	11	26	57	2	11	8	...
spp.	...	...	...	...	...	...	...	...	...	...	...	...	...	...	...	...	...	...	...	...
Rhoicosphenia																				
curvata	...	...	...	...	...	...	...	...	...	...	...	...	...	...	...	...	x	...	...	...
Surirella																				
angusta	...	...	...	x	...	...	x	x	...	1	x	1	...	4	4	13	...	x	3	...
brightwellii	...	...	...	...	x	...	...	...	x	2	...	x	...	2	2	x	...	...	1	...
ovata	...	...	x	x	...	...	x	x	1	2	x	x	7	3	x	2	x	5	3	...
Synedra																				
fasciculata	...	...	...	...	...	...	...	...	...	...	...	...	...	...	...	...	x	...	3	...
ulna	...	...	...	...	...	...	...	...	x	...	...	...	3	6	...	...	1	4	2	...
Unidentified pennates	x	...	...	...	...	...	...	...	...	...	...	...	...	...	...	...	...	...	...	...

[a] x = less than 1 percent.
[b] No sample.

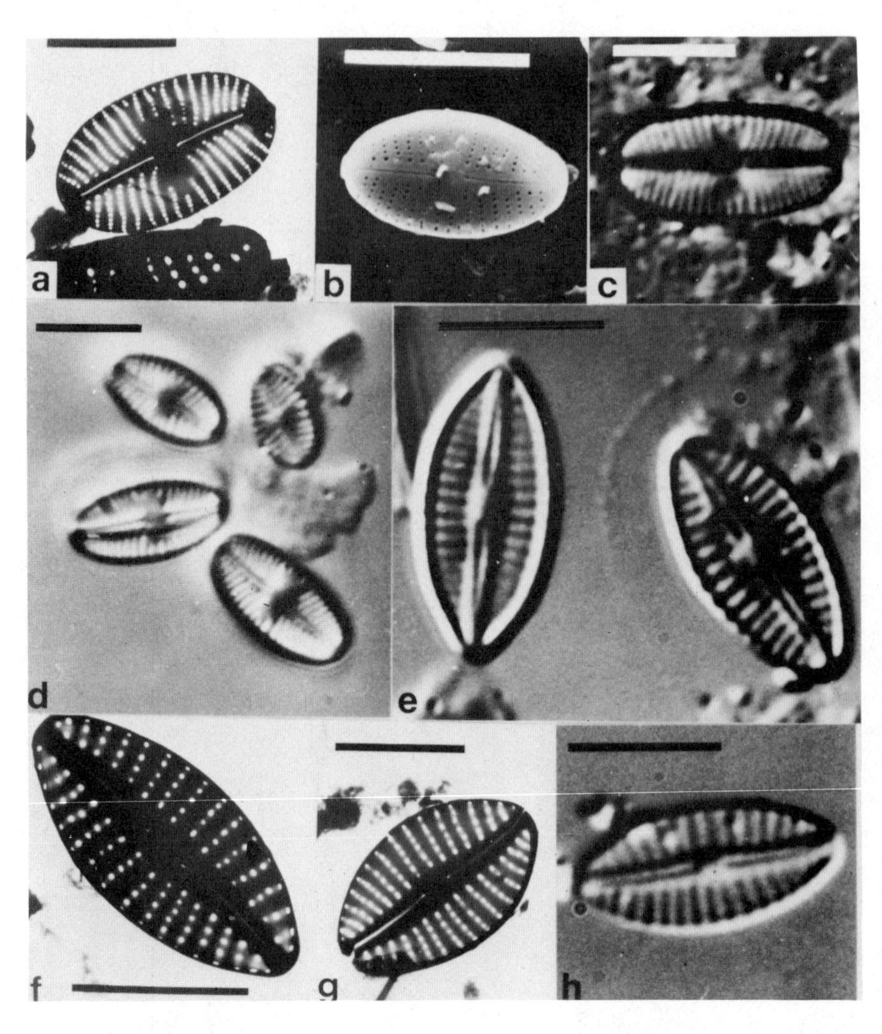

(*a–d*) *N. nigrii.*

(*a*) TEM of specimen from untreated riffle (Station 1).

(*b*) SEM of specimen labelled *Navicula* sp., collected by Blum from the Saline River just above Saline, Mich., outfall (Station 14), 6 Jul. 1952. The station was located below the outfall from a metal plating plant. The taxon was identified as *Navicula seminuloides* in Blum, 1953, and as *N. atomus* in Blum, 1957.

(*c*) PM of specimen from treated riffle (Station 4).

(*d*) PM of specimen from same sample as (*b*). (*e–h*) *N. frugalis.*

(*e*) PM of two specimens from untreated riffle (Station 1).

(*f* and *g*) TEMs of same sample as (*e*).

(*h*) PM of specimen from the Ille River, Colman, France, obtained from Lange-Bertalot, who identified this taxon as *N. frugalis* (Lange-Bertalot and Bonik, 1976). This material appears to be identical to *N. luzonensis* Hustedt, described by Patrick and Reimer (1966).

Bar equals 5 μm.

FIG. 4—Navicula nigrii *De Not.* (= N. minima *Grun.*) *and* N. frugalis *Hustedt (=* ? N. luzonensis *Hustedt). Specimens from Shayler Run except b, d, and h.*

TABLE 7—*Percent frequencies of occurrence and mean percent abundances of diatoms on glass slides for the period, 16 Feb. 1970–21 Dec. 1970. Taxa are generally ranked within each category in order of percent abundance.*

Taxon	Frequency of Occurrence		Mean Percent Abundance	
	Untreated Stations	Treated Stations	Untreated Stations	Treated Stations
Copper sensitive				
Cocconeis placentula var. *euglypta* (Ehr.) Cl.	70	52	21	<1
Gomphonema parvulum Kütz.	100	58	15	6
Gomphonema olivaceum (L.) Kütz.	50	47	12	5
Nitzschia amphibia Grun.	70	68	11	3
Nitzschia frustulum (Kütz.) Grun.	70	58	8	1
Amphora veneta (Kütz.) Hust.	60	42	2	<1
Achnanthes lanceolata (Bréb.) Grun. var. *dubia* Grun.	60	37	1	<1
Copper tolerant				
Nitzschia palea (Kütz.) W. Smith	80	89	4	31
Navicula seminulum var. *hustedtii* Pat.	80	84	2	21
Navicula nigrii De Not.	55	68	3	10
Surirella ovata Kütz.	65	74	4	5
Achnanthes minutissima Kütz.	20	58	<1	3
Navicula permitis Hust.	20[a]	32	<1[a]	3
Surirella angusta Kütz.	15	47	<1	1
Indifferent[b]				
Navicula frugalis Hust.	60	63	7	3
Navicula accomoda Hust.	40	47	<1	<1
Navicula confervacea (Kütz.) Grun.	30	26	2	<1
Navicula secreta var. *apiculata* Pat.	30	26	1	<1
Surirella brightwellii W. Smith	20	21	<1	<1

[a] Includes *Navicula saprophila* Lange-Bertalot and Bonik [*19*].
[b] Show no strong preference between the treated and untreated reaches.

TABLE 8—*Frequency of occurrence (percent of samples) of nondiatoms on glass slides during the period, 16 Feb. 1970–21 Dec. 1970.*

Taxon	Untreated Reach	Treated Reach
Schizothrix calcicola (Ag.) Gomont	30	63
Chlorococcum sp[a]	40	58
Cosmarium granatum Bréb.	5	53
Cosmarium subprotumidum Nordst.	0	32
Cladophora glomerata Kütz.[b]	21	0

[a] Believed to be principally *Chlorococcum*, but may include other coccoid green algae with similar morphology.
[b] Data from samples taken from the natural substrate. This species grew profusely on the natural substrate in the control reach, but did not develop on the glass slides because of the relatively short exposure period.

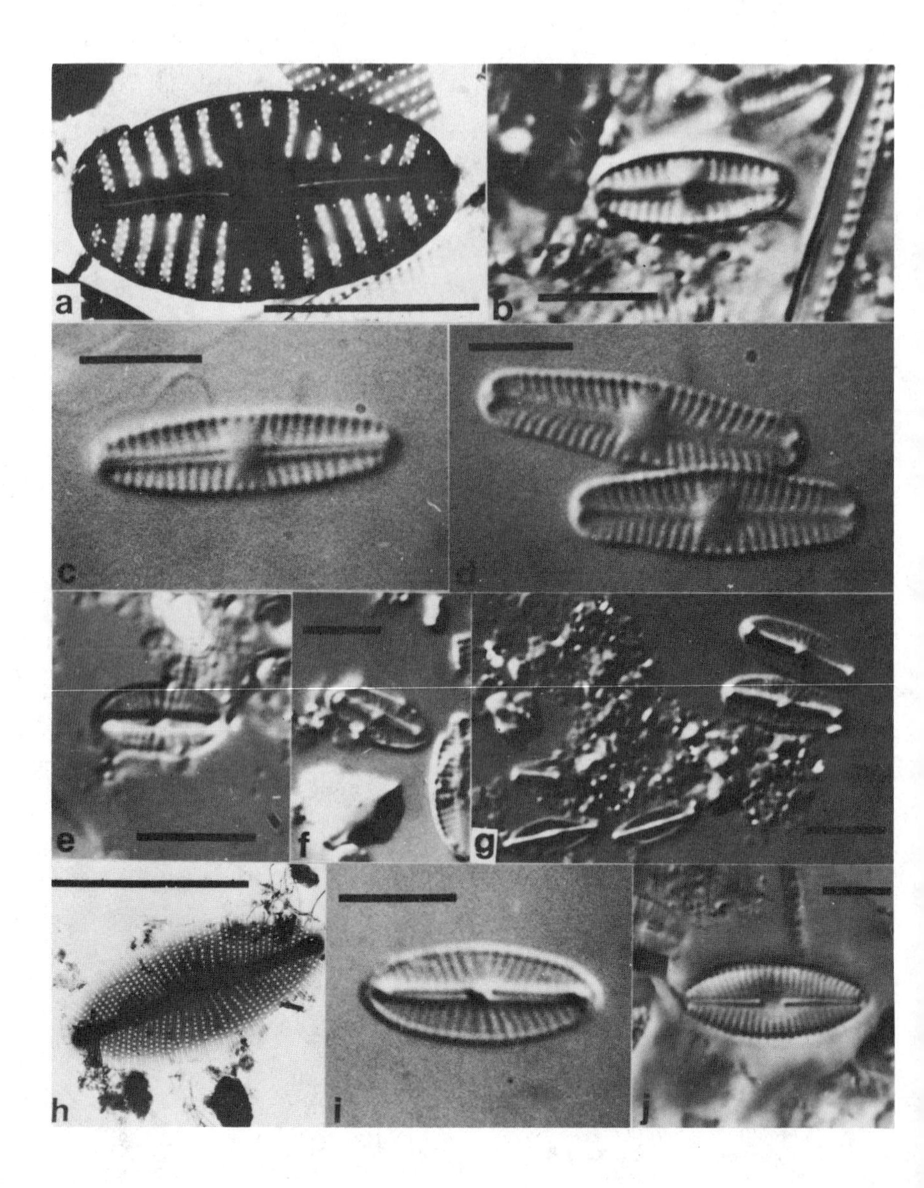

(*See caption page 123*)

Three species of the *N. permitis* association were very abundant in the copper-treated reach of Shayler Run: *N. nigrii* De Not., *N. permitis* Hust., and *N. frugalis* Hust.(= ? *N. luzonensis* Hust.) (Figs. 4 and 5). *N. saprophila* (Fig. 5) was present in the control reach of the stream, but was very sparce in the copper-treated reach.

Diatoms very similar to *N. pelliculosa* (Breb.) Hilse were also present. However, this taxon was ruled out because striae could sometimes be seen in the central area of the cells when viewed under the optical microscope. To confirm this decision, we examined the *N. pelliculosa* type material in the Van Heurck collection, "Types du Synopsis des Diatomées de Belgique," slide No. 145, France (Hodge Collection, Ohio State University, Columbus), and found that indeed no striae were visible under the light microscope and that the cells were larger than the unknown specimens from Shayler Run. When we examined the Shayler Run material with the transmission electron microscope, we found that those specimens in which striae had been visible under the optical microscope were identical to *N. permitis* Hust. (Fig. 5), as described by Lange-Bertalot and Bonik [*19*]. The specimens from Shayler

---

(*a–c*) *N. seminulum* Grun. var. *hustedtii* Patr.

(*a*) TEM of specimen from same sample as Fig. 4*e*. Note that each stria consists of a double row of punctae.

(*b*) PM from same Hyrax slide as Fig. 4*c*.

(*c*) PM from Van Heurck "Types du Synopsis des Diatomées de Belgique," slide No. 142, from the Hodge Collection; labelled "*N. minima* Grun. var." Considered to be *N. seminulum* by Schoeman and Archibald (1977) and *N. seminulum* var. *hustedtii* by Patrick and Reimer (1966).

(*d*) *N. seminulum* Grun. PM from Hodge (Van Heurck) Collection, slide No. 141; illustrated to show lateral bulge in central area.

(*e–g*) *N. permitis* Hustedt.

(*e*) PM of specimen from Shayler Run, treated pool (Station 3). Note that the striae are distinct only in the central area. The structure is finer than in *N. atomus* (Fig. 5*i*).

(*f*) PM of specimen from same slide as Fig. 4*h*, showing this species as described by Lange-Bertalot and Bonik (1976). Note that as in the Shayler Run material, the striae in this specimen are distinct only in the central area, and are too fine at the ends of the valves to be distinguished with the optical microscope.

(*g*) PM showing two cells of *N. permitis* (upper right corner) from the same Hyrax mount as Fig. 4*e*. The three valves in the lower left corner of this PM are *N. saprophila* Lange-Bertalot and Bonik.

(*h*) *N. saprophila.* TEM from same sample as Fig. 4*e*. The striae cannot be resolved with the optical microscope. This taxon is very closely related to *N. pelliculosa* (Bréb.) Hilse, but is smaller and the angle between the striae and the raphe is different. The valves of both taxa commonly appear eroded at the margins.

(*i*) *N. atomus* Kütz. PM from Hodge (Van Heurck) Collection, slide No. 149. This diatom is very similar to *N. permitis,* except that it has coarser striae, and the punctae in the striae are usually visible with the optical microscope. Specimens of this taxon in Shayler Run were intermediate between *N. permitis* and *N. atomus,* but fit the description of *N. permitis* more closely.

(*j*) *N. confervacea* Kütz. Sample from Shayler Run, untreated pool (Station 2). Commonly formed long chains of cells.

Bar equals 5 μm.

FIG. 5—*Species of* Navicula *in Shayler Run.*

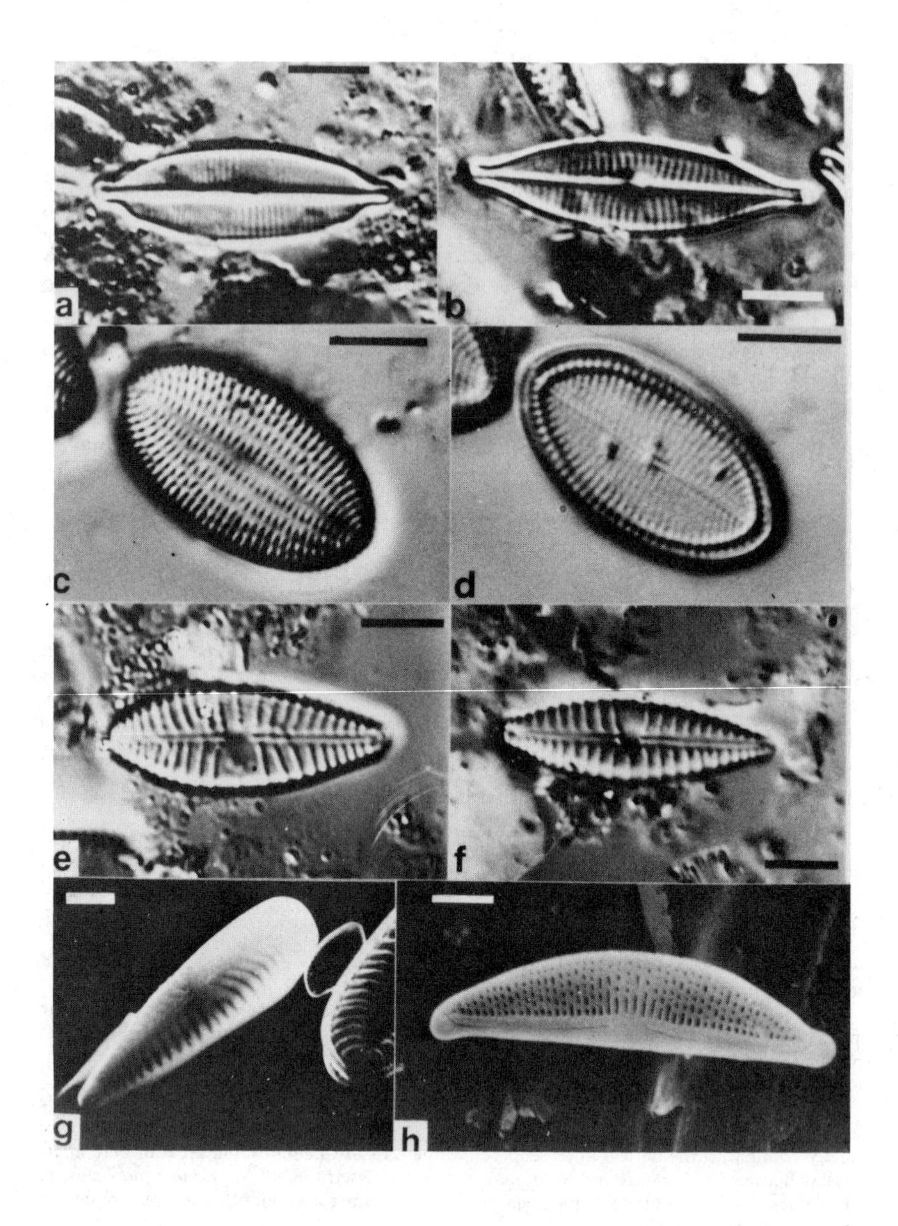

(*See caption page 125*)

Run in which striae could not be resolved with the optical microscope fit their description of *N. saprophila* (Fig. 5). The most significant difference between *N. saprophila* and *N. pelliculosa* is that in the former, the striae at the ends of the cell radiate from the raphe at a 70-deg angle, whereas in the latter, the angle is about 45 deg. This difference is visible only under the electron microscope.

Assuming that the interpretation of *N. pelliculosa* by Lange-Bertalot and Bonik [*19*] is correct, most reports of this species in the literature are probably incorrect. For example, Strain No. 668 of the Indiana Culture Collection of Algae, which is labelled *N. pelliculosa,* is instead *N. permitis.* Recently this culture was examined by Schoeman, Archibald and Barlow [*20*], using transmission electron microscopy, and was (erroneously) identified as *N. muralis* Grun. (their Figs. 24 to 27). Furthermore, in the same report, Figs. 14 to 18 are *N. saprophila* rather than *N. pelliculosa.* However, their Figs. 21 to 23, from Rabenhorst material and Hauck and Richter material, are correctly identified as *N. pelliculosa.* The probability of confusing *N. pelliculosa* and *N. saprophila* in field collections is reduced by the fact that the former is not commonly found in polluted waters.

Another interesting observation made in this study was the discovery that one of the taxa found in Shayler Run is generally identified as *N. frugalis* by Europeans and as *N. luzonensis* by Americans. We happened upon this fact when we observed that the material identified as *N. frugalis* by Lange-Bertalot and Bonik was identical to *N. luzonensis* Hust. described in Patrick and Reimer [*21*] (p. 492, pl. 46, Fig. 24). We compared material from Shayler Run that we had identified as *N. luzonensis* (Fig. 4), using Patrick and Reimer [*21*], with material identified as *N. frugalis* by Lange-Bertalot and Bonik, and have little doubt that they are identical. When similar material from the Cincinnati area was sent to the staff of the Philadelphia Academy of Natural Sciences and to the Institute of Meeresforschung at

---

(*a*) *N. accomoda* Hustedt. PM of valve collected from untreated riffle (Station 1).
(*b*) *N. secreta* var. *apiculata* Pat. PM from same slide as Fig. 6*a*.
(*c*) *Cocconeis placentula* var. *euglypta* (Ehr.) Cl. PM of pseudoraphe valve from same slide as Fig. 5*e*.
(*d*) Raphe valve of same frustule shown in Fig. 6*c*.
(*e*) PM of *Gomphonema parvulum* Kütz. Sample collected from untreated pool (Station 2).
(*f*) Same as (*e*), except collected from the treated riffle (Station 4).
(*g*) *Gomphonema olivaceum* (L.) Kütz. SEM at 45-degree angle. Collected from untreated riffle (Station 1).
(*h*) *Amphora veneta* (Kütz.) Hustedt. SEM of specimen collected from untreated riffle (Station 1).
Bar equals 5 μm.

FIG. 6—*Species of* Navicula, Cocconeis, Gomphonema *and* Amphora. *All material was collected from Shayler Run.*

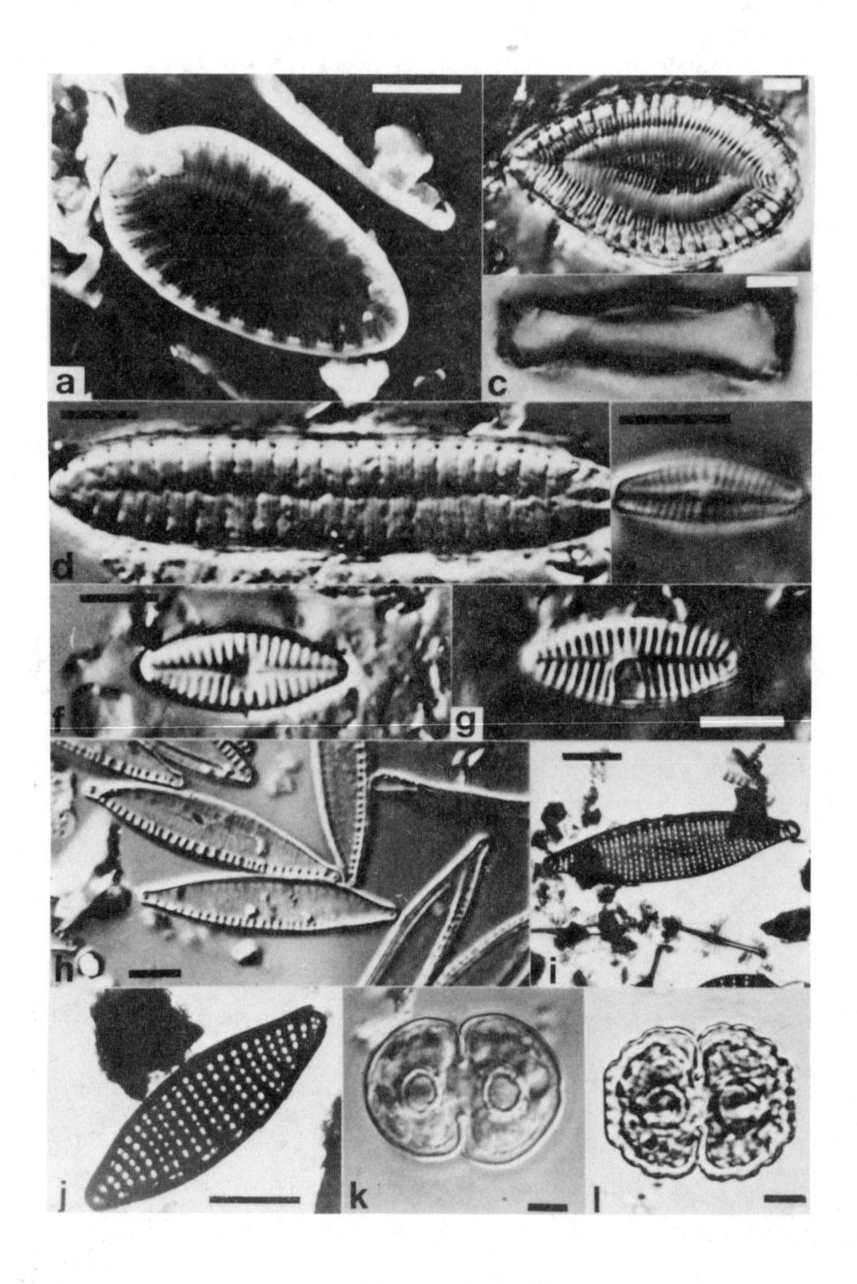

(*See caption page 127*)

Bremerhaven, it was identified as *N. luzonensis* by the former and *N. frugalis* by the latter. Lange-Bertalot and Bonik [*19*] also noted that the finer structured forms of *N. frugalis* intergrade with *N. muralis* Grun.

The small *N. seminulum* var. *hustedtii* was the most abundant species of *Navicula* in the copper-treated reach of Shayler Run, and could also be included in the *N. permitis* association, especially since it is almost always present in waste water along with *N. nigrii* (=*N. minima*). The type variety of this species was not found in the stream, but was present in small numbers in a stream-side, glass, flow-through fish bioassay chamber treated with 160 μg copper/litre. The most detailed descriptions of *N. seminulum* and the variety *hustedtii* were found in Patrick and Reimer [*21*] and in Schoeman and Archibald [*22*], but the later do not acknowledge *hustedtii* as a distinct variety. We have shown both forms from the Van Huerck collection in Fig. 5.

Of the parameters evaluated in this study, the species composition was found to be the most sensitive and informative measure of the effects of copper on the periphyton. It has been frequently reported that water pollution results in a decrease in the diversity of diatoms and other aquatic organisms, and there has been an increasing tendency to report the biological effects of pollution in terms of numerical species diversity indexes [*7*]. It may even be tempting to the nonbiologist to derive and utilize these indexes without determining the species composition of the organisms [*23*]. In Shayler Run, however, the effect of the copper was contrary to the prevailing pattern. The replacement of *Cocconeis placentula* by *Nitzschia palea, Navicula nigrii,* and *N. seminulum* resulted in an increase rather than a decrease in the diatom species diversity index in the treated riffle (Table 10). It is evident, therefore, that the knowledge of the species present and their relative abundances is

---

(*a*) *Surirella ovata* Kütz. SEM from same collection as Fig. 6*h*.
(*b*) *S. brightwellii* W. Sm. Collected in the treated reach. Notice highly undulate nature of valve face.
(*c*) PM of *S. brightwellii* in girdle view.
(*d*) PM of *S. angusta* Kütz. from same Hyrax mount as Fig. 5*e*.
(*e*) *Achnanthes minutissima* Kütz. from untreated riffle (Station 1).
(*f*) *A. lanceolata* var. *dubia* Grun. PM of raphe valve from same Hyrax mount as Fig. 5*e*.
(*g*) PM of pseudoraphe valve of same frustule shown in Fig. 7*f*.
(*h*) PM of *Nitzschia palea* (Kütz.) W. Sm. from a nearly pure growth, which formed a ring at water line on rocks in the treated riffle (Station 4).
(*i*) TEM of *N. frustulum* (Kütz.) Grun. from same sample as Fig. 4*e*.
(*j*) TEM of *N. amphibia* Grun. from same sample as in Fig. 4*e*.
(*k*) PM of *Cosmarium subprotumidum* from formalin-preserved wet mount. Collected from treated riffle (Station 4).
(*l*) PM of *C. granatum* Breb. from same sample as Fig. 7*k*.
Bar equals 5 μm.

FIG. 7—*Species of* Surirella, Achnanthes, Nitzschia *and* Cosmarium. *All material was collected from Shayler Run.*

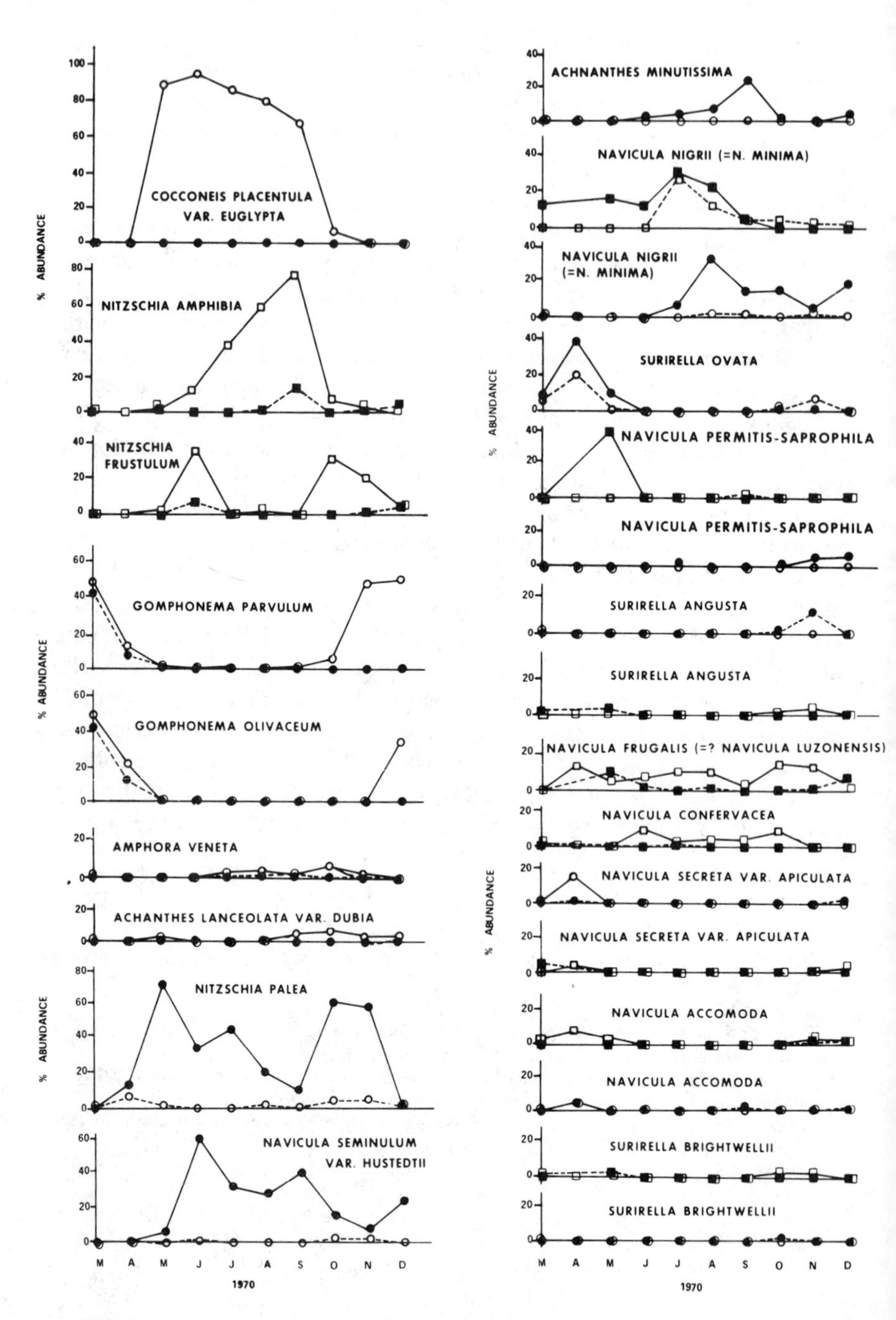

FIG. 8—*Percent relative abundances of dominant diatoms in Shayler Run: upper (control) riffle (○); lower (treated) riffle (●); upper (control) pool (□); lower (treated) pool (■).*

TABLE 9—*Algal periphyton tolerant of copper.*

Investigator (year)	Species	Copper Concentration Tolerated	Remarks
Schroeder (1939)	*Achnanthes affinis*	1.5 mg/litre	...
	Cymbella naviculiformis		
	Cymbella ventricosa		
	Fragilaria virescens		
	Gomphonema parvulum		
	Navicula viridula		
	Neidium bisculcatum		
	Nitzschia palea		
	Achnanthes affinis	2.1 mg/litre	...
	Cymbella ventricosa		
	Nitzschia palea		
Blum (1953, 1957)	*Navicula nigrii*[a]	(copper present)	dominant algae below discharge containing Cu, Cr, CN
	Nitzschia linearis		
	Nitzschia palea		
	Stigeoclonium tenue		
Butcher (1955)	*Achnanthes affinis*	(copper present)	first algae to appear below factory discharge containing copper
	Chlorococcum sp.		
	Stigeoclonium tenue		
Fjerdingstad (1965)	*Navicula viridula*	...	copper resistant
	Nitzschia palea		
	Stigeoclonium tenue		

[a] Reported as *Navicula seminuloides* [10] and later as *Navicula atomus* [11].

TABLE 10—*Species diversity of Diatoms (D-Bar calculated according to Wilhm* [7]).

Date	Station				Sum	Mean
	Control Riffle	Control Pool	Treated Pool	Treated Riffle		
16 Feb.-16 Mar. 1970	1.373	2.270[a]	1.700	1.629	6.972	1.743
7 Apr.-5 May 1970	2.983	2.745	2.685[a]	2.544	10.957	2.739
5 May-2 June 1970	0.750	3.566	1.679	1.498	7.493	1.873
2 June-1 July 1970	0.374	2.804	1.453	1.335	5.966	1.492
1 July-31 July 1970	0.792	2.185	1.506	2.045	6.528	1.632
31 July-30 Aug. 1970	1.249	1.919	1.697	2.321	7.186	1.796
30 Aug.-29 Sept. 1970	1.861	1.295	2.323	2.279	7.758	1.940
29 Sept.-27 Oct. 1970	2.822	3.104	1.293	1.732	8.951	2.238
27 Oct.-24 Nov. 1970	2.641	3.369	2.473	2.164	10.647	2.662
24 Nov.-21 Dec. 1970	1.897	3.002	3.632	2.953	11.484	2.871
Sum	16.742	26.259	20.441	20.500	83.942	
Mean	1.674	2.626	2.044	2.050		2.099

Analysis of variance[b]

Source	*df*	Sum of Squares	Mean Square	*F*	F_{05}
Stations	3	4.2609	1.4203	3.462	2.99
Dates	9	8.2976	0.9220	2.248	2.28[c]
Error	25	10.2552	0.4102		

[a]Missing item determined according to Snedecor [*24*].
[b]Analysis corrected for missing item.
[c]Significant at the 95 percent probability level.

basic to an understanding of the effects of pollution on aquatic communities. After this information has been derived from the samples, it may be compared with published information relating to associations of species that are indicative of environmental conditions, or used to calculate numerical species-diversity indexes.

Acknowledgments

We are grateful to the following persons for their assistance in this study; Michael Weisner, U.S. Environmental Protection Agency, Cincinnati, for providing slide mounts of diatoms from the Ille River, France, and the Rhine River, Germany, which he received from Dr. H. Lange-Bertalot; Dr. Hanna Croasdale, Dartmouth College, Hanover, New Hampshire, for identifying the desmids found in our samples; Dr. Charles Reimer, Academy of Natural Sciences of Philadelphia, for identifying *Navicula luzonensis* Hustedt; Dr. Reimer Simonsen, Institut für Meeresforschung, Bremerhaven, for identifying *Navicula frugalis* Hustedt; Dr. J. L. Blum, Department of Botany, University of Wisconsin, Milwaukee, for providing material from his collections from the Saline River, Michigan; and to Donald Moore of our laboratory, for providing valuable assistance in carrying out the Sedgwick-Rafter counts and other analyses.

References

[1] Geckler, J. R., Horning, W. B., Neiheisel, T. M., Pickering, Q. H., Robinson, E. L., and Stephan, C. E., "Validity of Laboratory Tests for Predicting Copper Toxicity in Streams," EPA 600/8-76-116, U.S. Environmental Protection Agency, Duluth, Minn., 1976.

[2] Butcher, R. W., *Verh. Int. Ver. Limnol.*, Vol. 12, 1955, pp. 8823-8827.

[3] Fjerdingstad, E., *International Revue ges. Hydrobiologie*, Vol. 50, No. 4, 1965, pp. 475-604.

[4] Kolkwitz, R. and Marsson, M., *Ber. Deut. Bot. Gesell.*, Vol. 26a, 1908, pp. 505-519.

[5] Weber, C. I., "Methods of Collection and Analysis of Plankton and Periphyton Samples in the Water Pollution Surveillance System," Application and Development Report No. 19, Federal Water Pollution Control Administration, Division of Pollution Surveillance, Cincinnati, Ohio, 1966.

[6] Weber, C. I. and Raschke, R. L., "Use of a Floating Periphyton Sampler for Water Pollution Surveillance," Application and Development Report No. 20, Federal Water Pollution Control Administration, Division of Pollution Surveillance, Cincinnati, Ohio, 1966, pp. 1-22.

[7] Wilhm, J. L., *Journal of the Water Pollution Control Federation*, Vol. 42, No. 5, 1970, pp. R221-224.

[8] Weber, C. I. in *Bioassay Techniques and Environmental Chemistry*, G. Glass, Ed., Ann Arbor Science Publishers, Ann Arbor, Mich., 1973, pp. 119-138.

[9] Schroeder, H., *Pflanzenf.*, Vol. 21, 1939, p. 1088.

[10] Blum, J. L., "The Ecology of Algae Growing in the Saline River, Michigan, with Special Reference to Water Pollution, Ph.D. thesis, University of Michigan, Ann Arbor, Mich., 1953.

[11] Blum, J. L., *Hydrobiology*, Vol. 9, 1957, pp. 361-408.

[12] McFarland, B. and Weber, C. I., "The Systematics and Pollution Tolerance of the Diatoms *Navicula minima* and *Navicula seminulm* var. *hustedtii*," U.S. Environmental Protection Agency, National Environmental Research Center, Cincinnati, Ohio, 1973. Presented at the 21st Annual Meeting, Midwest Benthological Society, Michigan State University, East Lansing, Mich., 21-23 March 1973.

[13] Granetti, B., *Giorn. Bot. Ital.*, Vol. 102, 1968, pp. 427-437.

[14] Van Heurck, H., *Synopsis des Diatomeés de Belgique, Anvers*, 2 Vols., text and atlas, 1880-1885.

[15] Thomas, E. A., *Archiv für Mikrobiologie*, Vol. 42, 1962, p. 246.

[16] Whitton, B. A., *Archiv für Mikrobiologie*, Vol. 58, 1967, pp. 21-29.

[17] Drouet, F., "Revision of the Classification of the Oscillatoriaceae," Monograph 15, Academy of Natural Sciences, Philadelphia, 1968, pp. 1-370.

[18] Brook, A. J., *Limnology and Oceanography*, Vol. 10, 1965, pp. 403-411.

[19] Lange-Bertalot, H. and Bonik, K., Algological Studies 16, *Archiv für Hydrobiologie* Suppl. Bd., Vol. 49, No. 3, 1976, pp. 303-332.

[20] Schoeman, F. R., Archibald, R. E. M., and Barlow, D. J., *British Phycological Journal*, Vol. 11, 1976, pp. 251-263.

[21] Patrick, R. and Reimer, C. W., "The Diatoms of the United States Exclusive of Alaska and Hawaii," Vol. 1, Monograph 13, Academy of Natural Sciences, Philadelphia, 1966.

[22] Schoeman, F. R. and Archibald, R. E. M., "The Diatom Flora of Southern Africa," No. 3, C.S.I.R. special report WAT 50, Pretoria, South Africa, 1977.

[23] McFarland, B. H. and Weber, C. I., "Seasonal Changes in the Periphyton of a Small Calcareous Stream," Federal Water Pollution Control Administration, Division of Pollution Surveillance, Cincinnati, Ohio, 1970. Presented at the 18th Annual Meeting, Midwest Benthological Society, St. Mary's College, Winona, Minn., 1-3 April 1970.

[24] Snedecor, G. W., *Statistical Methods*, Iowa State University Press, Ames, Iowa, 1956.

J. H. Sullivan, Jr.,[1] *H. D. Putnam,*[1] *J. T. McClave,*[2]
and D. R. Swift[3]

Statistical Techniques for Evaluating Procedures and Results for Periphyton Sampling

---

REFERENCE: Sullivan, J. H., Jr., Putnam, H. D., McClave, J. T., and Swift, D. R., **"Statistical Techniques for Evaluating Procedures and Results for Periphyton Sampling,"** *Ecological Assessments of Effluent Impacts on Communities of Indigenous Aquatic Organisms, ASTM STP 730,* J. M. Bates and C. I. Weber, Eds., American Society for Testing and Materials, 1981, pp. 132–141.

ABSTRACT: Statistical procedures for evaluating periphyton sampling techniques and results are described. The sampling technique utilized was glass slides incubated at various locations in a reservoir receiving industrial effluent. The density, diversity, and dominance of organisms were utilized as dependent variables. Initially, the sampling and processing technique was evaluated to determine the variability at each step. From this information, decisions were made regarding where in the process replication would be most important. Further, statistical procedures are demonstrated for relating sampling replication to the ability to detect differences between results obtained at different sampling locations. Techniques for relating processing or counting effort to the reliability of estimates of dominant species are discussed. The statistical techniques illustrated can be used to relate overall sampling or processing effort or costs to the reliability of the final results.

KEY WORDS: statistics, periphyton sampling, density, diversity, dominant species, environmental survey, impact assessment, diatom sampling, statistical evaluation techniques, ecology, effluents, aquatic organisms

In the spring of 1977 an integrated chemical-biological study was conducted in a portion of an impoundment in Chattanooga, Tenn., for the purpose of assessing the water quality impacts of an industrial effluent. As a part of this work, a more detailed sampling and statistical evaluation was con-

[1]Environmental engineer and aquatic biologist, respectively, Water and Air Research, Inc., Gainesville, Fla. 32602.

[2]Statistician, Water and Air Research, Inc., Gainesville, Fla. 32602; current address: Info Tech, Inc., Gainesville, Fla.

[3]Phycologist, Water and Air Research, Inc., Gainesville, Fla. 32602; current address: South Florida Water Management District, West Palm Beach, Fla.

ducted on the diatom community as sampled, utilizing artificial substrates. The purpose of this work was to define the key variables in the procedures utilized and to relate the degree of reliability of the results to the sampling and processing efforts. This latter purpose can often be achieved by conducting a pilot study prior to undertaking a major sampling program. Once one is able to estimate the degree of reliability that can be placed on the results as a function of effort, the most economical use of resources can be determined.

In the work reported here, the dependent variables selected were organism density, organism diversity, and organism dominance (that is, the relative fraction of the total organisms). These are, of course, not the only indicators of environmental quality but are used here to illustrate a rational approach to sampling plan design based on relating the degree of reliability in the results to the sampling and processing effort. The approach and techniques utilized herein can be applied to any type of environmental sampling that yields finite numerical estimates for the parameters of interest.

Sampling and Analysis Techniques

Standard 25 by 75-mm microscope glass slides were placed 25 mm below the water surface in slide racks at nine sampling locations in the bay receiving the industrial effluent and at three sampling locations in an adjacent unnamed "reference" bay. After a 30-day incubation period, the slides were collected and placed in individual glass bottles. The slides to be used for diatom enumeration were selected randomly.

Periphyton growth was scraped from the glass slides and oxidized with hydrogen peroxide and potassium dichromate. The solution was cooled, settled, decanted, and brought to a known volume. Permanent slides were prepared by pipetting a 0.4-ml aliquot of the "cleaned" material onto an 18 by 18-mm coverslip and allowing the sample to dry at 65°C (150°F) on a laboratory hotplate. The dried coverslip was placed on a standard microscope slide containing 1 drop of Hyrax mounting medium, and the slide was gently heated to drive off the toluene solvent. Under an oil immersion lens (Zeiss microscope, ×1000) diatoms were identified and enumerated to the species level, where possible.

A total of ten microscope fields were examined at each of six randomly selected positions for a total of 60 fields. The number of organisms found in the 60-field area was mathematically related back to the incubated field slide for determination of organism density.

In the statistical analyses the logarithmic transformation ln(density + 1) was used for density data to obtain a variable that more nearly satisfies the assumptions necessary for an analysis of variance. Shannon's [*1*][4] measure of diversity, $-\Sigma p \ln p$, was utilized with the following modification. Basharin

[4]The italic numbers in brackets refer to the list of references appended to this paper.

[2] has shown that Shannon's estimate of the true population diversity is biased and that the bias can be estimated by $(s - 1)/2N$, where s is the number of species, and N is the total number of organisms observed. Shannon's diversity (Base e) plus the adjustment is relatively stable as the sample size changes. Basharin has also derived an estimate of the standard error for Shannon's diversity. This allows one to measure the reliability of the diversity estimate and to compare diversities from different sampling locations.

Discussion of Results

Statistical analyses were conducted to consider the following factors:

1. The sources of variability in the technique, specifically: (*a*) variability between aliquots, (*b*) variability between fields, (*c*) variability between individual slides, and (*d*) variability between composites of five slides.
2. Comparisons between sampling locations and the reliability of such estimates as a function of sample size.
3. Determination of dominant species as a function of the number of cells identified.

More extensive sampling was done at four sampling locations representative of the range of conditions found during the survey. Much of the statistical evaluation was based on the results from these four locations.

Comparison of Diversity and Density Variability Between Field with That Between Aliquots

One slide was selected from each of four stations. Ten aliquots were taken from each slide, and 120 fields or 500 organisms, whichever came first (with a 60-field minimum), were counted from each aliquot. Counts were made in units of ten fields.

The variability between fields was compared with that between aliquots at each station. Bias-adjusted diversity values and logarithmic transforms of density were analyzed using a nested analysis of variance technique. The resulting ratios provide no evidence that the aliquot variability and the field within aliquot variability differ. (Statistical details may be obtained from the authors.)

The practical significance of this finding is that no additional reliability is gained by replicating aliquots as opposed to replicating fields.

Comparison of Diversity and Density Variability Between Slides to That Within Slides (Between Fields)

Nine slides were selected from each of four stations, and a count was obtained for 60 fields (ten fields at each of six random locations) on a single microscope slide prepared from each field-incubated slide. These slides were used to obtain an estimate of between-slide variability. An independent estimate of the within-slide variability was obtained from the previous experiment, which was performed on a different slide.

The F ratios comparing these two sources of variability indicate that there is slightly more variability between slides than within slides for both diversity and density. The implication is that one should obtain replication by sampling several slides rather than aliquoting a single slide.

Comparison of Diversity and Density Variability When One Slide is Used with That When Five Slides are Composited

Growth from five slides was composited. A permanent mount was prepared from this composite, and 60 fields were counted. Five replications of this experiment were performed at each of four stations. The variance for the five slide composites was compared with the single-slide variance found previously using a two-way analysis of variance.

The results showed that the compositing of five slides does not seem to reduce the variability of either diversity of density estimates. Since more replication (and thus better variance estimates) can be obtained by analyzing the slides separately, the single-slide method is probably preferable to compositing slides.

Comparisons Between Sampling Locations

Since the previous results indicated that a replicated, single-slide experiment at each location was preferred, this experiment was performed at seven other locations, five replicates at each location.

An analysis of variance comparison was deemed inappropriate due to the probable inequality of the diversity variances. Thus, the stations were compared in a pairwise fashion. The lines on the left of the station numbers in Table 1 reflect the results of comparing the diversities by using the two-sample Z statistic, connecting stations that cannot be declared significantly different. The formula for Z is

$$Z = \frac{(\text{Station } i \text{ diversity} - \text{Station } j \text{ diversity})}{\sqrt{(\text{standard error } i)^2 + (\text{standard error } j)^2}}$$

Since Basharin shows that Shannon's diversity estimate is asymptotically normal, this statistic should approximate a standard normal random variable when the true population diversities are the same, and it does not necessitate equal variances. When the value of Z is large (negative or positive), the implication is that the true station diversities differ. A conser-

TABLE 1—*Organism diversities for all stations.*

Station	Diversity	Standard Error
6	0.716	0.035
7	1.189	0.033
5	1.608	0.043
8	2.045	0.025
9	2.175	0.032
4	2.440	0.074
11	2.557	0.020
1	2.573	0.074
12	2.790	0.019
2	2.918	0.089
3	2.929	0.061

vative 0.01 significance level, $Z = 2.58$, was used since so many comparisons were made.

The logarithmic density transforms were subjected to a standard analysis of variance. The station comparison yielded a highly significant F value of 319.81, leaving no doubt that the station densities differ. The transformed densities were then subjected to Duncan's multiple comparison procedure [*3*] to determine which stations differ. The estimated densities (untransformed) are shown in Table 2, with the solid lines on the left of the station numbers connecting those stations that one cannot conclude to be different at the 0.05 significance level.

Reliability as a Function of Sample Size

Perhaps the most crucial decision that must be made in planning an ex-

TABLE 2—*Population densities for all stations.*[a]

Station	Density (number/mm^2)
11	37 293
8	27 803
9	22 462
12	19 698
7	18 936
6	14 911
5	750
4	174
1	144
3	85
2	28

[a]The conclusion is that the station densities are aligned as follows: 11 > 9, 12, 7, 6, 5, 4, 1, 3, 2; 8 > 6, 5, 4, 1, 3, 2; 9, 12, 7, 6 > 5 > 4 > 1, 3 > 2.

periment is the sample size, or number of replicates to be processed. One can use information from previous experiments to plan more carefully.

Previously, it was concluded that replication of slides was more desirable than replication of aliquots from a single slide. The question then becomes, How many slides should be sampled? Since diversity is analyzed on an untransformed scale, it is easier to work in absolute rather than percentage differences. The following equation is for the absolute difference between two stations' diversities that can be detected with 90 percent confidence for n replicate slides:

$$D = 2(1.645)\sigma_1\sqrt{2/n} = 4.653\sigma_1/\sqrt{n}$$

where σ_1 is the standard deviation of the diversity based on single-slide estimates. This assumes that the stations being compared have the same standard deviation σ_1. The probability of falsely concluding a difference exists is set at $\alpha = 0.10$.

A plot of D versus n for the four experimental stations is shown in Fig. 1. Note that for all but Station 3, where the density of organisms was very low, one can detect a diversity difference of slightly less than 0.3 with 90 percent confidence using five replicate slides.

Since the statistical analysis of density is performed on a logarithmic scale, the differences one can detect between station densities will be expressed in

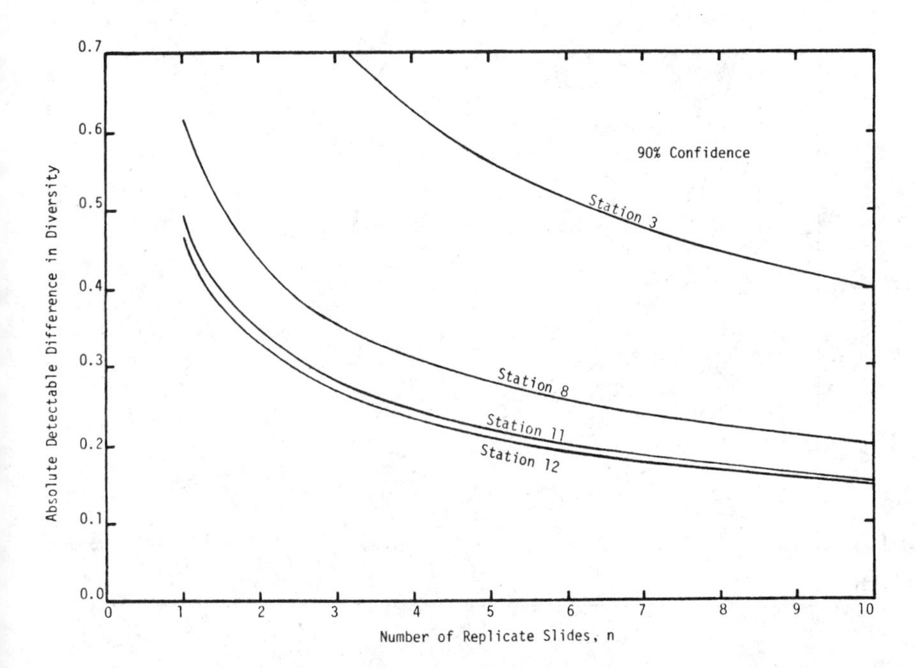

FIG. 1—*Sample size to meet diversity criteria.*

percentages. The absolute percentage difference which can be detected with 90 percent confidence, with $\alpha = 0.10$, is given by

$$P = 100[\exp(4.653\sigma_1/\sqrt{n}) - 1]$$

where σ_1 is the standard deviation of the *transformed* density for single-slide estimates. A plot of p versus n for the four experimental stations is shown in Fig. 2. Note that one must have at least five replicate slides before one can be 90 percent sure of detecting density differences of 100 percent. It is obvious that density comparisons require more replication than do diversity comparisons to achieve similar reliability.

In order to determine the necessary sample size for estimating and comparing diversities or densities, some preliminary estimate of the variance σ_1^2 must be obtained.

Determination of Dominant Species

An analysis of dominant species was conducted using nine replicate slides at a single station. The basis for dominant species selection was that each species make up at least 3 percent of the pooled sample at the station. Six species satisfied this criteria. The data are shown in Table 3.

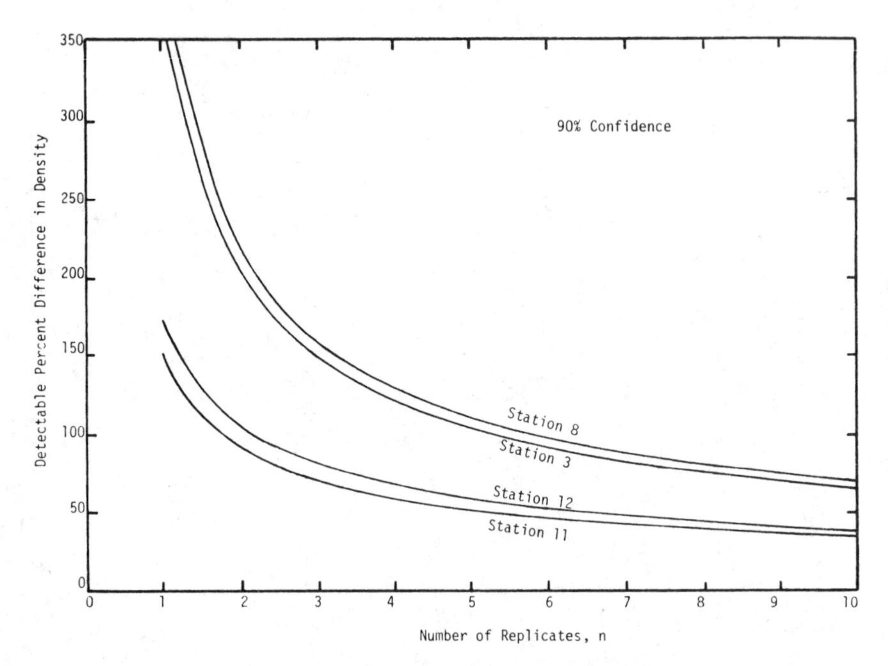

FIG. 2—*Sample size to meet density criteria.*

TABLE 3—*Dominant species for Station 8—periphyton, artificial substrate.*

	Achnanthes minutissima		*Cymbella prostrate* v. *auerswaldii*		*Diatoma tenue* v. *elongatum*		*Fragilaria vaucheriae*		*Melosira varians*		*Synedra rumpens*		All other species		
Slide	No.	%	No.	%	No.	%	No.	%	No.	%	No.	%	No.	%	Total
1	56	8.0	23	3.3	284	40.8	138	19.8	33	4.7	81	11.6	82	11.8	697
2	35	7.9	19	4.3	179	40.1	77	17.3	23	5.2	56	12.6	57	12.8	446
3	12	5.1	6	2.6	120	51.1	40	17.0	8	3.4	26	11.1	23	9.8	235
4	27	7.5	12	3.3	131	36.2	100	27.6	8	2.2	47	13.0	37	10.2	362
5	27	8.1	16	4.8	150	44.9	54	16.2	7	2.1	38	11.4	42	12.6	334
6	40	12.5	5	1.6	95	29.7	64	20.0	15	4.7	44	13.8	57	17.8	320
7	27	7.2	10	2.7	145	38.8	51	13.6	13	3.5	75	20.1	53	14.2	374
8	13	6.1	10	4.7	83	39.0	36	16.9	10	4.7	26	12.2	35	16.4	213
9	21	7.4	6	2.1	130	45.9	47	16.6	8	2.8	29	10.3	42	14.8	283
Total	258	7.9	107	3.3	1317	40.4	607	18.6	125	3.8	422	12.9	428	13.1	3264

Binomial theory indicates that one could pool the results of all replicates to obtain estimates and confidence intervals for the true proportion of each species of interest. In order to test the applicability of this theory, a χ^2 contingency table test was conducted to determine whether there is significant difference between replicates. This analysis showed that a significant difference exists between the replicate estimates of percent composition for the six species. Since the replicates are all taken from the same population (location), the percent composition of each replicate estimates the true composition at the location. The fact that the χ^2 test shows a significant difference in replicate percentages warns that the pooling and use of binomial theory is inadvisable.

The variability observed between the replicate proportions is significantly greater than that predicted by binomial theory. Thus, in determining the sample size necessary to estimate the percent composition for a species with a specific accuracy, the binomial formula will underestimate the necessary sample size.

The number of organisms necessary to estimate the true percentage of a species to a desired bound with 95 percent confidence is given by the formula

$$n = \frac{4(\text{estimated variance})}{(\text{bound})^2} \times 10^4$$

As noted previously, the estimated variance is obtained from the observed replicate percentages. Figure 3 shows the number of organisms to be counted versus the bound of the estimate for several of the species in Table 3.

In summary, if slide replicates are to be used for estimating percent compositions, the variability of the estimates may be considerably more than is expected assuming a binomial distribution. More realistic estimates of variability can be calculated from the actual sample results, and these in turn can be used to estimate the sample sizes necessary for a specified accuracy of estimation.

In this paper, procedures for examining the variability associated with the various steps in a sampling and analytical technique have been demonstrated. Through such an effort, one can identify the optimum points to employ replication in order to reduce the confidence limits around the final result. Techniques have been demonstrated that relate sample replication to the ability to detect differences between results obtained at different sampling locations. Finally, a technique for relating processing or counting effort to the reliability of estimates of dominant species has been demonstrated. Through the use of statistical techniques such as are illustrated here, one can relate overall sampling and processing effort or costs to the reliability of the final results.

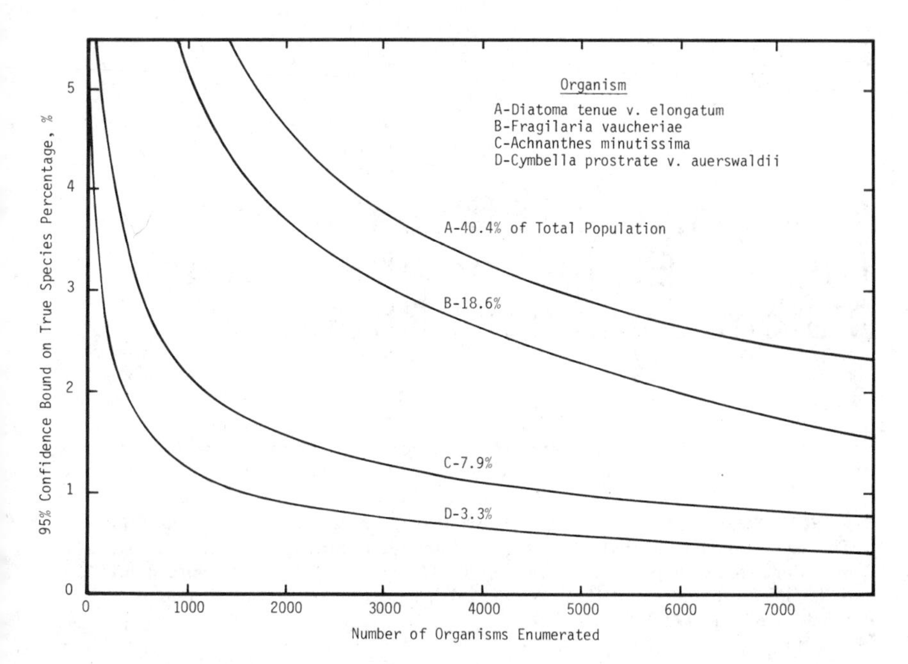

FIG. 3—*Confidence bound for dominant species as a function of the number of organisms counted.*

Acknowledgments

This research was supported by the U.S. Army Medical Research and Development Command, Ft. Detrick, Frederick, Md., and the U.S. Army Armament Research and Development Command, Ecological Research Office, Aberdeen Proving Ground, Md., under Contract No. DAMD-17-75-C-5049.

An appreciation is also extended to the Ecological Research Office at Edgewood Arsenal for assistance in the field study.

The authors especially wish to thank J. Gareth Pearson and Edward Bender for helpful criticism and encouragement throughout the entire investigation.

References

[1] Shannon, C. E., *The Mathematical Theory of Communication*, University of Illinois Press, Urbana, Ill., 1948, pp. 3-89.

[2] Basharin, G. P., *Theory of Probability and Its Applications*, Vol. 4, 1959, pp. 333-336.

[3] Duncan, D. B., *Biometrics*, Vol. 31, 1975, pp. 339-359.

R. L. Weitzel[1] *and J. M. Bates*[2]

Assessment of Effluent Impacts Through Evaluation of Periphyton Diatom Community Structure

---

REFERENCE: Weitzel, R. L. and Bates, J. M., **"Assessment of Effluent Impacts Through Evaluation of Periphyton Diatom Community Structure,"** *Ecological Assessments of Effluent Impacts on Communities of Indigenous Aquatic Organisms, ASTM STP 730,* J. M. Bates and C. I. Weber, Eds., American Society for Testing and Materials, 1981, pp. 142-165.

ABSTRACT: Evaluation of changes in periphyton diatom community structure proved to be a useful means of assessing the impact of an electroplating waste discharge on the biota of the Muskingum River in Ohio. The copper waste discharge decreased species richness and species diversity (as indicated by the Shannon-Weaver and Brillouin indexes) of the diatom community collected from artifical substrates. Corresponding evenness and redundancy components also indicated stressed conditions.

Short-count intervals of detailed readings were compared to determine the ability and sensitivity of these shorter efforts of enumeration to respond to community changes. Counts of 500 diatom cells, while accurately responding to gross perturbations when applied in the form of biotic indexes, were somewhat inconsistent and apparently insensitive to minor shifts in diatom community structure. Counts of 1000 diatom cells more closely paralleled the trends developed by the 5000-cell long counts. Extension of the diatom counts beyond 2000 cells yielded less species information for the man-hours required, and it was at this level that the species curve began to approach an asymptote.

KEY WORDS: periphyton, diatoms, species diversity, artificial substrates, biomonitoring, copper pollution, ecology, effluents, aquatic organisms

Historically, various methods of biological monitoring have been employed to assess water quality and to evaluate environmental conditions. Basically, three approaches have been used [*1*]:[3]

(*a*) measurements that utilize the types and quantities of indicator organisms [*1-9*],

[1] Regional marketing coordinator, Ecological Analysts, Inc., Northbrook, Ill. 60062.
[2] President, Ecological Consultants, Inc., Ann Arbor, Mich. 48103.
[3] The italic numbers in brackets refer to the list of references appended to this paper.

(*b*) mathematical indexes summarizing information on community structure [*10–19*], and

(*c*) measurements of physiological and behavioral responses of select organisms to stress [*1*].

Patrick [*7*] has concluded that diatom community structure is one of the better monitors of water quality. Patrick, Hohn, and Wallace [*10*] have shown that various kinds of pollution cause shifts in the structure of diatom communities and that diatom communities fit a truncated normal curve of distribution. The concept and application of the truncated normal curve is further developed by Patrick and Strawbridge [*11*] and by Cairns et al [*12*]. Although the most reliable procedure statistically, application of the truncated normal distribution has been criticized because of the level of effort required to generate the necessary species data to assure accurate and meaningful interpretations. This method requires detailed sample readings and may be prohibitively expensive in studies where one is looking for only gross differences in periphyton community structure.

Sufficiently meaningful information regarding periphyton diatom community structure may be achieved from what has become termed "short counts" and the application of other means of community structure analysis. Numerous studies have been published describing data generated from short-count sample analyses. Counts of 500 cells have been reported by Collingsworth et al [*20*], Stoermer et al [*21*], and Weitzel et al [*22*]. As few as 200 cells were enumerated by Barron [*23*], while Archibald [*24*] counted 400 cells, and Bradbury [*25*] reported on counts of 400 to 1000 cells. Anderson [*26*] counted 2500 cells, while 500 to 2000 cells were enumerated by Evans and Stockner [*27*]. Tippett [*28*] reported counts of 500 to 1000 valves, and 400 to 500 valves were considered statistically reliable by Schoeman [*29*]. Sreenivasa and Duthie [*30*] applied the total number of individuals encountered in three transects of a prepared slide. Castenholz [*31*] applied counts of 300 cells from random grids to determine the diatom abundance and concluded that the data from counts of 300 cells compared well with those from counts of 1000 cells. A counting procedure that directly relates the relative abundance of algal species to biomass was developed by McIntire [*32*]. In this procedure, no set number of individuals are enumerated, but several slide preparations and 15 to 30 random fields on different slides are observed at magnifications of $\times 200$, $\times 500$, and $\times 1250$.

Patrick and Strawbridge [*11*], however, suggest that the application of diatom counts of less than 7000 cells is often inaccurate or questionable. This is due to the probability of high variance caused by the degree of domination by only a few species. This can be further complicated by poor techniques of slide preparation. For the results to be accurate, it is necessary that the organisms on the slides to be analyzed have a random distribution. Clumping of specimens, which can be caused by a number of factors, such as drying the material too rapidly, species morphology, and sample debris, serves significantly to increase the total number of cells enumerated while only adding incidental species information. This is especially true of *Cocconeis* or *Achnanthes* species, which seem to have greater susceptibility to clumping. When clumped organisms are enumerated, the total number of cells enumerated increases rather rapidly, while the probability of encountering new species is reduced. Using a good slide preparation that has a randomized distribution of specimens provides an equal opportunity for the organisms (species) to be enumerated.

This study was designed: (*a*) to assess the impact of electroplating waste on a river, using periphytic diatoms as a biological monitor, (*b*) to compare the effectiveness of different biological indexes in measuring the response of periphyton communities to the discharge, and (*c*) to compare the results of biological indexes, based on different levels of sampling or effort.

Background

This study was part of a characterization and biological monitoring program conducted on the Muskingum River in southeastern Ohio, which originates at the confluence of the Walhonding and Tuscarawas Rivers and flows approximately 160 km south-southeast to the Ohio River at Marietta, Ohio. The Muskingum River is characterized by a general uniformity of conditions along its course created by ten navigation dams spaced at approximately 15-km intervals [*33-37*].

The biological monitoring program was initiated in 1972 when high mortalities of freshwater mussels were observed immediately downstream of McConnelsville, Ohio. This mortality was attributed to an electroplating plant effluent containing copper originating from Gould, Inc., Engine Parts and Foil Division (GEP), located in McConnelsville [*37*]. As part of this study, the impact of the copper effluent on periphyton colonizing artificial substrates was assessed during 1973 [*38*]. Following abatement of the discharge, a follow-up study was conducted in 1976 in the Muskingum River in the areas upstream and downstream of the GEP. Results of analyses of periphyton community structure based on samples collected on 1 July 1973, 8 Sept. 1973, 21 May 1976, and 28 May 1976 are reported here.

Methods and Materials

Sampling locations were established upstream and downstream of the GEP discharge. Data from the locations are reported in Table 1, using the GEP discharge as a "zero point" (Fig. 1).

Artificial substrate samplers, each comprised of five 12.5 by 51.8 by 0.6-cm

TABLE 1—*Sampling locations on the Muskingum River.*

Approximate Location	Description
−11 km	downstream of Rokeby Lock, beyond Dam No. 8
−1 km	immediate area upstream of GEP outfall
GEP	immediate area downstream of GEP outfall
+5 km	downstream of McConnelsville, beyond Dam No. 7
+21 km	downstream of Stockport, beyond Dam No. 6
+42 km	downstream beyond Dam No. 4 and Beverly, Ohio

Plexiglas plates attached to an aluminum frame (Fig. 2) [*38*], were placed at each location under visually similar conditions of flow, depth, and exposure to sunlight. The samplers were attached to floats that held them at an exposure depth of 46 cm. The samples were collected after 14 and 21-day exposure periods. In the laboratory, periphytic growth was scraped from the plates, and permanent slides were made by a digestion and Hyrax medium procedure [*9*]. The sampling locations and procedures remained similar between the 1973 and 1976 studies, including the use of the same sampler type.

The slides were examined at ×1250 magnification, and the species were enumerated. Counts of approximately 500 diatom cells were made on the 1973 material to establish the species occurrence and relative abundance by percent. Counts of at least 5000 diatom cells were made on the 1976 material. To permit comparison with 1973 data, the species counts were tallied and recorded in sets of 500 cells, which were referred to as "intervals" of the long count. Ten sets (or intervals) of 500 cells each were equivalent to the long counts of 5000 cells.

Diversity indexes were computed using the following equations [*13,14*]:

$$H' = -\Sigma \frac{n_i}{N} \log_2 \frac{n_i}{N}$$

$$\overline{H} = \left(\frac{1}{N}\right) (\log N! - \sum_{1}^{s} \log N_i$$

where

N = the total number of organisms,
n_i = the number of individuals per species, and
S = the number of species.

For the 1976 data, diversity indexes were computed using data from each 500-cell data set and combination of sets of the 5000-cell long counts.

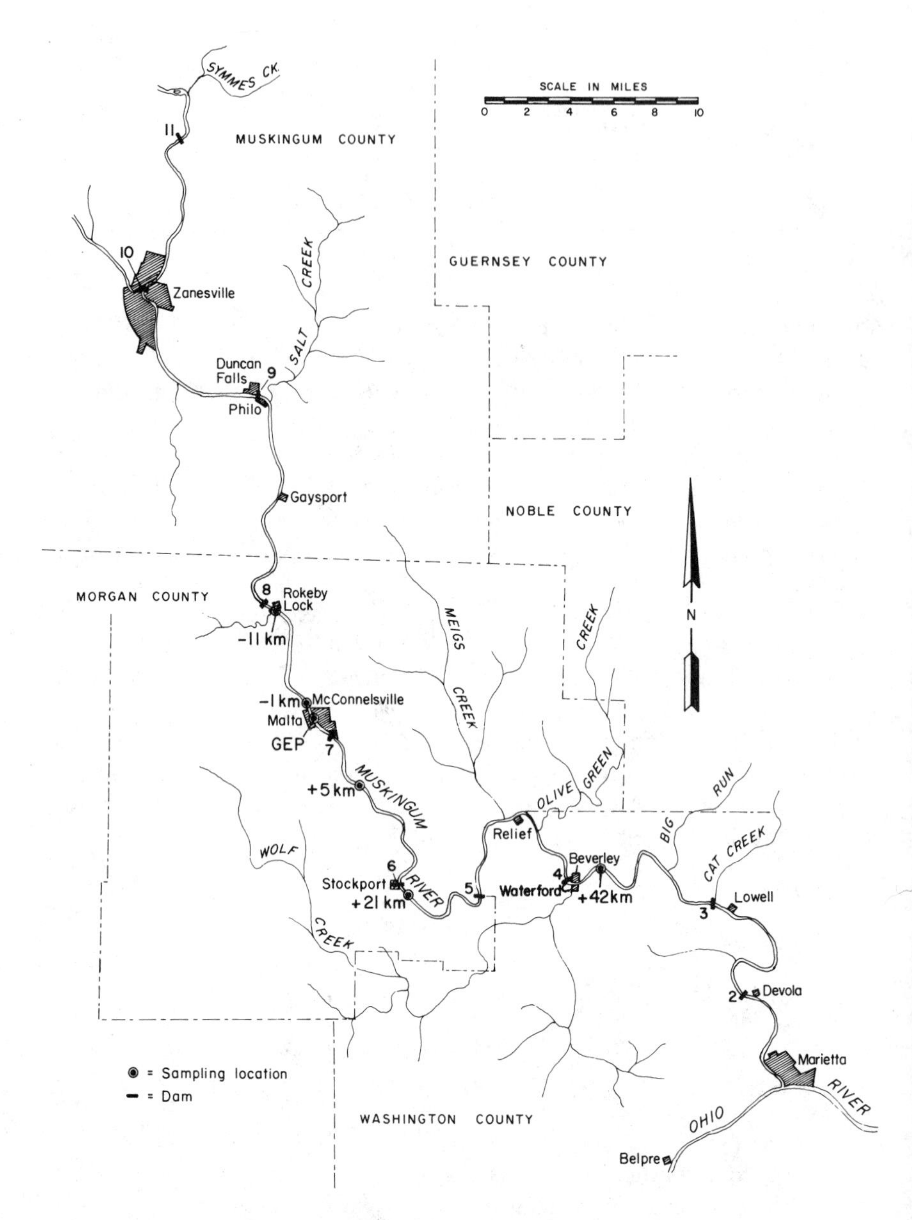

FIG. 1—*Location of periphyton sampling points in relation to industrial discharge, Muskingum River, Ohio.*

FIG. 2—*Acrylic plastic artificial substrate samplers.*

Results

1973 Survey

After 21 days of incubation, samples were collected from four of six locations on 1 July 1973: −1 km, GEP, +5 km, and +42 km (Fig. 1). The samples from the other two locations were lost. Good periphyton growth was observed on substrates from Locations −1 km, +5 km, and +42 km, whereas little growth was observed on GEP substrates.

In relation to Station −1 km, the species richness was lower downstream of GEP, with the lowest value immediately below the GEP discharge point (Table 2). At Location −1 km, 69 taxa were recorded, 84 percent of which occurred at a relative abundance of ≤2.0 percent. Of 17 species at the GEP location, about 50 percent occurred at a relative abundance of ≤2.0 percent. Between −1 km and GEP, *Navicula graciloides* A. Mayer var. *graciloides* decreased in relative abundance from 17 to 4 percent, while

TABLE 2—*Periphyton species richness in the vicinity of GEP copper electroplating effluent, Muskingum River, Ohio, 1973.*

	1 July 1973[a]			8 Sept. 1973[a]		
Sampling Location	Number of Species	Number of Individuals	Species Richness ($S/N^{1/2}$)	Number of Species	Number of Individuals	Species Richness ($S/N^{1/2}$)
−11 km	NS	NS	NS	31	483	1.41
−1 km	69	610	2.70	26	505	1.16
GEP	17	51	2.38	18	300	1.04
+5 km	53	401	2.65	30	741	1.10
+21 km	NS	NS	NS	20	205	1.40
+42 km	44	300	2.54	18	425	0.50

[a]NS = no sample collected.

Achnanthes species and *Nitzschia fonticola* Grunow increased by 13 and 11 percent, respectively. Periphyton at the GEP location was composed primarily of the coccoid green alga *Chlorella vulgaris* Beyerinck. Diatom species richness increased downstream of GEP (Table 2). The species association at +5 km was similar to that at −1 km, with the major difference being the absence of a few species at +5 km that were considered occasional (≤1.0 percent occurrence). Also at +5 km, *Nitzschia palea* (Kuetzing) W. Smith was present at 22.0 percent relative abundance, which was three times its abundance at −1 km and four times its abundance at GEP. Diatom species richness was slightly depressed at +42 km (Table 2), which may have been due to luxurious growths of *Cladophora glomerata* (Linnaeus) Kuetzing. *Navicula graciloides* dominated (30 percent) the assemblage at +42 km.

Samples were collected from all six sampling locations on 8 Sept. 1973 following a 21-day exposure period. As in July, the substrates at +42 km were covered with growths of *Cladophora glomerata.* Because this was not observed at other locations, these samples were considered not comparable, having been influenced by factors other than copper. The greatest change in diatom species richness occurred between −11 km and −1 km and between +5 km and +21 km (excluding +42 km) (Table 2). At −1 km, GEP, and +5 km, the species richness values were nearly equal.

Cyclotella meneghiniana Kuetzing dominated (48 to 65 percent) the assemblages at all the locations but was most abundant at GEP. As observed in July, the green alga *Chlorella vulgaris* was abundant at GEP, not only on the artificial substrates but also on the surrounding natural substrates. The shifts in species richness between locations were due to the addition or loss of diatom species that normally occurred in low numbers (<2.0 to 3.0 percent relative abundance).

1976 Survey

As in the 1973 surveys, trends determined from the indexes of community structure during 1976 suggested perturbation at GEP. Similar trends were observed for count lengths of 500, 1000, and 5000 cells. Because it was assumed that 5000-cell counts most accurately describe the natural situation, the results described here are based on the 5000-cell long counts.

21 May 1976

Species richness—Margalef's richness index ($S/N^{1/2}$) [*13*] showed a decrease at the GEP outfall, with the value decreasing to a level of 0.55 from values of 0.70 and 0.67 at Locations −11 km and −1 km, respectively (Table 3). At locations further downstream, the richness index suggested recovery by rising values that became equal to the upstream values.

The 500 and 1000-cell counts effectively responded to the GEP perturbation, although the specific trends differed. Trends determined from counts of 1000 cells more closely mimicked the trend shown by the longer, 5000 cell count (Table 3). The trend of the 500-cell counts was that values generally increased in these directions: from −11 km to −1 km, +5 km to +21 km, and +21 km to +42 km, which is the opposite of the 1000 and 5000-cell count trends (Table 3). Furthermore, richness values derived from the 500 and 1000-cell counts were generally higher than those for the 5000-cell count. This resulted because the majority of the species found in the longer counts were always recorded within the first 500 cells tallied.

Shannon-Weaver diversity—The expression of species diversity derived from the Shannon-Weaver equation indicated a pronounced effect at the GEP outfall. This was apparent in the 5000-cell count as well as in the 1000 and 500-cell counts (Fig. 3*a, b,* and *c*, respectively). For the 5000-cell count, the species diversity decreased about 1.4 points between −11 km and GEP (Table 4). At the remaining downstream locations (+5 km, +21 km and +42 km) the species diversity increased to above 3.0.

The 1000-cell short count trend closely mimicked that of the 5000-cell count (Fig. 3). Trends from the 500-cell counts seemed more variable than those from the 1000-cell and 5000-cell counts. Although the species diversity for each of the count intervals was within the same range at each respective location (Table 4), the data from the 500-cell short counts tended to be more variable.

Brillouin's diversity—The Brillouin equation yielded results and trends in species diversity similar to those determined by the Shannon-Weaver diversity equation. On 21 May 1976, the diversity was severely depressed at GEP. This held true for the 5000-cell count as well as the 1000 and 500-cell short counts (Fig. 4*a, b,* and *c*, respectively). For the 5000-cell count, the diversity decreased about 1.3 points between −11 km and GEP. The lowest

TABLE 3—*Summary of species counts and species richness, Muskingum River, Ohio, 21 May 1976.*

		Enumeration Interval (cells)						
Location	Index	500	500	1000	1000	1000	1000	5000
−11 km	total individuals	533	576	1131	1123	1080	1071	5514
	number of species	26	26	35	36	35	26	52
	richness	1.13	1.08	1.04	1.07	1.07	0.79	0.70
−1 km	total individuals	538	568	994	1066	1032	1005	5203
	number of species	30	25	33	35	32	28	48
	richness	1.29	1.05	1.05	1.07	1.00	0.88	0.67
GEP	total individuals	501	498	991	970	...	...	2960
	number of species	18	23	27	26	...	...	30
	richness	0.80	1.03	0.86	0.83	...	...	0.55
+5 km	total individuals	538	529	1111	1105	1176	1045	5504
	number of species	25	23	32	34	33	40	54
	richness	1.08	1.00	0.96	1.02	0.96	1.24	0.73
+21 km	total individuals	496	536	1069	1254	1056	1121	5532
	number of species	24	24	30	35	30	33	50
	richness	1.08	1.04	0.92	0.99	0.92	0.99	0.67
+42 km	total individuals	561	647	1211	965	950	973	5307
	number of species	24	29	35	29	27	32	48
	richness	1.01	1.14	1.01	0.93	0.88	1.03	0.66

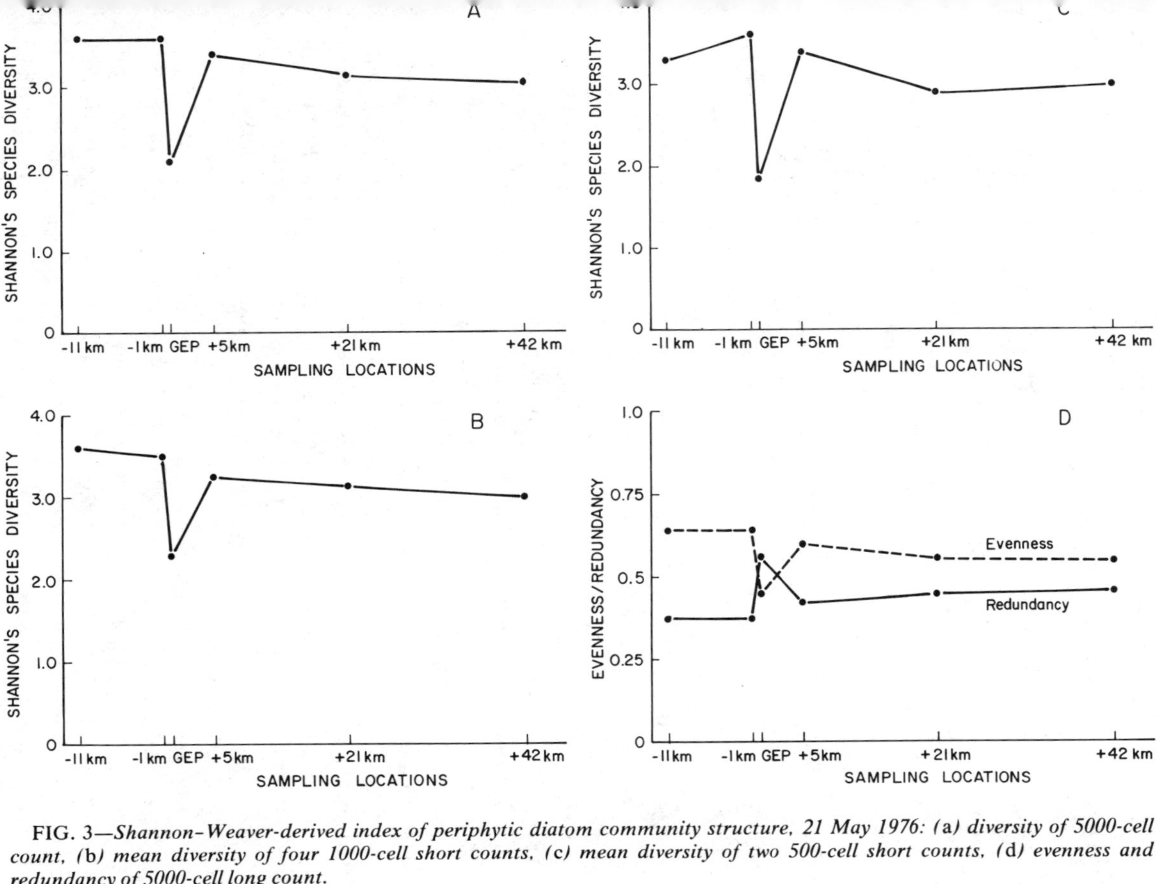

FIG. 3—*Shannon–Weaver-derived index of periphytic diatom community structure, 21 May 1976: (a) diversity of 5000-cell count, (b) mean diversity of four 1000-cell short counts, (c) mean diversity of two 500-cell short counts, (d) evenness and redundancy of 5000-cell long count.*

TABLE 4—*Summary of species diversity in the vicinity of GEP electroplating effluent, Muskingum River, Ohio, 21 May 1976.*

		Enumeration Interval (cells)						
Location	Index	500	500	1000	1000	1000	1000	5000
−11 km	Shannon-Weaver	3.27	3.37	3.68	3.68	3.57	3.45	3.62
	Brillouin	3.15	3.25	3.60	3.59	3.48	3.38	3.59
−1 km	Shannon-Weaver	3.75	3.48	3.51	3.65	3.51	3.37	3.59
	Brillouin	3.61	3.36	3.42	3.56	3.43	3.30	3.56
GEP	Shannon-Weaver	1.71	1.99	2.33	2.30	...	...	2.21
	Brillouin	1.63	1.88	2.25	2.23	...	...	2.18
+5 km	Shannon-Weaver	3.38	3.39	3.27	3 11	3.19	3.33	3.39
	Brillouin	3.26	3.28	3.29	3.03	3.11	3.23	3.36
+21 km	Shannon-Weaver	2.76	2.99	3.11	3.15	3.14	3.13	3.16
	Brillouin	2.65	2.88	3.03	3.07	3.07	3.05	3.13
+42 km	Shannon-Weaver	2.96	2.98	3.25	2.86	2.93	2.88	3.06
	Brillouin	2.85	2.86	3.17	2.78	2.85	2.80	3.03

species diversity (2.2) was observed at GEP, with increases to above 3.0 at the remaining downstream locations (Table 4).

The 1000-cell counts more closely resembled the trend developed by the 5000-cell count (Fig. 4). Although the 500-cell count trend responded as expected to the GEP stress, it did not closely follow the trend of more extended levels of enumeration. This was apparent at −1 km, which had a species diversity nearly equal to that of −11 km by the 1000 and 5000-cell counts, but the 500-cell count trend showed a diversity increase between −11 km and −1 km, which was opposite to the trend of the longer counts. The diversity based on 5000-cell count data was usually higher in value than that based on the shorter counts, with the 500-count yielding the lowest species diversity (Table 4). Furthermore, utilizing the same data, Brillouin's diversity index tended to be at least 0.1 points less than the Shannon-Weaver diversity index for 500 and 1000-cell counts, while the two indexes were almost equal at the 5000-cell level (Table 4).

Evenness—Shannon-Weaver-derived evenness responded to the GEP stress as anticipated on 21 May 1976. A decrease of almost 0.2 points with a recovery of 0.13 points was observed between −1 km and GEP and between GEP and +5 km, respectively (Fig. 3*d*), The evenness calculated from the 500-cell counts was more variable but did respond as expected to the GEP stress by decreasing in value. The values calculated from the 1000-cell intervals were a mimic of those from the 5000-cell trend. Short counts routinely yielded higher evenness values than the 5000-cell count, with the 500-cell short count having the highest values.

The evenness derived from Brillouin's diversity index responded with

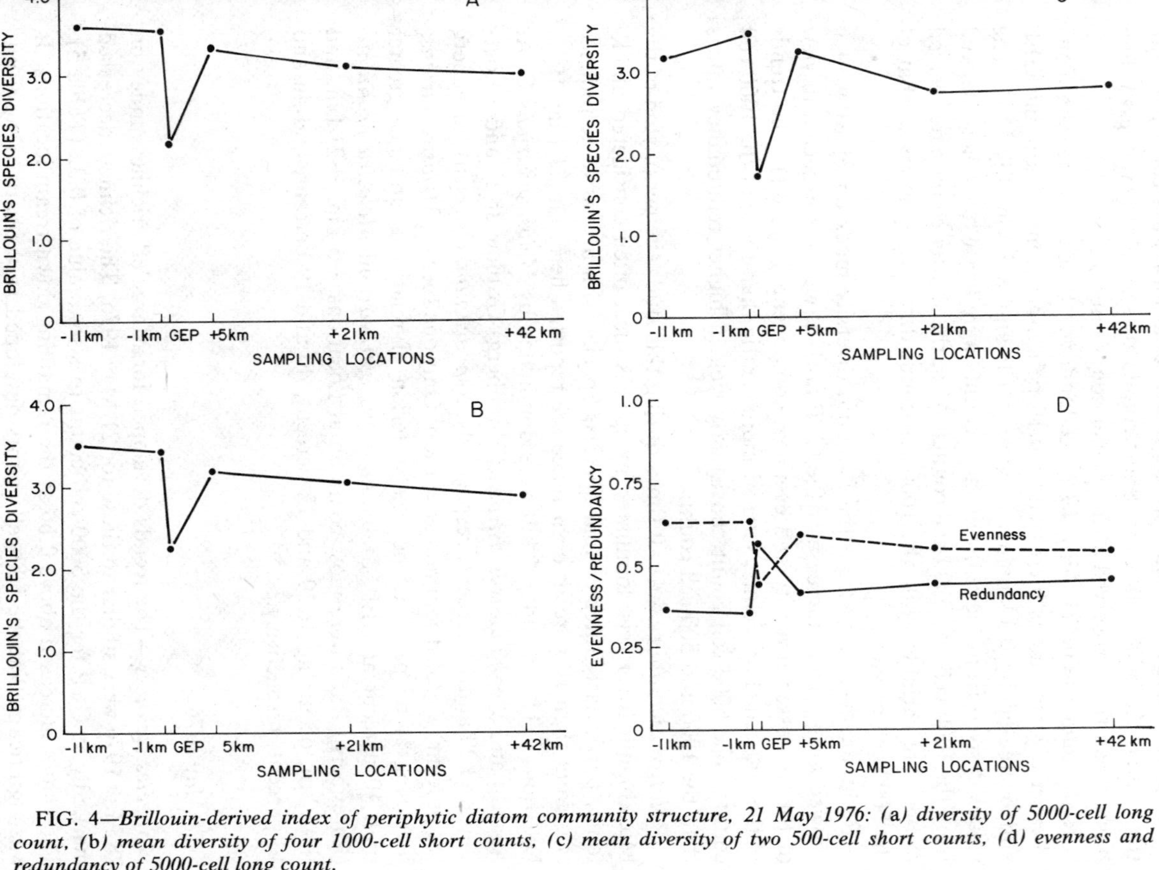

FIG. 4—*Brillouin-derived index of periphytic diatom community structure, 21 May 1976:* (a) *diversity of 5000-cell long count,* (b) *mean diversity of four 1000-cell short counts,* (c) *mean diversity of two 500-cell short counts,* (d) *evenness and redundancy of 5000-cell long count.*

depressed values to the GEP stress on 21 May 1976. A decrease of about 0.19 points with a recovery of 0.14 points was observed between Locations −1 km and GEP, and GEP and +5 km, respectively (Fig. 4*d*). The short-count-derived evenness values were variable and not always consistent with the trend of the 5000-cell count. Short counts always yielded evenness values higher than those observed at the longer count intervals.

Redundancy—On 21 May 1976, redundancy based on the Shannon-Weaver diversity index increased by 0.19 points between −1 km and GEP and decreased by 0.14 points between GEP and +5 km (Fig. 3*d*). The values derived from the 500-cell short counts were lower than those derived from the 1000-cell and 5000-cell counts. Also, the shorter count data seemed to be more variable while the 1000-cell-count data closely followed that of the 5000-cell count.

Redundancy derived from the Brillouin diversity index on 21 May 1976 showed a significant increase at the GEP outfall, which is expected in view of the depressed diversity and evenness components (Fig. 4*d*). The trends of the 500-cell short counts seemed more erratic than those of the 1000-cell counts. The 5000-cell count consistently yielded higher redundancy values than the 1000 and 500-cell counts.

Relative abundance—At all times during this study the periphyton of the Muskingum River was dominated by species most often considered planktonic, particularly three or four species. Species with a relative abundance of greater than 10 percent occurrence are reported here. On 21 May 1976, Locations −11 km and −1 km were dominated by *Melosira varians* C. A. Agardh and *Stephanodisus astraea* (Ehrenberg) Grunow at 27 and 19 percent and 29 and 12 percent, respectively. *Synedra ulna* (Nitzsch) Ehrenberg (11 percent) was also common at −1 km. Only one species, *Melosira varians,* occurring at 67 percent, met the criterion of relative abundance greater than 10 percent at GEP. *Stephanodiscus astraea* and *Melosira granulata* (Ehrenberg) Ralfs were most common at Locations +5 km, +21 km, and +42 km, occurring at 39 and 12 percent, 40 and 19 percent, and 46 and 13 percent, respectively.

28 May 1976

Species richness—The trends developed for Margalef's richness index on 28 May 1976 were similar to those for 21 May 1976. The richness decreased markedly at GEP for the 5000-cell count to a low value of 0.32 (Table 5). The values increased above 0.66 at the remaining downstream locations. It was significant that the 500-cell counts developed very erratic trends and, in one case, failed to detect the GEP outfall, where the counts showed a high richness value and an increasing trend rather than a depression. Trends developed from the 1000-cell counts were more similar to those from the 5000-cell count, although somewhat erratic at −1 km, +21 km, and

TABLE 5—*Summary of species counts and species richness, Muskingum River, Ohio, 28 May 1976.*

Location	Index	Enumeration Interval (cells)						
		500	500	1000	1000	1000	1000	5000
−11 km	total individuals	633	582	1078	1021	1064	1207	5585
	number of species	24	29	28	33	30	36	49
	richness	0.95	1.20	0.85	1.03	0.92	1.04	0.66
−1 km	total individuals	558	544	1016	1088	1176	1115	5497
	number of species	20	20	30	28	33	30	46
	richness	0.85	0.86	0.94	0.85	0.96	0.90	0.62
GEP	total individuals	553	613	1059	1000	986	1092	5303
	number of species	23	20	19	18	19	17	23
	richness	0.98	0.81	0.58	0.57	0.61	0.51	0.32
+5 km	total individuals	523	558	1026	1023	1059	1188	5377
	number of species	25	29	30	34	32	31	54
	richness	1.09	1.23	0.94	1.06	0.98	0.90	0.74
+21 km	total individuals	581	614	1017	1104	1030	1247	5592
	number of species	25	25	28	27	25	34	48
	richness	1.04	1.01	0.88	0.81	0.78	0.96	0.64
+42 km	total individuals	513	508	1106	1043	1032	1037	5239
	number of species	29	24	33	29	33	27	48
	richness	1.28	1.06	0.99	0.90	1.00	0.84	0.66

+42 km (Table 5). When the mean trend of four 1000-counts is considered, it closely mimics that of the 5000-cell counts.

As shown previously, the trends of the shorter counts were artifically higher in value than that of the 5000-cell count, because most of the species were recorded in the first 500 cells observed, with proportionately fewer species being added per number of cells tallied in the extended counts. The 500-cell count generated 20 to 30 species, even at the stressed location, while only about 10 species were usually added in the next interval of 500 tallies, that is, the 1000-cell count. A doubling of the number of species between the 500 and 5000-cell counts was common (Table 5), except for the stressed location (GEP), at which a maximum number of species was reached in the first 500-count interval.

Shannon-Weaver diversity—Periphyton diatom samples collected on 28 May 1976 indicated stress at the GEP outfall through depression of the Shannon-Weaver species diversity index, although not as dramatically as was seen on 21 May 1976. The 5000-cell count indicated a decrease of only 0.22 points between −1 km and GEP, with subsequent decreases at the remaining downstream locations (Table 6). The trends for the 5000 and 1000-cell counts were quite similar (Fig. 5*a* and *b*), while the 500-cell count trend was in some cases opposite (Fig. 5*c*). The 1000-cell counts showed the same trends as the 5000-cell count when the change in diversity was most dramatic, such as between −11 km and GEP, where species diversity decreased by 0.62 points (Table 6). Under conditions where the change in diversity was only slight, such as between GEP and +42 km, the 1000-cell count trends were less consistent with the 5000-cell count. It is noted that 500-cell counts were less reliable for measuring stressed conditions than

TABLE 6—*Summary of species diversity in the vicinity of GEP electroplating effluent, Muskingum River, Ohio, 28 May 1976.*

		Enumeration Interval (cells)						
Location	Index	500	500	1000	1000	1000	1000	5000
−11 km	Shannon-Weaver	3.44	3.65	3.54	3.50	3.61	3.48	3.61
	Brillouin	3.34	3.52	3.46	3.41	3.52	3.40	3.58
−1 km	Shannon-Weaver	3.08	2.81	3.17	3.16	3.27	3.14	3.21
	Brillouin	2.98	2.72	3.09	3.09	3.20	3.06	3.19
GEP	Shannon-Weaver	3.18	3.00	2.75	2.86	3.06	2.78	2.99
	Brillouin	3.07	2.91	2.70	2.81	3.00	2.73	2.97
+5 km	Shannon-Weaver	3.17	3.13	2.73	2.67	2.73	2.91	2.95
	Brillouin	3.05	3.01	2.66	2.59	2.65	2.84	2.92
+21 km	Shannon-Weaver	2.92	2.97	2.85	2.81	2.75	2.71	2.88
	Brillouin	2.81	2.87	2.77	2.74	2.68	2.64	2.85
+42 km	Shannon-Weaver	2.94	2.56	2.80	2.63	2.81	2.69	2.80
	Brillouin	2.81	2.45	2.72	2.56	2.73	2.62	2.78

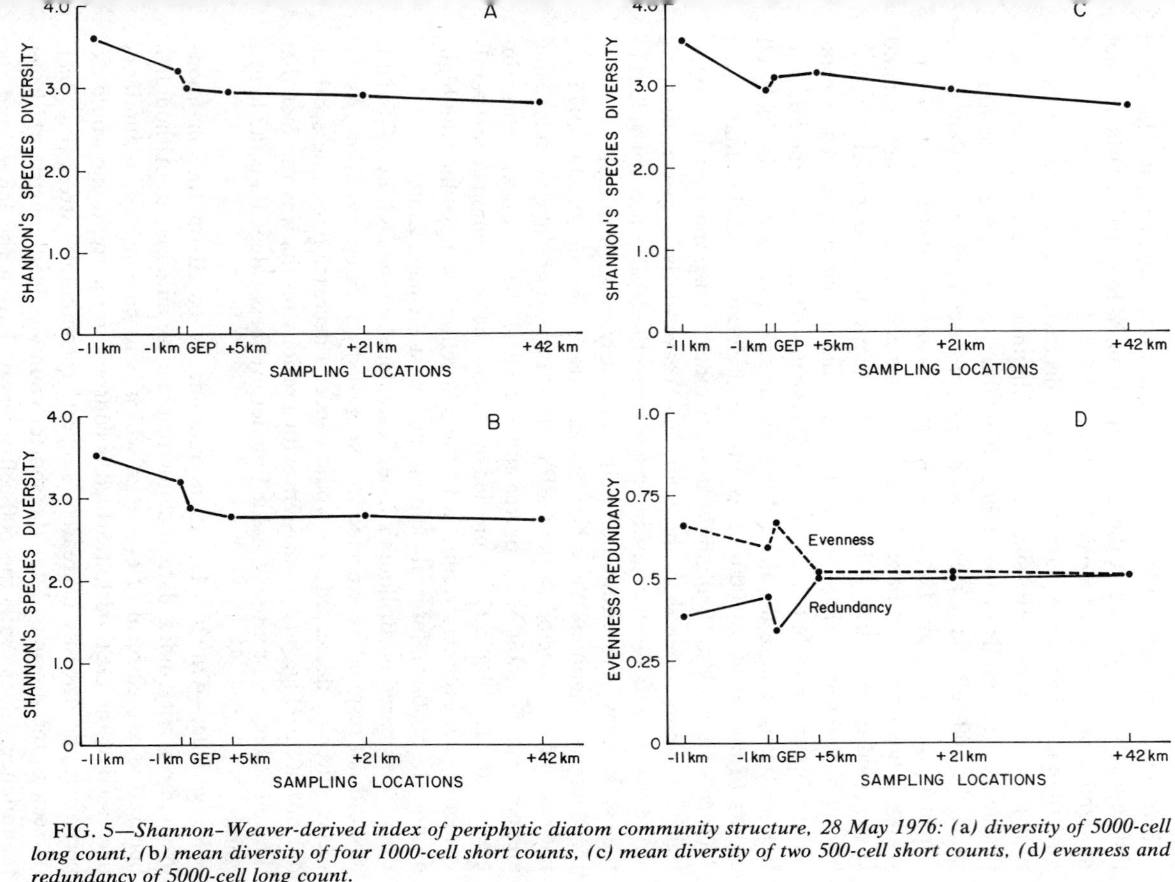

FIG. 5—*Shannon–Weaver-derived index of periphytic diatom community structure, 28 May 1976: (a) diversity of 5000-cell long count, (b) mean diversity of four 1000-cell short counts, (c) mean diversity of two 500-cell short counts, (d) evenness and redundancy of 5000-cell long count.*

1000-cell intervals. The 500-cell counts yielded inconsistent trends and did not respond to the GEP outfall as expected, since they actually showed an increase in diversity (Fig. 5*c*) rather than the expected decrease due to stress.

Brillouin's diversity—The diversity of the diatom community based on this index, was depressed at the GEP outfall on 28 May 1976, although not as significantly as was observed on 21 May 1976. The 5000-cell count based on Brillouin's diversity equation showed a decrease of only 0.22 points between −1 km and GEP, similar to the Shannon–Weaver diversity index (Table 6). The trends of the 5000 and 1000-cell counts were quite similar, while the 500-cell count trend was in several cases opposite to that of the longer counts (Fig. 6). The 500-cell count yielded inconsistent species diversity values when compared with the longer counts and did not respond as expected to the GEP outfall, which was shown by this shorter count to have a high diversity (Fig. 6*c*), rather than the anticipated decrease in diversity due to perturbation. The diversity values derived from the 500 and 1000-cell counts were usually lower than those derived from the 5000-cell counts (Table 6). The 1000-cell count trend closely resembled that of the 5000-cell count. Also, Brillouin's diversity index tended most often to be at least 0.1 points lower than the Shannon–Weaver-derived species diversity at the 500 and 1000-cell count intervals, while the 5000-cell count values of the two indexes were within 0.05 points or nearly equal (Table 6).

Evenness—Shannon–Weaver-index-derived evenness on 28 May 1976 did not respond as expected to the GEP outfall. Instead of showing decreasing evenness concomitant with decreasing diversity, the evenness values increased at GEP (Fig. 5*d*). As previously shown, the short counts consistently yielded higher evenness values, with the 500-cell interval yielding the highest and the 5000-cell interval yielding the lowest evenness values.

Evenness based on Brillouin's index, calculated on the 28 May 1976 data, also did not respond as expected to a supposedly stressed condition. As the species diversity decreased, one would expect a decrease in evenness, which did not occur (Fig. 6*d*). Consistent with previous results was the fact that the 500-cell counts routinely yielded higher evenness values than the longer counts.

Redundancy—On 28 May 1976, redundancy based on the Shannon–Weaver diversity index data did not respond as anticipated. Although it behaved as a reciprocal of evenness, which is to be expected, redundancy decreased at the GEP outfall, which is contrary to the trend one would expect since the diversity decreased, suggesting community stress (Fig. 5*d*). As seen in the previous data, the shorter counts yielded consistently lower redundancy values, with the 500-cell intervals having the lowest and the 5000-cell intervals having the highest values.

Calculations of redundancy based on Brillouin's diversity index on 28 May 1976 yielded unusual results. Again, redundancy behaved as a recipro-

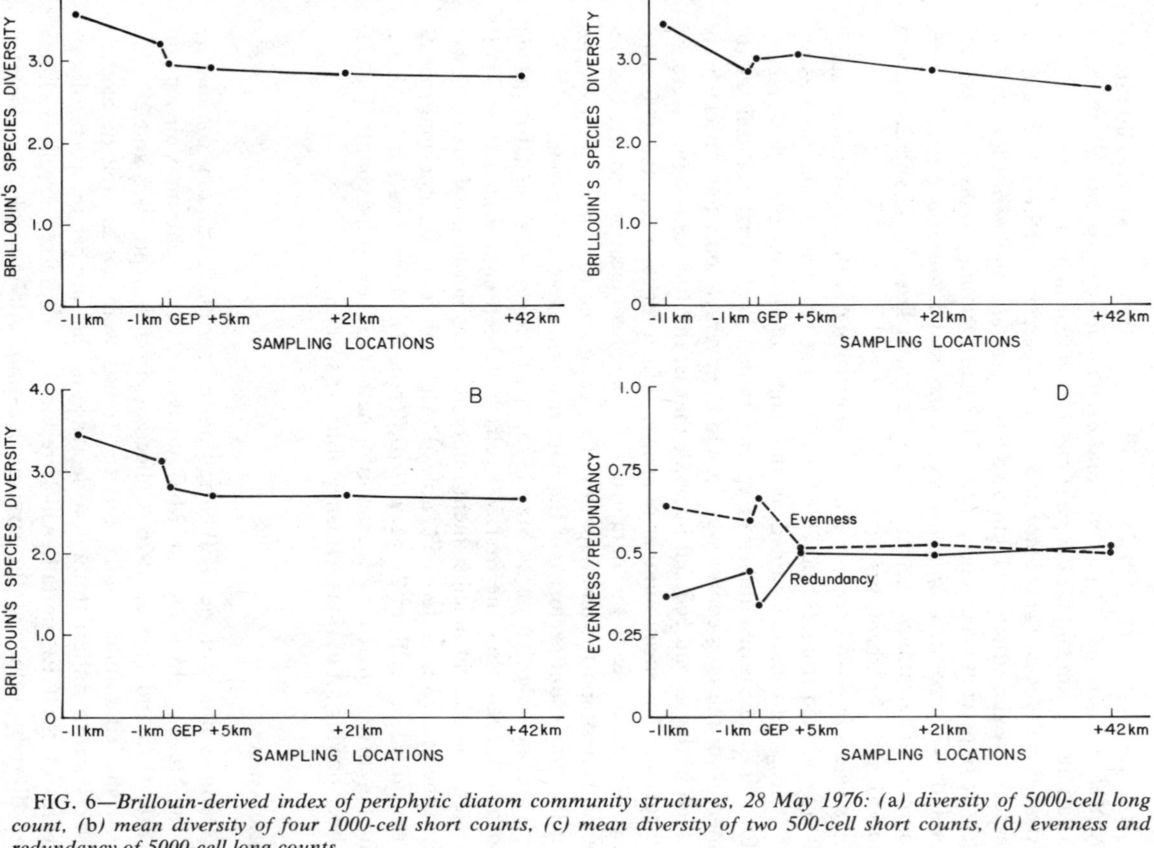

FIG. 6—*Brillouin-derived index of periphytic diatom community structures, 28 May 1976: (a) diversity of 5000-cell long count, (b) mean diversity of four 1000-cell short counts, (c) mean diversity of two 500-cell short counts, (d) evenness and redundancy of 5000-cell long counts.*

cal of evenness, which is to be expected, however, redundancy decreased rather than increased at the suspected stress location (Fig. 6*d*). As shown previously, the values derived from the shorter counts were consistently lower than those of the 5000-cell long count.

Relative abundance—On 28 May 1976, Location −11 km was dominated by *Melosira varians* (27 percent) and *Stephanodiscus astraea* (20 percent). Locations −1 km and GEP were somewhat similar in having *Stephanodiscus astraea* and *Melosira granulata* at 33 and 18 percent and 35 and 15 percent, respectively. Location −11 km also had *Cyclotella stelligera* Cleve et Grunow at 10 percent occurrence, while GEP had *Melosira varians* occurring at a level of 17 percent. *Stephanodiscus astraea* and *Melosira granulata* were common at Locations +21 km and +42 km at levels of 43 and 25 percent and 51 and 14 percent, respectively.

Species curve—A species curve was constructed with the combined count data generated from 21 and 28 May 1976. The curve shows the progression of species information gained as the length of diatom count is extended (Fig. 7). The mean number of species tallied in the first 500-cell set (or interval) of the long counts was 26. On the average, only four species were added in the second 500-cell interval, that is, the 1000-cell counts averaged 30 species. The mean number of species tallied for long counts (5000 cells) was 46, inclusive of the data from the stressed location, which characteristically supported fewer species. This compared to a mean of 50 species for the long count exclusive of the data from the stressed location.

It was notable that on 28 May 1976, the data generated at GEP yielded no new species after the first 500 cells had been enumerated (Fig. 7). It was also noted that in subsequent intervals it was not unusual to tally fewer than the first 23 taxa derived (Table 5). Although fewer than normal species were tallied at GEP on 21 May 1976 (Table 3), these data still yielded a similar curve of increasing species information with extended counts (Fig. 7), while showing an obvious tendency of fewer species.

Discussion

Definite effects of the GEP outfall on the periphyton diatom community of the Muskingum River near McConnelsville, Ohio, were most obvious on 1 July 1973 and 21 May 1976. Collections from 8 Sept. 1973 and 28 May 1976 showed similar results, but the trends were not as well developed. Effects were manifest through a decrease in the number of species downstream of the GEP outfall, which resulted in depressed species richness and diversity values at the stressed location. All the indexes calculated (species richness, Shannon-Weaver and Brillouin's diversity, evenness, and redundancy) documented the effect of the waste outfall.

Results based on short counts were more variable than results based on longer counts. This was especially true of 500-cell short counts. Although

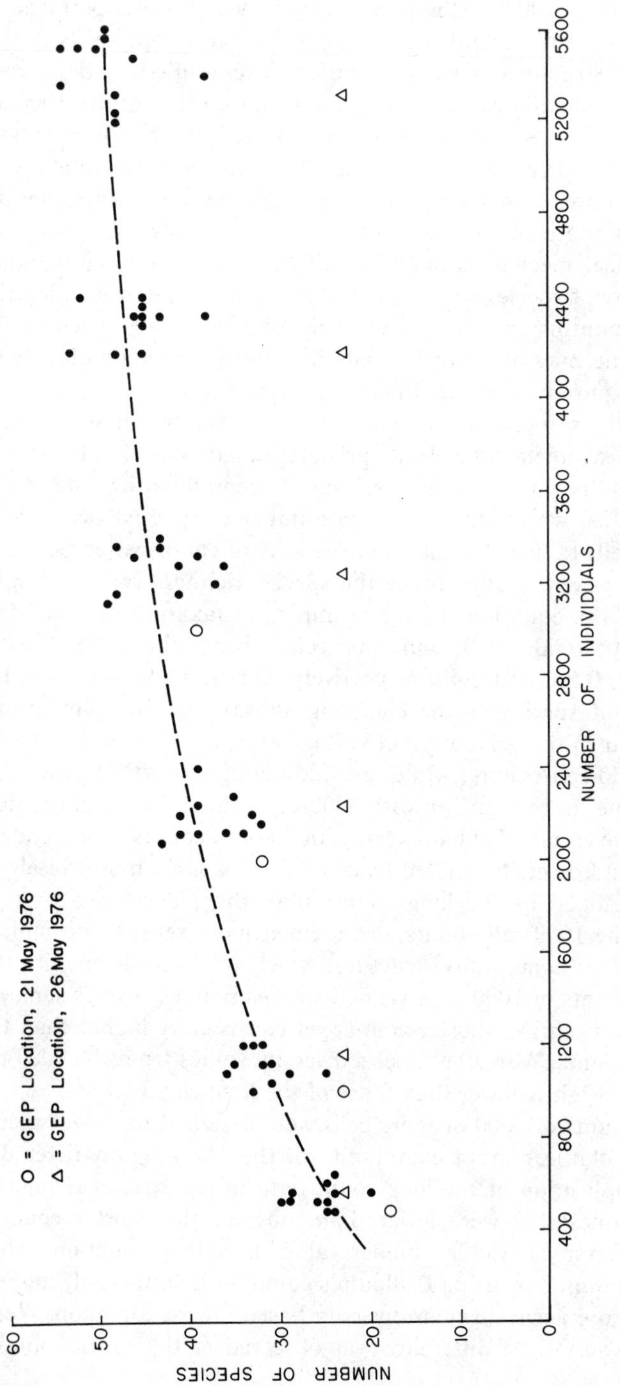

FIG. 7—Species curve of combined diatom count data from 21 and 28 May 1976.

short counts of 500 cells effectively responded to gross perturbations, this level of enumeration tended to be insensitive to minor shifts in the diatom community structure. This situation was exemplified by those cases where the trend of the 500-cell counts was opposite to that of the 1000 and 5000-cell counts. The 500-cell short counts yielded species richness values that were always higher than those of the 1000 and 5000-cell counts. Conversely, the short-count data usually produced species diversity values that were lower than those of the 1000 and 5000-cell counts. This was due to the mathematical mechanics of the equation and was most pronounced when the majority of species were tallied in the lowest level of enumeration.

When counting to a predetermined number of cells, such as 500, 1000, or 5000, one may mask real trends if there is a paucity of cells and extra effort is required to achieve the desired number of cells. In doing so, one is changing the sample size, because twice the number of slide transects may need to be scanned to reach the predetermined count level from stressed as opposed to "normal" conditions. This is exemplified by data from GEP on 28 May 1976, where the maximum number of species was achieved in the first 500 cells tallied. Extension of the level of enumeration beyond 500 cells served to systematically reduce the species richness value through the mechanics of the equation. If, for example, 25 taxa are recorded from intervals of 500, 1000, 2000, and 4000 cells, then richness ($S/N^{1/2}$) would be 1.12, 0.79, 0.56, and 0.40, respectively. Due to variations in count length, clumping of specimens, or changing subsample size, similar differences could occur in the generation of biological monitoring data.

The 1000-cell counts, while not adding significantly to the sample processing time, in comparison with 5000-cell counts, long counts, significantly reduced the variability characteristic of 500-cell counts. The trends of species richness and diversity derived from 1000-cell counts more closely paralleled those developed by the long counts than those developed by the 500-cell counts. The 1000-cell counts also seemed more sensitive to minor changes in the diatom community structure, which was a problem with the shorter counts. Counts of 1000 cells yielded species richness values somewhat lower than those from the shorter counts yet consistently higher than those from the long counts. Also, the species diversity values from the 1000-cell counts were often slightly lower than those of the long count.

Long counts of 5000 or more cells were assumed to reflect accurately the dynamics of the diatom community in the Muskingum River during this study. Application of the long-count data in the species richness equation yielded values that were lower than those of the shorter counts, but for species diversity it yielded higher values by either equation. Although diversity computed by using Brillouin's equation is supposedly more appropriate and more accurate than diversity based on the Shannon–Weaver equation [*39*], very little difference was observed in the results obtained using these equations. Small differences between the indexes were observed at all levels of sample enumeration.

To reduce the probability of artifacts in diatom community analyses, the following precautions should be taken: (*a*) use an index that is relatively independent of sample size, (*b*) count a predetermined area of a sample slide to maintain consistency in sample and subsample size, or (*c*) convert the number of diatom cells enumerated to cells per unit area rather than apply raw abundance data. The last approach will be more accurate, unless one can guarantee that enumeration of a predetermined number of cells can be achieved through nearly equal analytical effort and sample size.

In this study, it was found that, given the same number of species but a difference of less than 100 individuals (10 percent) in two 1000-cell counts of the same sample, differences of as much as 0.3 points (≤ 10 percent) in the Shannon-Weaver diversity measurement could occur, whereas using Brillouin's expression, the difference was 0.1 points or less (≤ 3 percent). A difference of fewer than 50 cells (10 percent) in two 500-cell counts of the same sample with equal species numbers resulted in differences of 0.1 to 0.3 points (≤ 10 percent) in both diversity expressions. Such variations are, of course, dependent on the species association and the dominance, evenness, and redundancy of the species. This effect may be insignificant when measuring acute or gross pollution effects, but, when one is trying to detect minor changes in community structure, the patterns may be masked by characteristics (artifacts) of the analytical methods applied.

Results of this study suggest that after enumeration of 1000 cells, and definitely after 2000 cells, the amount of species information gained diminishes in relation to the effort required to complete the longer counts. From review of the species curve generated in this study, it is apparent that 50 percent of the total species tallied in counts of 5000 to 5600 cells were recorded within the first 500-cell interval. It was common to record 20 to 30 ($\overline{X} = 26$) species in this first 500 cells tallied. At a maximum, an additional ten species were added within the next succeeding 500-cell interval, with the average being only four additional species. The mean species recorded in the first 1000 cells enumerated was 30, accounting for 60 percent of the total species recorded in the typical long count. An average of about four new species was added when the counts were extended from 1000 to 2000 cells, after which the curve begins to approach an asymptote.

Conclusions

In using periphyton diatoms as a biological monitoring tool and in assessing impact or perturbation through analysis of changes in the diatom community structure (diversity), it is not necessary to complete detailed sample analysis to assess gross impacts. Short counts of 500 cells effectively isolated gross stress but were inconsistent and insensitive to secondary or minor changes in community structure. Counts of 1000 cells were more accurate and sensitive to community structure shifts and more closely mimicked the trends of 5000-cell long counts.

When using raw diatom abundance data in the calculation of expressions of diversity, a 10 percent difference in total cells enumerated in a short count can have the effect of ≤ 10 percent variation in the resultant species diversity values. Though seemingly insignificant, this could mask other, naturally occurring changes in community structure. Artifacts introduced by methods of community analysis may lead to false descriptions of baseline conditions or limit the value of these expressions to the detection and isolation of only gross perturbations.

To reduce such artifacts, the authors suggest that the following considerations be given to one's study plan: (*a*) use an index that is relatively independent of the sample size, (*b*) maintain equal sample and subsample sizes between replicates and between study locations, and (*c*) convert the enumeration data to quantitative estimates of cells per unit area.

References

[*1*] Ott, W. R., *Environmental Indices, Theory and Practice,* Ann Arbor Science, Ann Arbor, Mich., 1978, p. 371.

[*2*] Kolkwitz, R. and Marsson, M., *Biology of Water Pollution,* U.S. Department of the Interior, Washington, D.C., 1967, pp. 47-52.

[*3*] Fjerdingstad, E., *Folia Limnologica Scandinavica,* No. 5, 1950, p. 123.

[*4*] Cholonoky, B. J., *Council for Scientific and Industrial Research,* Reprint R. W. No. 129, 1960, pp. 215-225.

[*5*] Lowe, R. L., "Environmental Requirements and Pollution Tolerance of Freshwater Diatoms," U.S. Environmental Protection Agency, Cincinnati, EPA-670/4-74-005, 1974, p. 334.

[*6*] Patrick, R. in *The Biology of Diatoms,* Dietrich Werner, Ed., University of California Press, Berkeley, Calif., 1977, pp. 284-332.

[*7*] Patrick, R., *Proceedings of the Academy of Natural Sciences of Philadelphia,* Vol. 101, 1949, pp. 277-341.

[*8*] Patrick, R. and Hohn, M. H., *Proceedings of the American Petroleum Institute, Section III, Refining,* Vol. 36, No. 3, 1956, pp. 332-339.

[*9*] Patrick, R. and Reimer, C. W., "Diatoms of the United States," *Monographs of the Academy of Natural Sciences of Philadelphia,* 1966, p. 688.

[*10*] Patrick, R., Hohn, M. H., and Wallace, J. H., *Notulae Naturae, Academy of Natural Sciences of Philadelphia,* No. 259, 1954, p. 12.

[*11*] Patrick, R. and Strawbridge, D., *The American Naturalist,* Vol. 97, No. 892, 1963, pp. 51-57.

[*12*] Cairns, J., Jr., Lanza, G. R., and Parker, B. C., *Proceedings of the Academy of Natural Sciences of Philadelphia,* Vol. 124, No. 5, 1972, pp. 79-127.

[*13*] Wilhm, J. L., *Journal of the Water Pollution Control Federation,* Vol. 39, No. 10, 1967, pp. 1673-1683.

[*14*] Wilhm, J. and Dorris, T. C., *Bioscience,* Vol. 18, No. 6, 1968, pp. 477-481.

[*15*] Cairns, J., Jr. and Dickson, K. L., *Journal of the Water Pollution Control Federation,* Vol. 43, No. 5, 1971, pp. 755-772.

[*16*] Hurlbert, S. H., *Ecology,* Vol. 52, No. 4, 1971, pp. 577-586.

[17] Patrick, R. in *Biological Methods for the Assessment of Water Quality, ASTM STP 528,* American Society for Testing and Materials, Philadelphia, 1973, pp. 76-95.
[18] Pielou, E. C., *Ecological Diversity,* Wiley, New York, 1975, p. 165.
[19] Pielou, E. C., *Mathematical Ecology,* Wiley, New York, 1977.
[20] Collingsworth, R. F., Hohn, M. F., and Collins, G. B., *Michigan Academy of Science, Arts and Letters,* Vol. 52, 1967, pp. 19-30.
[21] Stoermer, E. F., Taylor, S. M., and Callender, E., *Transactions of the American Microscopical Society,* Vol. 90, No. 2, 1971, pp. 195-206.
[22] Weitzel, R. L., Sanocki, S. L., and Holecek, H. in *Methods and Measurements of Periphyton: A Review, ASTM STP 690,* American Society for Testing and Materials, Philadelphia, 1979, pp. 90-115.
[23] Barron, J. A., *Journal of Paleontology,* Vol. 49, No. 4, 1975, pp. 619-632.
[24] Archibald, R. E. M., *Water Research,* Vol. 6, 1972, pp. 1229-1238.
[25] Bradbury, J. P., *Geological Society of America,* Paper No. 171, 1975, p. 147.
[26] Anderson, N., "Recent Diatoms from Douglas Lake, Cheboygan County, Michigan," Ph.D. dissertation, University of Michigan, Ann Arbor, Mich., 1976.
[27] Evans, D. and Stockner, J. G., *Journal of the Fisheries Research Board of Canada,* Vol. 29, No. 1, 1972, pp. 31-44.
[28] Tippett, R., *British Phycology Journal,* Vol. 5, No. 2, 1970, pp. 187-199.
[29] Schoeman, F. R., *Journal of the Limnological Society of Southern Africa,* Vol. 2, No. 1, 1976, pp. 21-24.
[30] Sreenivasa, M. and Duthie, H. C., *Canadian Journal of Botany,* Vol. 17, 1973, pp. 25-31.
[31] Castenholz, R. W., *Limnology and Oceanography,* Vol. 5, No. 1, 1960, pp. 1-28.
[32] McIntire, C. D., *Ecology,* Vol. 49, No. 3, 1968, pp. 520-537.
[33] Bates, J. M., "Part I—Mussel Studies," Ohio Mussel Fisheries Investigation, Center for Aquatic Biology, Eastern Michigan University, Ypsilanti, Mich., 1970, p. 108.
[34] Bates, J. M., "Part II—Water Chemistry and Sediment Analyses," Ohio Mussel Fisheries Investigation, Center for Aquatic Biology, Eastern Michigan University, Ypsilanti, Mich., 1970, p. 111.
[35] Bates, J. M., "Part III—Plankton Survey," Ohio Mussel Fisheries Investigation, Center for Aquatic Biology, Eastern Michigan University, Ypsilanti, Mich., 1970, p. 111.
[36] Weitzel, R. L., "An Investigation of Net Plankton from the Muskingum River, Ohio—Summer of 1969," M.S. thesis, Eastern Michigan University, Ypsilanti, Mich., 1971, p. 104.
[37] Foster, R. B. and Bates, J. M., *Environmental Science and Technology,* Vol. 12, No. 8, 1978, pp. 958-962.
[38] Bowling, J. W., "The Effects of Copper Effluent on Periphyton Communities in the Lower Muskingum River," M.S. thesis, Eastern Michigan University, Ypsilanti, Mich., 1974, p. 70.
[39] Kaesler, R. L., Herricks, E. E., and Crossman, J. S. in *Biological Data in Water Pollution Assessment, Quantitative and Statistical Analyses, ASTM STP 652,* American Society for Testing and Materials, Philadelphia, 1978, pp. 92-112.

C. I. Weber[1] *and B. H. McFarland*[1]

Effects of Exposure Time, Season, Substrate Type, and Planktonic Populations on the Taxonomic Composition of Algal Periphyton on Artificial Substrates in the Ohio and Little Miami Rivers, Ohio

---

REFERENCE: Weber, C. I. and McFarland, B. H., **"Effects of Exposure Time, Season, Substrate Type, and Planktonic Populations on the Taxonomic Composition of Algal Periphyton on Artificial Substrates in the Ohio and Little Miami Rivers, Ohio,"** *Ecological Assessments of Effluent Impacts on Communities of Indigenous Aquatic Organisms, ASTM STP 730.* J. M. Bates and C. I. Weber, Eds., American Society for Testing and Materials, 1981, pp. 166–219.

ABSTRACT: Standard glass microscope slides were exposed at the surfaces of the Ohio and Little Miami Rivers at Cincinnati, Ohio, for 1, 2, 4, and 8-week periods in June through November 1966. The maximum cell densities were attained in samples from the Little Miami River after an exposure of 1 or 2 weeks, except in November. In the Ohio River, the maximum counts were observed after exposures of 4 weeks in June and July, and 1 week in August and September. In October and November, the cell density continued to increase with increasing periods of exposure.

Distinct successional patterns in the composition of the periphyton populations were not found. The changes in the composition of the periphyton with increased substrate exposure time resulted principally from seasonal changes in environmental conditions and phytoplankton populations.

The maximum algal cell density, 29 000/mm^2, was observed in a 1-week sample collected in the Ohio River. The periphyton populations in both rivers were generally dominated by diatoms, which averaged 68 percent of the cell count. The most important genera were *Achnanthes, Cyclotella, Gomphonema, Melosira, Navicula, Nitzschia, Stephanodiscus,* and *Synedra*. Other important groups were the green flagellates and filamentous blue-green, coccoid green, and filamentous green algae.

The biomasses on five sets of glass and Plexiglas slides exposed for 2 weeks in the Little Miami River during the period of August through November did not differ significantly.

[1]Chief, Aquatic Biology Section, and aquatic biologist, respectively, Biological Methods Branch, Environmental Monitoring and Support Laboratory, U.S. Environmental Protection Agency, Cincinnati, Ohio 45268.

KEY WORDS: ecology, effluents, aquatic organisms, diatoms, biomass, periphyton, phytoplankton

Kolkwitz and Marsson [*1*],[2] Butcher [*2-4*], Blum [5], Fjerdingstad [*6-8*], Sladeckova and Sladecek [*9*], and others have demonstrated a close relationship between water quality and the composition of attached microorganisms, and have described a large number of indicator species and species associations. Full realization of the potential usefulness of this parameter in long-term monitoring of water quality will not be achieved, however, until the effects of community succession, seasonal environmental changes, phytoplankton, and basic water chemistry on the standing crop and taxonomic composition of periphyton communities have been characterized.

Artificial substrates, such as glass slides, are often preferred to natural surfaces for sampling periphyton because they permit greater control over experimental conditions by providing a uniform surface, and offer greater flexibility in the selection of station location and exposure time. The slides are commonly exposed for periods of 2 to 4 weeks [*10–12*], but few studies have been made of the effects of factors such as exposure time, substrate type, and season on the properties of the periphyton collected.

An extensive study of the factors affecting substrate colonization, undertaken in 1966, during preliminary investigations of the use of periphyton populations in the Water Pollution Surveillance System (WPSS) [*13-16*], operated by the Federal Water Pollution Control Administration (FWPCA), was designed to acquire data on the effects of the phytoplankton density and species composition, the substrate type, the length of exposure time, and the season of the year on the cell density, biomass, and species composition of the periphyton.

Study Sites

Little Miami River

The Little Miami River is a small calcareous stream that rises in southwestern Ohio and flows 170 km through agricultural land to its confluence with the Ohio River at Cincinnati, Ohio. The water chemistry is determined principally by agricultural drainage and, to a lesser extent, by domestic wastes discharged into the river [*17*].

The average annual flow rate is approximately 49 m^3/s. During this study, flow rates ranged from 2.6 m^3/s on 13 Sept. to 170 m^3/s on 11 Nov. (Fig. 1). The annual means for selected water quality parameters are listed in Table 1 [*17*].

[2] The italic numbers in brackets refer to the list of references appended to this paper.

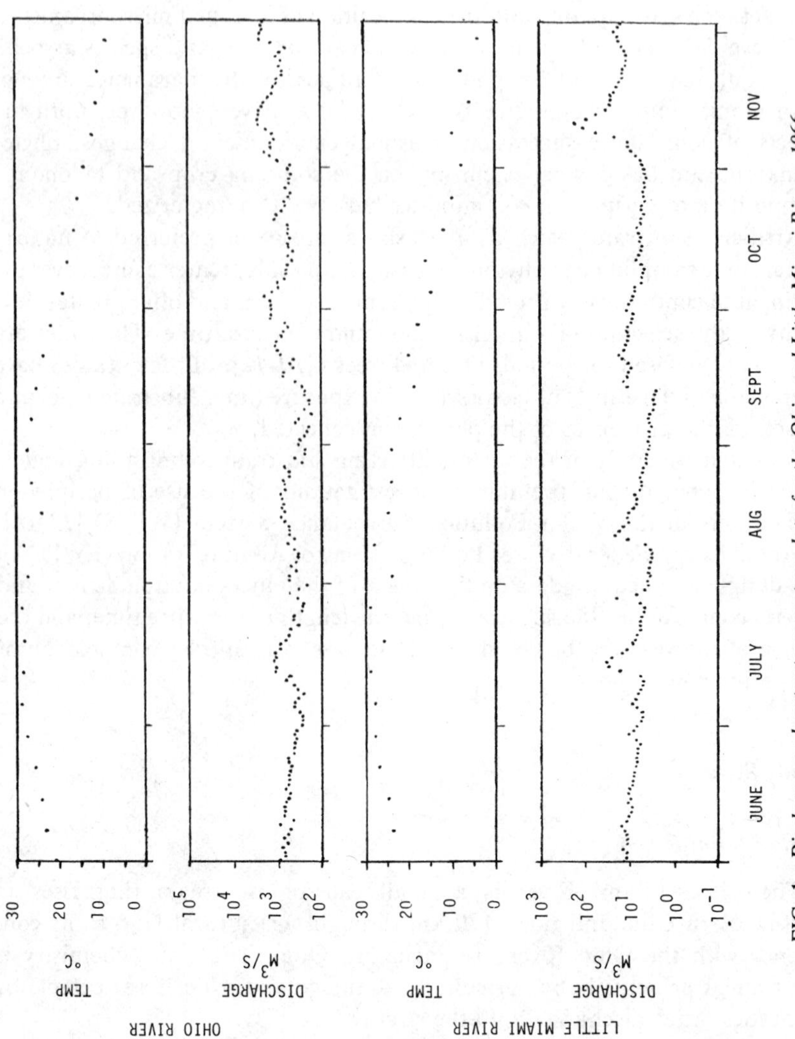

FIG. 1—*Discharge and water temperature data for the Ohio and Little Miami Rivers, 1966.*

TABLE 1—*Selected water quality parameters (annual means), calculated from weekly measurements.*

Parameter	Little Miami River	Ohio River
pH	7.9	7.3
Alkalinity, mg/litre as $CaCO_3$	196	38
Total hardness, mg/litre as $CaCO_3$	253	128
Chloride, mg Cl/litre	21	32
Sulfate, mg SO_4/litre	73	99
Diss orthophosphate, μg PO_4/litre	500	15
Total dissolved solids, mg/litre	318	271

Ohio River

The Ohio River at Cincinnati, Ohio, has an average annual flow rate of approximately 2700 m^3/s. During this study the flow rates ranged from 170 m^3/s on 8 Sept. to 3000 m^3/s on 12 Nov. (Fig. 1). The water chemistry differs greatly from that of the Little Miami River [*17*]. The concentrations of dissolved oxygen and phosphorus and the hardness, alkalinity, and pH are generally much lower, and the sulfate concentrations are higher (Table 1).

The upper Ohio River receives large amounts of industrial and domestic wastes, as well as acid mine drainage. Industrial wastes probably exert the dominant influence on the density of the phytoplankton in the river. Phytoplankton counts are generally much lower in the Ohio River at Cincinnati than in the Little Miami River. Although this may be due in part to lower nutrient concentrations and alkalinity, toxic substances in the river at Cincinnati frequently inhibit algal growth [*18,19*].

Methods

Periphyton

The periphyton sampler employed was a modification of the Catherwood Diatometer [*20*], consisting of a Styrofoam float, approximately 30 by 30 by 30 cm, supporting a central Plexiglas-cradle with slots for eight 25 by 75-mm plain glass microscope slides. The slides were held 1 cm apart, with their longitudinal axes parallel to the current and their transverse axes perpendicular to the surface. The upper edges of the slides were held 5 mm below the water surface (Fig. 2). Samplers were placed 6.4 km above the mouth of the Little Miami River, near WPSS Station 90, and in the Ohio River, approximately 1 km upstream from the Cincinnati municipal water treatment plant (WPSS 37). There were no significant local waste discharges to the river above the samplers at either station.

Six exposure series were established at each station, starting on 5 June 1966. At the beginning of each following month, through November, a

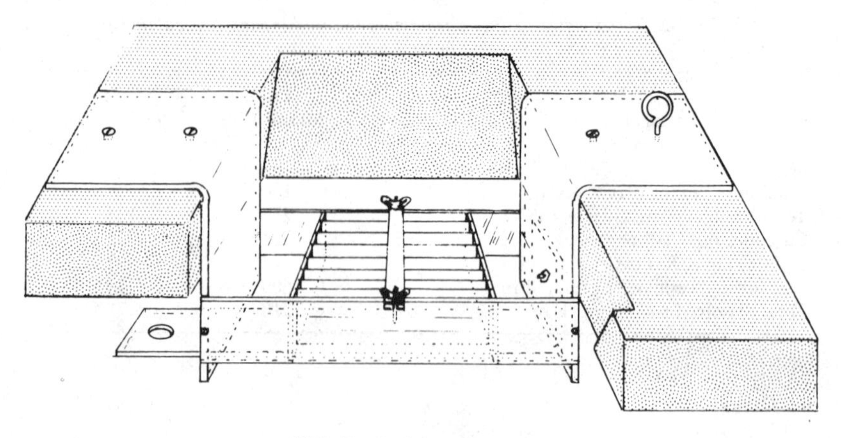

FIG. 2—*Periphyton sampler.*

sampler with eight glass slides was placed at each station. One pair of slides was removed from each sampler at intervals of 1, 2, 4, and 8 weeks. Upon collection, the slides were placed in 5 percent formalin. The attached material was removed from each slide with a razor blade and dispersed in approximately 100 ml of preservative. The periphyton specimens on the paired slides were pooled and treated as a single sample. One millilitre of the suspension was transferred to a Sedgwick-Rafter chamber, and cell counts and identifications were made at ×200. The diatom slides were prepared by digesting an aliquot of the samples in 5 percent $K_2S_2O_8$ for 30 min at 95°C. The cleaned frustules were rinsed with three changes of distilled water and mounted in Hyrax. The diatom species proportional counts were made of 200 to 500 cells at ×1000. Aliquots of the samples were dried to constant weight at 105°C, and ashed at 500°C for 1 h [*13*].

On five occasions in August, September, and October, paired Plexiglas slides were exposed for two weeks adjacent to paired glass slides to study differences in biomass accumulation.

Plankton

Weekly surface plankton grab samples were taken from the Little Miami River at midstream from the Highway 125 bridge (WPSS Station 90), using a DO sampler. Ohio River samples were taken monthly from the raw-water intake line at WPSS Station 37. All the plankton samples were preserved with thimerosol (Merthiolate) [*21*]. The phytoplankton were enumerated and identified at ×200 in a direct, Sedgwick-Rafter strip count. Each solitary cell or natural group of cells (colony) was counted as one unit, except for the diatoms, which were counted on a cell-by-cell basis [*13*]. The diatoms were concentrated by centrifuging 100 ml of sample 20 min at ×1000 g. The frustules were cleaned by incineration, mounted in Hyrax, and counted as described for periphyton [*13*].

Results and Discussion

Cell Density

Periphyton—Maximum cell densities were attained in 1 or 2 weeks at both stations and generally remained in the range of 1 000 to 10 000 cells/mm² (Fig. 3, Tables 2 and 3). The cell counts observed at both stations were similar to those found by Butcher [*2,3*] in polluted waters. The highest density obtained in this study, 29 292/mm², occurred in a 1-week sample from

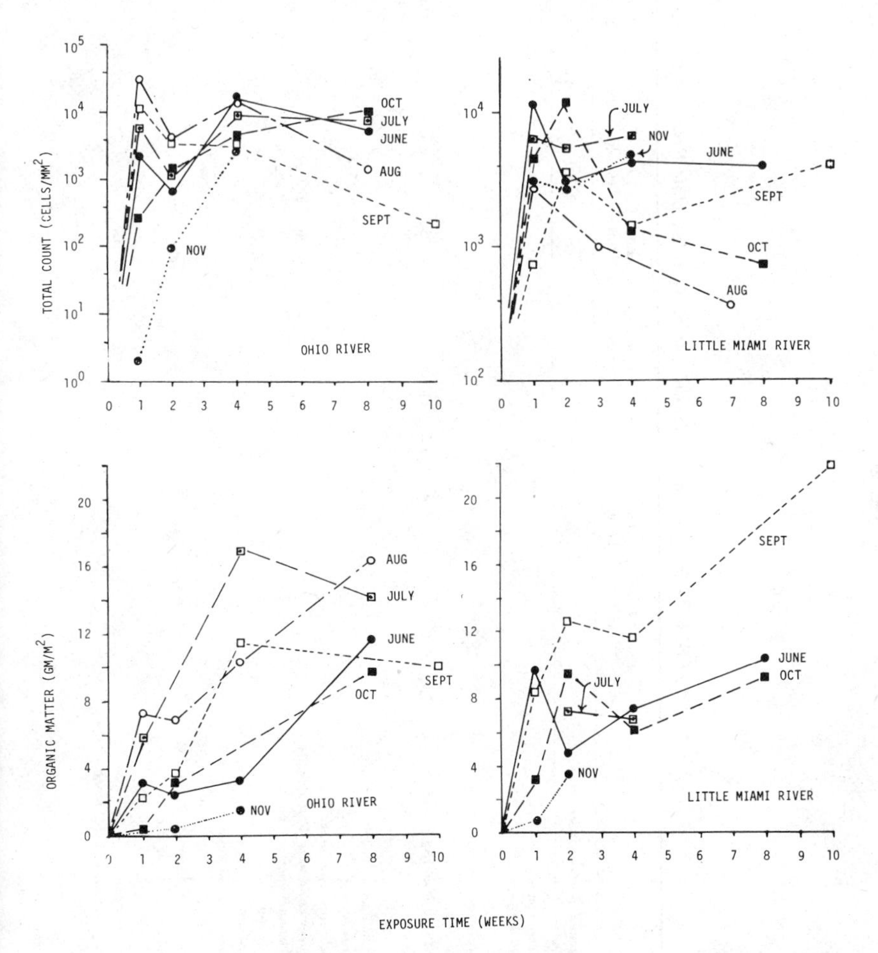

FIG. 3—*Periphyton cell counts and biomass (organic matter), Ohio and Little Miami Rivers, 1966.*

TABLE 2—*Periphyton cell densities on glass slides, Ohio River, 1966 (cells/mm²).*

	5 June–12 June	5 June–17 June	5 June–1 July	5 June–28 July	1 July–9 July	1 July–15 July	1 July–28 July	1 July–26 Aug.	28 July–4 Aug.	28 July–12 Aug.	28 July–26 Aug.	28 July–3 Sept.
Number of weeks slides exposed	1	2	4	8	1	2	4	8	1	2	4	8
Coccoid blue-green algae												
Agmenellum spp.	...	...	...	...	...	...	...	...	432	...	...	...
Filamentous blue-green algae												
Oscillatoria spp.	...	...	...	...	...	...	...	...	...	...	11	...
Schizothrix calcicola (Ag.) Gomont	...	34	5 203	...	1 305	...	...	1 652	...	1 224	12 057	...
Unidentified	...	...	...	...	1 279	...	...	1 101	21 037	...	...	...
Coccoid green algae												
Actinastrum spp.	...	...	...	...	26	...	...	...	...	...	...	...
Ankistrodesmus convolutus Corda	...	...	...	...	...	...	...	...	54	...	...	...
Ankistrodesmus falcatus (Corda) Ralfs	...	...	...	61	26	...	...	...	162	...	...	...
Ankistrodesmus setigerus Lemm.	...	...	...	...	52	...	...	...	...	...	...	...
Characium spp.	...	...	...	...	78	12	...	...	...	...	...	...
Closterium spp.	...	...	...	61	...	...	...	...	...	9	...	...
Cosmarium circulare Reinsch.	...	...	...	...	...	...	...	...	...	...	...	...
Cosmarium spp.	...	...	...	61	...	...	60	...	108	...	...	...
Crucigenia rectangularis (A. Braun) Gay	...	...	...	...	...	...	...	...	216	...	...	...
Crucigenia tetrapedia (Kirch.) W. and W.	...	...	...	...	...	...	...	...	...	5	...	...
Crucigenia spp.	...	...	...	61	...	...	...	...	...	...	...	...
Dictyosphaerium spp.	...	...	...	...	104	...	...	...	...	...	...	...
Kirchneriella lunaris (Kirch.) Moeb.	...	...	...	...	...	...	...	...	54	...	...	...
Kirchneriella spp.	...	...	...	...	...	3	...	...	...	...	...	...
Netrium spp.	...	...	...	...	...	...	...	55	...	...	...	...
Oocystis spp.	...	...	...	...	209	...	...	...	...	...	...	...

Pediastrum simplex (Meyen) Lemm.	...	...	...	...	...	...	...	...	...	...	...	...
Scenedesmus abundans (Kirch.) Chod.	...	...	...	...	...	...	...	...	...	14	...	...
Scenedesmus denticulatus Lager.	...	...	...	...	104	...	...	...	...	...	...	...
Scenedesmus quadricauda (Turp.) Breb.	...	...	...	...	104	...	...	...	...	...	6	...
Scenedesmus spp.	...	...	...	...	104	...	...	...	108	5	16	35
Tetraedron minimum (A. Braun) Hans.	...	...	...	...	...	...	...	...	...	...	...	6
Tetrastrum staurogeniaeforme (Schr.) Lemm	...	...	...	...	...	12	...	...	...	...	...	...
Unidentified	...	...	17	...	157	12	...	...	216	...	...	...
Filamentous green algae												
Mougeotia spp.	678	...	...	184	...	16	...	...	...	...	...	...
Rhizoclonium spp.	...	...	...	...	...	...	...	...	...	...	...	12
Spirogyra spp.	...	...	...	...	...	...	...	...	...	...	...	...
Stigeoclonium spp.	...	51	...	...	...	...	...	...	...	...	...	...
Ulothrix spp.	...	...	...	...	...	462	...	...	...	...	...	...
Uronema spp.	...	...	...	...	...	...	...	...	...	...	...	...
Unidentified	...	...	...	...	...	...	...	165	...	9	...	...
Green flagellated algae												
Chlamydomonas spp.	...	177	...	61	...	...	...	...	54	...	...	6
Pandorina spp.	...	...	...	...	...	...	...	...	...	...	5	...
Trachelomonas spp.	...	...	17	122	...	...	60	...	...	...	...	...
Unidentified	...	...	17	...	78	326	...	...	...	9	15	...
Other pigmented flagellates												
Chrysococcus spp.	...	...	...	...	...	...	...	...	...	5	...	...
Unidentified	...	...	...	...	...	...	60	...	...	...	...	...
Centric diatoms	203	...	...	2 387	679	9	3 957	606	1 996	1 145	54	463
Pennate diatoms	1 311	401	10 370	2 326	1 259	264	4 557	3 578	4 855	1 404	2 949	796
Total live algae	2 192	663	15 624	5 324	5 564	1 116	8 694	7 157	29 292	3 829	15 113	1 318

TABLE 2—*Continued.*

	26 Aug.- 2 Sept.	26 Aug.- 9 Sept.	26 Aug.- 23 Sept.	26 Aug.- 5 Nov.	7 Oct.- 14 Oct.	7 Oct.- 21 Oct.	7 Oct.- 5 Nov.	7 Oct.- 2 Dec.	5 Nov.- 12 Nov.	5 Nov.- 18 Nov.	5 Nov.- 2 Dec.
Number of weeks slides exposed	1	2	4	10	1	2	4	8	1	2	4
Coccoid blue-green algae											
Agmenellum spp.	...	...	...	...	...	...	...	...	...	...	...
Filamentous blue-green algae											
Oscillatoria spp.	...	...	...	...	...	...	...	...	...	...	...
Schizothrix calcicola (Ag.) Gomont	1 684	2 344	...	...	...	...	...	...	...	...	...
Unidentified	...	...	...	...	...	...	...	...	...	...	...
Coccoid green algae											
Actinastrum spp.	...	...	...	...	...	...	...	...	...	...	...
Ankistrodesmus convolutus Corda	...	...	...	...	...	...	...	...	...	...	...
Ankistrodesmus falcatus (Corda) Ralfs	...	...	...	...	...	3	...	...	...	...	...
Ankistrodesmus setigerus Lemm.	...	...	...	...	...	...	...	...	...	...	...
Characium spp.	168	5	172	...	...	...	...	...	...	...	...
Closterium spp.	...	...	...	...	...	...	...	...	...	...	...
Cosmarium circulare Reinsch.	112	...	...	...	...	...	...	...	...	...	...
Cosmarium spp.	...	...	...	...	...	...	...	...	...	...	...
Crucigenia rectangularis (A. Braun) Gay	...	...	...	...	...	...	...	...	...	...	...
Crucigenia tetrapedia (Kirch.) W. and W.	...	...	...	...	...	...	...	...	...	...	...
Crucigenia spp.	...	...	...	...	...	...	...	...	...	...	...
Dictyosphaerium spp.	...	...	...	...	...	...	...	...	...	...	...
Kirchneriella lunaris (Kirch.) Moeb.	...	...	...	...	...	...	...	...	...	...	...
Kirchneriella spp.	...	...	...	...	1	...	...	...	...	...	...
Netrium spp.	...	...	...	...	...	...	...	...	...	...	...
Oocystis spp.	...	...	...	...	...	...	...	...	...	...	...
Pediastrum simplex (Meyen) Lemm.	898	...	...	...	...	...	...	...	...	...	...

Scenedesmus abundans (Kirch.) Chod.	...	...	...	...	...	...	...	...	...	...	...
Scenedesmus denticulatus Lager.	...	...	...	...	...	...	...	...	...	...	...
Scenedesmus quadricauda (Turp.) Breb.	224	...	...	...	...	...	...	...	...	...	...
Scenedesmus spp.	...	...	...	...	...	...	60	...	...	...	...
Tetraedron minimum (A. Braun) Hans.	...	...	...	...	...	...	...	...	...	...	...
Tetrastrum staurogeniaeforme (Schr.) Lemm	...	...	...	...	...	...	...	...	...	...	...
Unidentified	...	118	...	...	...	...	...	...	...	...	...
Filamentous green algae											
Mougeotia spp.	...	...	...	...	13	44	...	...	...	...	...
Rhizoclonium spp.	...	...	...	...	...	...	...	...	...	...	...
Spirogyra spp.	393	...	...	...	...	...	...	...	...	...	...
Stigeoclonium spp.	...	...	...	...	...	...	...	360	...	...	...
Ulothrix spp.	4 602	...	...	...	...	...	...	...	...	...	...
Uronema spp.	...	...	...	...	14	...	...	...	...	...	...
Unidentified	...	75	...	...	...	...	...	...	...	...	...
Green flagellated algae											
Chlamydomonas spp.	...	...	...	...	3	...	...	...	...	...	...
Pandorina spp.	...	...	...	...	...	...	...	...	...	...	...
Trachelomonas spp.	...	5	...	...	...	...	...	...	...	...	...
Unidentified	...	16	86	...	...	...	...	...	...	...	...
Other pigmented flagellates											
Chrysococcus spp.	...	5	...	24	...	...	...	...	...	...	...
Unidentified	...	...	...	...	...	...	...	...	...	...	...
Centric diatoms	561	900	1 935	129	68	389	1 079	540	...	16	120
Pennate diatoms	2 469	291	1 075	55	172	688	3 628	9 115	2	79	2 518
Total live algae	11 111	3 759	3 268	208	271	1 124	4 767	10 015	2	95	2 638

TABLE 3—*Periphyton cell densities on glass slides, Little Miami River, 1966 (cells/mm²).*

	5 June–12 June	5 June–17 June	5 June–1 July	5 June–28 July	1 July–9 July	1 July–15 July	1 July–28 July	4 Aug.–12 Aug.	4 Aug.–26 Aug.	4 Aug.–23 Sept.
Number of weeks slides exposed	1	2	4	8	1	3	7	1	3	7
Filamentous blue-green algae										
Oscillatoria spp.	...	1 218	...	...	...	...	93	...	...	...
Schizothrix calcicola (Ag.) Gomont	...	...	...	...	...	...	5 529	...	114	...
Unidentified	...	...	792	...	...	...	...	...	...	...
Coccoid green algae										
Actinastrum spp.	...	420	66	...	...	...	...	...	...	...
Ankistrodesmus convolutus Corda	...	...	...	...	201	396	...	...	...	...
Ankistrodesmus falcatus (Corda) Ralfs	...	...	99	...	25	...	133	162	...	6
Characium spp.	...	...	...	...	25	...	...	...	23	6
Coelastrum spp.	...	...	33	...	377	...	...	...	...	...
Crucigenia tetrapedia (Kirch.) W. and W.	...	...	...	...	...	...	...	...	...	96
Crucigenia spp.	...	...	...	...	...	...	9	...	...	...
Dictyosphaerium spp.	282	34	...	...	...	...	...	...	...	...
Kirchneriella obesa (W. West) Schmidle	...	...	...	...	...	...	2	...	...	...
Kirchneriella spp.	...	...	...	...	50	...	...	...	...	...
Micractinium spp.	...	...	33	...	...	...	...	...	...	...
Oocystis spp.	...	...	...	...	...	330	...	...	...	...
Pediastrum duplex Meyen	...	...	...	...	...	...	...	...	...	...
Pediastrum spp.	...	...	...	...	...	...	...	...	...	...
Scenedesmus abundans (Kirch.) Chod.	...	50	66	...	...	...	...	108	...	...

Scenedesmus bijuga (Turp.) Lager.	...	...	...	...	...	...	14	...	...	...
Scenedesmus intermedius Chod.	...	34	...	...	...	...	...	...	...	...
Scenedesmus quadricauda (Turp.) Breb.	...	...	264	...	...	264	19	...	...	...
Scenedesmus spp.	...	...	66	...	100	792	28	...	...	...
Tetraedron caudatum (Corda) Hansg.	...	...	...	...	...	...	2	...	...	...
Tetrastrum spp.	...	67	99	...	...	...	...	...	...	...
Unidentified	282	8	33	...	1 556	66	107	...	...	...
Filamentous green algae										
Rhizoclonium hieroglyphicum (C. A. Ag.) Kuetz.	...	...	...	...	...	...	9	...	...	...
Rhizoclonium spp.	...	...	...	...	...	...	...	...	...	...
Stigeoclonium spp.	7 432	17	...	...	...	...	...	...	400	48
Green flagellated algae										
Chlamydomonas spp.	47	8	33	56	101	...	...	108	6	...
Euglena spp.	...	8	...	...	...	...	...	...	...	...
Pandorina spp.	...	...	...	391	...	...	...	...	...	...
Pteromonas spp.	...	8	...	...	...	...	...	...	...	...
Trachelomonas spp.	47	...	...	...	...	...	...	...	...	...
Unidentified	94	118	66	...	...	132	9	...	...	...
Other pigmented flagellates										
Kephyrion spp.	...	...	33	...	...	...	...	...	...	...
Centric diatoms	1 693	227	363	2 622	753	594	655	1 187	269	36
Pennate diatoms	1 788	680	2 310	893	3 263	3 036	189	1 187	160	155
Total live algae	11 665	2 897	4 356	3 962	6 451	5 610	6 798	2 752	972	347

TABLE 3—*Continued.*

	26 Aug.- 2 Sept.	26 Aug.- 9 Sept.	26 Aug.- 23 Sept.	26 Aug.- 5 Nov.	7 Oct.- 14 Oct.	7 Oct.- 21 Oct.	7 Oct.- 5 Nov.	7 Oct.- 2 Dec.	5 Nov.- 12 Nov.	5 Nov.- 18 Nov.	5 Nov.- 2 Dec.
Number of weeks slides exposed	1	2	4	10	1	2	4	8	1	2	4
Filamentous blue-green algae											
Oscillatoria spp.	...	...	...	...	...	...	...	...	...	...	...
Schizothrix calcicola (Ag.) Gomont	...	...	...	...	...	...	72	...	...	...	1 889
Unidentified	...	...	...	...	...	...	...	...	2 778	...	...
Coccoid green algae											
Actinastrum spp.	...	...	...	...	...	...	...	...	...	...	...
Ankistrodesmus convolutus Corda	...	...	...	...	...	...	...	...	...	...	...
Ankistrodesmus falcatus (Corda) Ralfs	62	31	28	64	...	...	36	3	...	...	...
Characium spp.	...	...	...	...	...	...	...	...	...	...	...
Coelastrum spp.	...	...	...	...	...	...	17	...	...	...	...
Crucigenia tetrapedia (Kirch.) W. and W.	...	...	...	...	...	...	11	24	...	...	...
Crucigenia spp.	...	...	...	...	...	...	...	...	...	...	...
Dictyosphaerium spp.	...	...	...	...	...	...	...	...	...	...	...
Kirchneriella obesa (W. West) Schmidle	...	...	...	...	...	...	...	...	...	...	...
Kirchneriella spp.	...	...	...	...	...	...	...	...	...	...	...
Micractinium spp.	...	...	...	...	...	...	...	...	...	...	...
Oocystis spp.	...	...	...	...	...	...	...	...	...	...	...
Pediastrum duplex Meyen	...	...	...	...	...	...	...	51	...	...	...
Pediastrum spp.	...	...	...	191	...	...	...	...	...	...	...
Scenedesmus abundans (Kirch.) Chod.	...	...	...	...	...	...	...	...	...	...	...
Scenedesmus bijuga (Turp.) Lager.	...	...	...	...	...	...	...	...	...	...	...

Scenedesmus intermedius Chod.	...	...	...	...	...	...	...	...	...	...	...
Scenedesmus quadricauda (Turp.) Breb.	...	...	...	...	...	...	...	...	...	...	...
Scenedesmus spp.	31	499	111	...	...	240	64	...	...	...	...
Tetraedron caudatum (Corda) Hansg.	...	...	...	...	...	...	...	...	...	...	...
Tetrastrum spp.	...	...	...	...	...	...	...	...	...	...	...
Unidentified	...	...	...	...	...	...	...	...	...	...	...
Filamentous green algae											
Rhizoclonium hieroglyphicum (C. A. Ag.) Kuetz.	...	...	...	...	...	...	...	...	...	...	...
Rhizoclonium spp.	...	...	138	...	...	...	...	...	...	...	...
Stigeoclonium spp.	...	...	...	...	...	...	...	...	...	...	...
Green flagellated algae											
Chlamydomonas spp.	...	...	...	...	...	...	102	3	...	60	54
Euglena spp.	...	...	...	...	...	...	...	3	...	...	...
Pandorina spp.	...	...	...	...	...	...	...	...	...	...	...
Pteromonas spp.	...	...	...	...	...	...	...	...	...	...	...
Trachelomonas spp.	...	...	...	...	...	...	...	3	...	...	...
Unidentified	...	...	...	...	...	...	3	3	3	...	...
Other pigmented flagellates											
Kephyrion spp.	...	...	...	...	...	...	11	...	...	...	...
Centric diatoms	499	2 276	1 026	1 590	2 248	7 556	388	75	...	180	252
Pennate diatoms	125	624	1 885	2 099	2 189	2 279	681	506	50	2 638	2 510
Total live algae	717	3 430	3 188	3 944	4 437	10 075	1 385	671	2 831	2 878	4 705

the Ohio River in August, and consisted largely of the filamentous blue-green alga, *Schizothrix calcicola* (Ag.) Gomont (Table 2). The frequent occurrence of *S. calcicola* and related blue-green algae appeared to be the principal reason for the generally higher cell densities at Station 37. The cell densities in samples exposed for 2 weeks in the Ohio River, beginning in June, July, August, and September, were significantly less than those in samples exposed for 1 week in the same series (Fig. 4). A similar pattern occurred in the ash-free dry weights. Since this phenomenon was not due to high flow rates or other known causes during the exposure period, it may have resulted from the natural sloughing of some of the attached material. In November, the rate of colonization declined significantly in the Ohio River, and the increase in the cell counts with increasing exposure time was nearly linear.

Plankton—Historically, the density of phytoplankton (units per millilitre) has been much higher in the Little Miami River at Cincinnati than in the Ohio River [*16*]. A similar pattern was observed during this study, in which the phytoplankton cell densities in the Little Miami River were greater on every sampling data during which data were obtained from both stations. The maximum density observed in the Little Miami River (122 703/ml) was

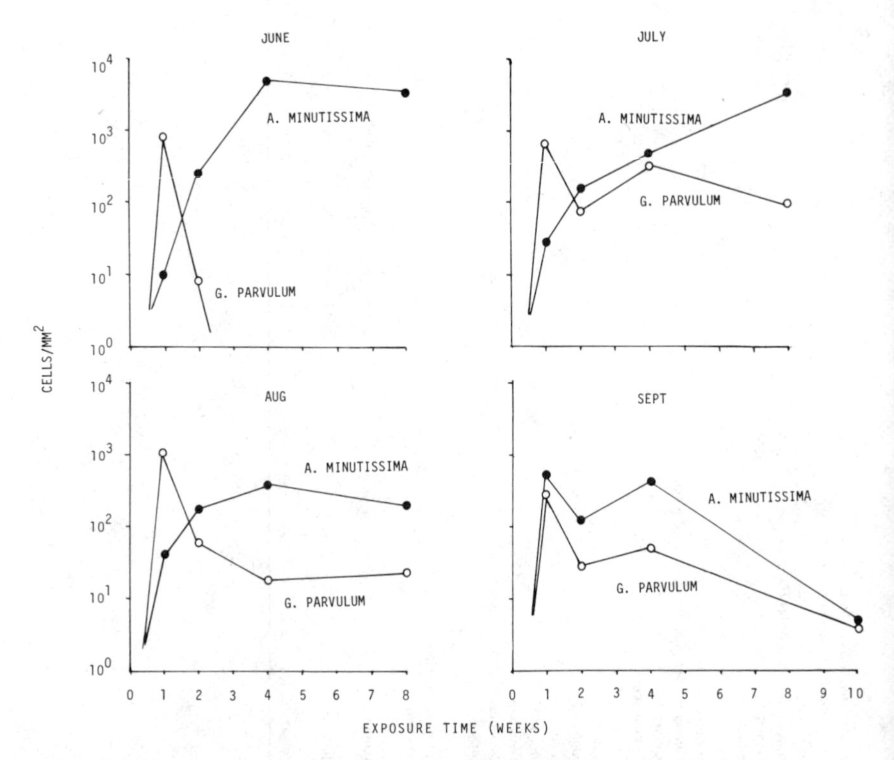

FIG. 4—*Abundances of* G. parvulum *and* A. minutissima *in samples of periphyton from the Ohio River, 1966.*

more than an order of magnitude greater than that in the Ohio River (8326/ml) (Tables 4 and 5).

Taxonomic Composition

Periphyton—Diatoms contributed an average of 69 percent of the total algal cell count in the Ohio River samples and 67 percent of the count in the samples from the Little Miami River. The pennate diatoms were generally the most abundant, constituting 72 percent of the diatoms in the Ohio River samples and 57 percent in the Little Miami River samples. The dominance of diatoms in the periphyton has been reported previously [*2,3,22,23*].

The blue-green and green algae of the families Chamesiphonaceae and Chaetophoraceae, which were abundant on glass slides exposed by Butcher in the English rivers, did not occur regularly in our samples.

The diatoms in samples from the Ohio River (Table 6) were dominated by *Achnanthes minutissima* Kuetz., *Gomphonema parvulum* Kuetz., *Melosira varians* Ag., *Navicula tripunctata* var. *schizonemoides* (V. Heur.) Patr., and *Nitzschia palea* (Kuetz.) W. Sm. Except for *Navicula*, all of these species are considered saprophilous (generally found in polluted waters) and typical of mesosaprobic conditions.

Centric diatoms, which were also abundant in samples from the Ohio River, included *Cyclotella meneghiniana* Kuetz., *Melosira ambigua* (Grun.) O. Muell., *M. distans* var. *alpigena* Grun. and *Thalassiosira pseudonana* (Hust.) Has. and Heim.

Filamentous blue-green algae were observed in nine of the 14 samples taken from the Ohio River between 5 June and 9 Sept., and contributed more than 30 percent of the counts in eight of these samples. Coccoid green algae occurred more often, but were usually not as abundant as the blue-green algae (Tables 2 and 3).

Although the diagnosis of the water quality at Station 37 as mesosaprobic is not strongly supported by the biochemical oxygen demand (BOD) (usually below 2.0), industrial waste at marginal levels of toxicity may be depressing the BOD and acting as a co-selecting agent in determining the composition of the diatom communities.

In samples of periphyton from the Little Miami River, the roles of the filamentous blue-green and coccoid green algae were the reverse of those observed in the Ohio River (Tables 3 and 7). Also, the centric diatoms were much more prominent. The most abundant centric diatoms were *Cyclotella atomus* Hust., *C. meneghiniana* Kuetz., *Melosira varians* Ag., *Stephanodiscus invisitatus* Hohn and Hell., and *Thalassiosira pseudonana* (Hust.) Has. and Heim. The most abundant pennate diatoms were *Achnanthes minutissima* Kuetz., *Amphora ovalis* Kuetz., the *Gomphonema angustatum/parvulum* complex, *G. olivaceum* (Lyngb.) Kuetz., *Navicula cryptocephala* Kuetz., *N. frugalis* Hust., *N. seminulum* var. *hustedtii* Patr.,

TABLE 4—*Sedgwick-Rafter counts of phytoplankton, Ohio River, 1966 (units/ml). Each solitary cell or natural colony of cells was counted as one unit, except for the diatoms, which were counted on a cell basis.*

	8 June	6 July	10 Aug.	9 Sept.	5 Oct.	9 Nov.
Filamentous blue-green algae						
Anabaena spp.	...	...	...	...	22	...
Coccoid green algae						
Actinastrum spp.	155	...	21	...	...	...
Ankistrodesmus falcatus (Corda) Ralfs	266	...	63	...	22	210
Closteriopsis spp.	...	...	...	...	...	21
Coelastrum microporum Naeg.	...	21	...	...	...	...
Coelastrum sp.	...	...	...	21	...	...
Cosmarium sp.	...	...	...	21	...	...
Dictyosphaerium spp.	133	84	105	63	45	21
Elakatothrix viridis (Snow) Printz	...	...	21	...	...	...
Eurastrum sp.	...	...	...	21	...	...
Golenkinia radiata (Chod.) Wille	...	21	...	...	...	...
Kirchneriella spp.	244	...	42	...	...	...
Micractinium spp.	...	...	21	...	22	...
Oocystis spp.	67	84	105	42	...	42
Pediastrum duplex Meyen	...	63	21	...	...	...
Scenedesmus abundans (Kirch.) Chod.	...	...	...	...	67	...
Scenedesmus bijuga (Turp.) Lager.	...	42	...	...	...	21
Scenedesmus dimorphous (Turp.) Kuetz.	67	...	...	...	...	21
Scenedesmus granulatus Hortob.	22	...	...	...	...	126
Scenedesmus intermedius Chod.	22	...	63	...	...	42
Scenedesmus quadricauda (Turp.) Breb.	111	42	42	21	22	42
Scenedesmus spp.	...	...	189	168	22	273

Schroederia setigera Lemm.	...	...	...	21	...	...
Tetrastrum staurogeniaeforme (Schr.) Lemm.	...	...	21	...	...	21
Tetrastrum spp.	733	...	...	21	...	...
Unknown	467	84	336	42	...	105
Green flagellated algae						
Chlamydomonas spp.	...	42	84	...	89	...
Euglena spp.	...	21	...	...	...	...
Pandorina spp.	...	...	21	...	...	...
Phacus spp.	...	...	...	21	...	...
Pteromonas spp.	...	...	21	...	...	...
Trachelomonas spp.	...	...	21	42	...	...
Unknown	111	63	63	126	...	63
Other pigmented flagellates						
Chrysococcus spp.	111	...	...	21	...	189
Peridinium spp.	...	42	21	...	...	...
Mallomonas spp.	22	...	...	...	...	...
Live centric diatoms	4 462	189	1 092	399	580	735
Live pennate diatoms	200	63	126	105	223	147
Percent dead centric diatoms[a]	11.1	18.2	46.4	36.7	39.6	39.7
Percent dead pennate diatoms	18.0	25.0	25.0	16.7	54.6	61.1
Total live algae	7 326	819	2 499	1 160	1 114	2 079

[a]Empty centric frustules ×100/(empty centric frustules + live centrics).

TABLE 5—*Sedgwick-Rafter counts of phytoplankton, Little Miami River, 1966 (units/ml).*

	1 June	8 June	15 June	22 June	29 June	7 July	13 July	20 July	27 July
Coccoid blue-green algae									
Agmenellum spp.	...	...	...	...	93	...	...	...	...
Anacystis spp.	...	...	...	...	...	...	...	...	...
Unidentified	...	...	...	...	...	...	...	...	...
Filamentous blue-green algae									
Anabaena sp.	...	...	44	...	...	...	...	...	...
Unidentified	42	...	...	...	...	...	...	...	...
Coccoid green algae									
Actinastrum spp.	168	266	7 974	4 107	10 345	532	532	84	
Ankistrodesmus convolutus Corda	1 431	44	133	333	373	354	399	168	
Ankistrodesmus falcatus (Corda) Ralfs	294	753	1 728	777	932	797	753	84	
Closterium spp.	...	...	...	...	...	...	...	...	...
Coelastrum microporum Naeg.	...	...	...	...	...	133	...	...	...
Coelastrum sphaericum Naeg.	...	...	...	111	...	266	...	...	...
Coelastrum spp.	...	...	...	...	466	...	133	...	...
Crucigenia crucifera (Wolle) Collins	...	...	...	...	...	...	...	...	...
Crucigenia quadrata Morren	...	...	...	...	...	...	...	...	...
Crucigenia tetrapedia (Kirch.) W. and W.	42	...	44	...	93	222	89	42	...
Crucigenia spp.	...	...	...	...	...	...	...	...	...
Dictyosphaerium spp.	379	576	1 683	1 554	559	709	310	...	...
Franceia spp.	...	...	...	...	...	...	...	...	...
Gloeoactinium spp.	...	...	...	...	...	399	...	...	...
Gloeocystis spp.	...	...	...	...	...	...	...	...	...
Golenkinia radiata (Chod.) Wille	...	44	...	...	...	...	...	...	...
Kirchneriella lunaris (Kirch.) Moeb.	...	...	...	...	...	...	...	...	...
Kirchneriella spp.	168	1 418	399	333	746	487	266	42	...
Lagerheimia quadriseta (Lemm.) G. M. Sm.	42	89	88	111	...	...	...	...	122
Lagerheimia spp.	...	...	...	...	...	...	...	...	...
Micractinium spp.	...	...	88	...	186	44	89	42	22
Oocystis spp.	337	133	1 639	1 554	2 982	487	354	42	...

Pediastrum boryanum (Turp.) Menegh.	...	...	...	...	...	...	...	...	...
Pediastrum duplex Meyen	...	...	...	...	93	...	44	...	...
Scenedesmus abundans (Kirch.) Chod.	...	...	44	...	93	...	...	...	...
Scenedesmus acuminatus (Lag.) Chod.	...	...	...	...	...	...	...	...	...
Scenedesmus dimorphus (Turp.) Kuetz.	...	...	...	...	466	222	177	...	...
Scenedesmus intermedius Chod.	42	133	133	111	373	133	133	...	...
Scenedesmus longus Meyen	...	...	...	...	...	44	...	...	...
Scenedesmus quadricauda (Turp.) Breb.	42	89	266	222	1 398	620	89	211	...
Scenedesmus spp.	168	133	88	1 110	559	620	310	42	245
Tetraedron muticum (A. Braun) Hansg.	...	...	...	...	...	...	...	...	...
Tetraedron spp.	...	...	...	...	...	...	...	...	...
Tetrastrum elegans Playf.	...	...	...	...	...	44	...	...	...
Tetrastrum spp.	...	...	177	888	186	88	...	...	...
Treubaria triappendiculata Bern	...	...	44	...	186	88	...	...	...
Unidentified	1 389	310	1 905	3 219	3 355	2 082	1 639	252	3 304
Green flagellated algae									
Chlamydomonas spp.	1 347	354	620	2 331	...	797	...	42	367
Euglena spp.	...	...	...	111	...	177	...	...	...
Phacus spp.	...	...	...	...	...	...	...	...	...
Strombomonas spp.	...	...	88	...	...	...	...	126	...
Trachelomonas spp.	42	...	...	111	93	...	...	...	...
Unidentified	84	1 097	619	888	653	367	177	210	...
Other pigmented flagellates									
Bicoeca spp.	...	...	...	...	...	...	...	...	...
Chrysococcus spp.	126	89	44	...	93	44	44	...	...
Kephyrion spp.	...	...	44	...	...	...	...	...	...
Peridinium spp.	...	...	...	...	...	...	...	...	...
Unidentified	...	...	88	...	93	...	...	...	612
Live centric diatoms	6 568	2 327	12 050	40 293	71 912	109 890	12 227	7 831	9 543
Live pennate diatoms	2 610	709	709	1 221	2 423	3 057	753	589	367
Percent dead centric diatoms[a]	...	17.0	5.9	13.2	11.8	9.2	1.8	27.9	1.0
Percent dead pennate diatoms	...	30.4	48.4	38.9	36.6	28.9	37.0	56.3	0.0
Total live algae	15 321	8 564	30 827	59 385	98 751	122 703	18 518	9 807	14 582

TABLE 5—*Continued.*

	3 Aug.	10 Aug.	17 Aug.	24 Aug.	31 Aug.	7 Sept.	14 Sept.	21 Sept.	28 Sept.
Coccoid blue-green algae									
Agmenellum spp.	...	...	...	...	...	21	...	...	...
Anacystis spp.	...	...	...	...	...	...	...	...	...
Unidentified	...	...	...	...	...	...	...	...	...
Filamentous blue-green algae									
Anabaena sp.	...	...	...	...	...	...	...	...	...
Unidentified	...	...	...	...	...	...	...	...	...
Coccoid green algae									
Actinastrum spp.	...	...	89	...	192	42	44	177	...
Ankistrodesmus convolutus Corda	...	...	...	...	...	...	...	265	...
Ankistrodesmus falcatus (Corda) Ralfs	...	258	89	79	...	168	1 108	443	848
Closterium spp.	...	...	44	...	...	21	...	...	...
Coelastrum microporum Naeg.	...	...	...	...	...	...	...	...	...
Coelastrum sphaericum Naeg.	...	...	...	...	...	42	...	...	...
Coelastrum spp.	...	...	...	...	...	...	89	...	...
Crucigenia crucifera (Wolle) Collins	...	...	...	...	...	42	...	...	...
Crucigenia quadrata Morren	...	...	...	...	...	...	...	...	...
Crucigenia tetrapedia (Kirch.) W. and W.	...	...	...	79	...	...	...	89	1 060
Crucigenia spp.	...	...	...	...	...	...	...	89	...
Dictyosphaerium spp.	...	...	222	79	96	336	1 108	354	...
Franceia spp.	...	...	...	...	...	...	...	...	...
Gloeoactinium spp.	...	129	...	...	...	...	...	...	...
Gloeocystis spp.	...	...	...	...	...	21	...	...	...
Golenkinia radiata (Chod.) Wille	...	129	...	...	...	21	...	...	...
Kirchneriella lunaris (Kirch.) Moeb.	...	...	...	...	...	...	...	...	...
Kirchneriella spp.	...	...	44	158	192	...	310	266	...
Lagerheimia quadriseta (Lemm.) G. M. Sm.	...	...	...	...	...	...	44	...	...
Lagerheimia spp.	...	...	...	...	...	...	...	...	...
Micractinium spp.	...	...	...	...	96	...	89	...	...
Oocystis spp.	...	129	89	...	...	84	89	266	...

Pediastrum boryanum (Turp.) Menegh.	...	...	...	...	...	42	44	...	...
Pediastrum duplex Meyen	...	...	...	...	...	84	44	...	...
Scenedesmus abundans (Kirch.) Chod.	...	...	...	...	...	42	...	89	...
Scenedesmus acuminatus (Lag.) Chod.	...	...	...	...	...	...	...	...	106
Scenedesmus dimorphus (Turp.) Kuetz.	...	129	...	...	...	42	44	...	...
Scenedesmus intermedius Chod.	...	129	...	...	...	105	222	177	...
Scenedesmus longus Meyen	...	...	...	...	96	42	...	...	...
Scenedesmus quadricauda (Turp.) Breb.	...	129	44	...	96	294	532	177	106
Scenedesmus spp.	...	...	178	79	96	126	399	433	318
Tetraedron muticum (A. Braun) Hansg.	...	129	...	...	...	...	...	...	...
Tetraedron spp.	...	...	...	...	...	...	...	...	106
Tetrastrum elegans Playf.	...	...	...	...	...	...	...	...	...
Tetrastrum spp.	...	...	...	...	96	...	...	...	...
Treubaria triappendiculata Bern	...	...	...	...	...	21	...	...	...
Unidentified	396	1 288	399	552	1 248	420	1 905	1 595	106
Green flagellated algae									
Chlamydomonas spp.	132	...	...	316	...	21	89	89	...
Euglena spp.	...	...	44	...	...	...	...	...	...
Phacus spp.	132	...	...	...	...	...	...	...	...
Strombomonas spp.	...	...	44	...	...	...	...	...	...
Trachelomonas spp.	...	...	...	...	...	...	...	...	...
Unidentified	132	516	44	158	1 856	63	44	89	424
Other pigmented flagellates									
Bicoeca spp.	...	...	...	...	...	...	...	...	...
Chrysococcus spp.	...	...	...	...	...	...	89	...	...
Kephyrion spp.	...	...	...	...	96	...	...	...	...
Peridinium spp.	...	...	...	...	96	...	...	...	...
Unidentified	...	129	89	...	...	462	576	...	...
Live centric diatoms	3 702	13 642	7 520	12 230	28 032	77 070	11 075	7 974	6 042
Live pennate diatoms	397	129	975	373	672	357	487	1 329	1 166
Percent dead centric diatoms[a]	30.0	14.5	50.0	9.4	7.6	14.3	38.4	40.3	10.9
Percent dead pennate diatoms	25.0	0.0	47.6	25.0	12.5	26.1	31.3	31.8	21.4
Total live algae	4 891	16 865	9 914	14 103	32 960	79 989	18 431	13 901	10 282

TABLE 5—*Continued.*

	5 Oct	12 Oct.	19 Oct.	26 Oct.	2 Nov.	9 Nov.	16 Nov.	23 Nov.	30 Nov.
Coccoid blue-green algae									
Agmenellum spp.	...	...	...	...	...	...	...	...	...
Anacystis spp.	74	...	...	...	...	...	...	...	...
Unidentified	...	...	44	...	...	...	...	...	...
Filamentous blue-green algae									
Anabaena sp.	...	...	...	...	...	...	...	...	...
Unidentified	...	...	...	...	...	...	...	...	...
Coccoid green algae									
Actinastrum spp.	...	...	...	...	66	...	...	...	...
Ankistrodesmus convolutus Corda	...	...	...	...	...	...	...	...	...
Ankistrodesmus falcatus (Corda) Ralfs	1 113	223	177	354	265	210	...	177	...
Closterium spp.	...	...	...	...	...	...	...	...	...
Coelastrum microporum Naeg.	...	...	...	44	...	...	...	...	...
Coelastrum sphaericum Naeg.	...	...	...	...	...	...	...	...	...
Coelastrum spp.	...	...	...	...	22	...	...	...	...
Crucigenia crucifera (Wolle) Collins	...	89	...	...	...	...	...	...	...
Crucigenia quadrata Morren	...	...	...	...	...	...	...	...	21
Crucigenia tetrapedia (Kirch.) W. and W.	74	...	44	...	22	...	...	22	...
Crucigenia spp.	223	...	...	...	...	...	44	...	...
Dictyosphaerium spp.	...	134	354	177	88	...	44	44	...
Franceia spp.	...	...	89	...	...	...	...	...	...
Gloeoactinium spp.	...	...	...	...	...	...	...	...	...
Gloeocystis spp.	...	...	...	...	...	...	...	...	...
Golenkinia radiata (Chod.) Wille	...	...	...	...	...	...	...	...	...
Kirchneriella lunaris (Kirch.) Moeb.	223	...	...	...	...	...	...	...	...
Kirchneriella spp.	...	...	44	...	44	...	...	...	...
Lagerheimia quadriseta (Lemm.) G. M. Sm.	...	...	44	...	...	...	22	22	...
Lagerheimia spp.	148	...	...	...	22	...	...	...	...
Micractinium spp.	...	...	...	44	...	...	...	...	...
Oocystis spp.	74	45	...	44	66	...	...	44	...
Pediastrum boryanum (Turp.) Menegh.	...	...	...	...	...	...	...	...	...

Pediastrum duplex Meyen	...	45	...	...	...	...	...	...	...
Scenedesmus abundans (Kirch.) Chod.	296	89	...	...	...	105	...	...	...
Scenedesmus acuminatus (Lag.) Chod.	...	...	...	...	...	...	...	...	...
Scenedesmus dimorphus (Turp.) Kuetz.	74	45	44	...	...	...	...	22	...
Scenedesmus intermedius Chod.	...	...	...	...	...	...	...	...	...
Scenedesmus longus Meyen	...	...	...	...	...	...	...	...	...
Scenedesmus quadricauda (Turp.) Breb.	...	89	44	88	22	...	...	...	...
Scenedesmus spp.	...	...	...	88	...	105	...	...	21
Tetraedron muticum (A. Braun) Hansg.	...	...	...	...	...	...	...	...	...
Tetraedron spp.	...	...	...	...	...	...	...	...	...
Tetrastrum elegans Playf.	...	...	...	...	...	...	...	22	...
Tetrastrum spp.	74	...	...	...	...	...	...	22	...
Treubaria triappendiculata Bern	...	...	...	...	...	...	...	...	...
Unidentified	74	179	753	1 063	486	...	...	44	...
Green flagellated algae									
Chlamydomonas spp.	594	...	133	665	199	...	...	199	...
Euglena spp.	...	...	...	133	...	...	...	...	...
Phacus spp.	...	...	...	...	...	...	...	...	...
Strombomonas spp.	...	...	...	...	...	...	...	...	...
Trachelomonas spp.	...	134	...	...	22	...	...	...	...
Unidentified	...	...	133	88	22	...	...	66	...
Other pigmented flagellates									
Bicoeca spp.	...	...	133	44	22	...	...	22	...
Chrysococcus spp.	371	...	177	443	155	105	...	155	21
Kephyrion spp.	...	...	399	665	641	...	...	...	...
Peridinium spp.	...	...	...	...	...	...	...	...	...
Unidentified	297	45	...	...	...	...	...	...	...
Live centric diatoms	35 319	6 512	9 214	16 539	6 365	1 470	133	1 879	148
Live pennate diatoms	668	268	399	487	1 216	4 515	221	243	170
Percent dead centric diatoms[a]	8.1	20.7	28.3	18.1	16.5	54.8	39.8	14.1	30.2
Percent dead pennate diatoms	25.0	33.3	40.0	31.3	15.4	32.8	47.4	38.9	0.0
Total live algae	39 696	7 897	12 225	20 966	9 745	6 510	464	2 983	381

[a] Empty centric frustules $\times 100$/(empty centric frustules + live centrics).

TABLE 6—*Species of diatoms in periphyton*

	5 June–12 June	5 June–17 June	5 June–1 July	5 June–28 July
Centric diatoms				
Cyclotella atomus Hust.	...	...	...	16
Cyclotella meneghiniana Kuetz.	7	5	...	422
Cyclotella pseudostelligera Hust.	...	...	...	...
Cyclotella stelligera Cl. and Grun.	...	...	...	...
Cyclotella spp.	...	...	...	...
Melosira ambigua (Grun.) O. Muel.	...	11	...	509
Melosira distans var. *alpigena* Grun.	...	...	13	326
Melosira granulata Ralfs	...	5	...	382
Melosira italica subsp. *subartica* O. Müll.	4	...	...	...
Melosira varians Ag.	358	11	...	...
Microsiphona potamos Weber	...	...	...	...
Stephanodiscus astrea var. *minutula* (Kuetz.) Grun.	...	...	...	...
Stephanodiscus hantzschii Grun.	11	2	...	143
Stephanodiscus invisitatus Hohn and Hell.	7	...	...	64
Thalassiosira fluviatilis Hust.	...	...	...	...
Thalassiosira pseudonana Has. and Heim.	11	1	13	16
Pennate diatoms				
Achnanthes biasolettiana Kuetz.	...	...	...	...
Achnanthes lanceolata Breb.	...	...	...	...
Achnanthes minutissima Kuetz.	11	360	10 586	4 917
Amphipleura pellucida W. Sm.	...	...	...	...
Amphora montana Krass.	...	...	...	...
Amphora ovalis Kuetz.	...	...	...	...
Amphora submontana Hust.	...	...	...	...
Anomoeoneis exilis (Kuetz.) Cl.	...	...	...	...
Asterionella formosa Hass.	...	...	...	16
Bacillaria paradoxa Gmel.	...	...	...	...
Caloneis bacillum (Grun.) Meres.	...	...	...	...
Cocconeis placentula Ehr.	...	...	...	...
Cocconeis placentula var. *euglypta* (Ehr.) Cl.	...	...	...	...
Cocconeis placentula var. *lineata* (Ehr.) Cl.	...	...	...	...
Cymbella affinis Kuetz.	...	1	...	...
Cymbella delicatula Kuetz.	...	...	...	...
Cymbella prostrata (Berk.) Cl.	...	...	...	...
Cymbella sinuata Greg.	...	...	...	40
Cymbella tumida (Breb.) V. Heur.	...	...	...	...
Cymbella ventricosa Kuetz.	11	4	64	...
Cymbella spp.	...	...	...	...
Diatoma elongatum Ag.	...	...	...	...
Diatoma vulgare Bory	7	...	...	...
Eunotia spp.	...	...	...	...
Fragilaria capucina Desm.	139	...	...	223
Fragilaria construens (Ehr.) Grun.	...	...	...	...
Fragilaria crotonensis Kitt.	...	...	...	159
Fragilaria intermedia Grun.	...	...	...	...
Gomphoneis herculeana (Ehr.) Cl.	...	...	...	...
Gomphonema angustatum (Kuetz.) Rabh.	7	1	...	...
Gomphonema gracile Ehr.	...	...	...	16
Gomphonema lanceolatum Ehr.	...	...	...	...
Gomphonema olivaceum (Lyngb.) Kuetz.	...	...	...	...
Gomphonema parvulum Kuetz.	873	2	...	...
Gomphonema spp.	...	1	...	...
Gyrosigma kuetzingii (Grun.) Cl.	...	...	...	119
Gyrosigma scalproides (Rabh.) Cl.	...	...	...	...
Meridion circulare Ag.	...	...	...	...
Navicula cincta (Ehr.) Kuetz.	...	...	...	...
Navicula circumtexta Meist.	...	...	...	...
Navicula crytocephala Kuetz.	...	...	...	...
Navicula crytocephala var. *veneta* (Kuetz.) Grun.	...	...	...	40

from the Ohio River, 1966 (cells/mm^2).

1 July–9 July	1 July–15 July	1 July–28 July	1 July–26 Aug.	28 July–4 Aug.	28 July–12 Aug.	28 July–26 Aug.	28 July–3 Sept.
...	1	120	10	32	68	7	...
234	1	881	40	1 218	369	38	49
30	3	20	...	...	...	...	5
...	...	...	...	...	...	...	...
...	...	...	...	...	6	...	...
154	1	1 422	...	465	233	10	24
6	5	1 241	50	144	421	44	136
15	3	120	...	...	42	10	...
6	...	...	...	...	...	...	...
...	4	50	140	481	211	...	402
6	...	...	10	...	13	...	...
...	...	...	...	...	...	...	...
87	...	100	...	48	19	...	...
182	2	50	...	48	32	10	24
...	1	...	...	...	...	...	...
43	...	70	...	...	26	...	30
...	1	...	...	...	...	...	...
...	...	...	10	...	...	20	5
30	157	511	4 268	48	191	2 688	151
...	...	...	...	...	...	...	...
...	...	...	20	16	...	48	5
...	...	...	...	...	...	...	...
...	...	...	...	...	26	...	...
...	...	...	...	...	...	...	...
6	1	20	...	...	6	7	...
...	...	...	...	...	...	...	30
...	...	...	20	...	13	62	9
...	...	...	...	...	6	...	...
...	...	...	20	...	...	34	306
...	...	...	...	...	...	...	...
...	...	...	...	16	...	...	...
6	...	...	...	...	...	...	...
...	...	...	...	...	...	...	...
...	...	...	...	...	...	...	...
...	8	...	...	64	...	...	...
...	3	70	20	...	6	17	...
...	...	...	20	...	13	...	...
...	...	...	...	...	...	...	...
...	...	70	...	...	...	...	...
...	...	...	...	...	...	...	...
74	...	1 582	...	64	...	...	...
6	...	...	...	...	...	...	...
219	3	361	...	...	...	...	...
...	...	100	...	...	...	...	...
...	...	...	...	...	...	...	...
...	3	120	...	312	6	7	9
...	...	...	...	...	...	...	...
...	...	...	...	...	...	...	...
6	...	...	...	...	...	...	...
659	72	220	100	906	62	116	9
22	...	...	...	112	...	7	5
...	...	70	...	...	13	...	5
...	...	...	...	...	...	...	...
...	...	20	...	...	...	...	...
...	...	...	...	...	...	...	...
...	...	...	...	...	...	...	...
...	...	...	20	80	19	17	5
6	1	70	...	...	...	...	...

TABLE 6—

	5 June-12 June	5 June-17 June	5 June-1 July	5 June-28 July
Navicula debilissima Grun.	...	...	...	...
Navicula dicephala Ehr.	...	...	...	...
Navicula frugalis Hust.	...	...	...	...
Navicula hungarica Grun.	...	...	...	...
Navicula menisculus Schum.	...	...	...	...
Navicula mutica Kuetz.	...	...	...	...
Navicula pupula Kuetz.	...	...	...	...
Navicula pygmaea Kuetz.	...	...	...	...
Navicula schroeteri var. *escambia* Patr.	...	...	...	...
Navicula seminulum var. *hustedtii* Patr.	...	...	...	16
Navicula tenera Hust.	...	...	...	...
Navicula tripunctata (O. Muel.) Bory	...	...	...	...
Navicula tripunctata var. *schizonemoides* (V.Heur.) Patr.	...	...	...	80
Navicula viridula Kuetz.	...	...	...	...
Navicula spp.	...	...	...	16
Nitzschia acicularis W. Sm.	...	...	...	...
Nitzschia amphibia Grun.	...	...	...	...
Nitzschia baccata Hust.	...	...	...	...
Nitzschia capitellata Hust.	...	...	...	...
Nitzschia clausii Hantz.	...	...	...	...
Nitzschia closterium (Ehr.) W. Sm.	...	...	...	...
Nitzschia dissipata (Kuetz.) Grun.	4	...	...	...
Nitzschia fasciculata Grun.	...	...	...	...
Nitzschia filiformis (W. Sm.) Hust.	...	...	...	...
Nitzschia fonticola Grun.	...	...	...	...
Nitzschia frustulum Kuetz.	...	...	...	...
Nitzschia gandersheimiensis Krass.	...	...	...	9
Nitzschia kuetzingiana Hilse	...	...	...	...
Nitzschia linearis W. Sm.	...	...	...	...
Nitzschia microcephala Grun.	...	...	...	...
Nitzschia palea (Kuetz.) W. Sm.	...	1	26	183
Nitzschia sublinearis Hust.	...	...	...	103
Nitzschia subrostrata Hust.	...	...	...	...
Nitzschia tryblionella Hantz.	...	...	...	...
Nitzschia spp.	...	1	...	16
Pinnularia braunii (Cl.) Grun.	...	...	...	...
Pinnularia spp.	...	...	...	...
Rhoicosphenia curvata (Kuetz.) Grun.	...	...	...	...
Stauroneis crucicula (Grun. *ex* Cl.) Ross	...	...	...	...
Surirella ovata Kuetz.	...	...	...	...
Synedra acus Kuetz.	...	...	...	...
Synedra amphicephala Kuetz.	...	...	...	16
Synedra nana Meist.	...	...	...	...
Synedra pulchella Kuetz.	...	...	...	...
Synedra rumpens Kuetz.	...	...	...	...
Synedra tabulata (Ag.) Kuetz.	...	...	...	...
Synedra tenera W. Sm.	...	...	...	...
Synedra ulna (Nitz.) Ehr.	179	1	...	16
Synedra ulna var. *oxyrhynchus* (Kuetz.) V.Heur.	...	...	...	...
Synedra vaucheriae Kuetz.	...	...	...	40
Synedra spp.	...	...	...	...
Unidentified pennate diatoms	...	1	...	16
Total (live + dead) diatoms	1 629	408	10 702	7 926

Continued.

1 July–9 July	1 July–15 July	1 July–28 July	1 July–26 Aug.	28 July–4 Aug.	28 July–12 Aug.	28 July–26 Aug.	28 July–3 Sept.
...	...	...	...	...	19	7	...
...	...	...	...	...	...	...	...
...	...	...	...	16	...	...	5
...	...	20	...	...	...	...	...
...	...	150	...	...	6	...	...
...	...	...	...	16	...	...	...
...	...	...	...	...	6	...	...
...	...	20	...	...	...	...	...
...	...	...	...	...	32	...	...
...	1	...	40	...	6	10	...
...	...	...	10	...	...	...	...
...	...	...	...	...	...	...	...
33	4	240	30	2 637	408	110	43
...	1	...	...	...	6	...	...
15	...	220	30	64	...	10	30
30	...	50	...	...	...	...	...
...	...	...	...	...	...	...	...
...	...	...	...	16	...	...	...
...	...	50	...	...	...	...	...
...	...	20	...	16	...	...	...
...	...	...	...	16	...	...	...
6	...	20	...	...	...	...	5
...	...	...	...	...	...	...	5
6	...	...	...	16	...	27	...
...	...	20	...	...	...	...	...
...	3	...	...	16	13	...	5
...	...	...	...	...	...	...	...
...	...	...	...	...	26	7	5
6	...	50	...	32	32	10	...
...	...	...	...	128	6	...	...
111	3	1 122	...	128	253	96	5
...	...	120	...	...	...	...	...
...	...	...	...	...	...	...	...
6	...	20	...	...	6	...	5
22	2	70	...	32	360	...	...
...	...	...	...	...	...	...	...
...	...	40	...	...	...	...	...
...	...	...	...	...	13	...	...
...	...	...	...	16	...	...	...
...	...	...	...	16	...	...	...
6	...	120	...	32	...	7	9
...	...	20	...	...	...	...	...
6	...	...	...	...	6	...	...
...	...	...	...	...	...	...	...
65	1	20	...	...	...	...	...
6	...	20	...	64	32	...	5
...	...	...	...	...	...	...	...
...	...	50	10	729	136	...	9
...	...	...	...	...	...	...	...
...	...	20	...	...	6	...	...
...	...	...	...	...	...	...	...
37	1	90	50	...	39	...	9
2 152	286	9 860	4 918	8 044	3 212	3 426	1 349

TABLE 6—

	26 Aug.- 2 Sept.	26 Aug.- 9 Sept.	26 Aug.- 23 Sept.	26 Aug.- 5 Nov.
Centric diatoms				
Cyclotella atomus Hust.	34	5	23	2
Cyclotella meneghiniana Kuetz.	324	35	166	17
Cyclotella pseudostelligera Hust.	34	3	12	1
Cyclotella stelligera Cl. and Grun.	...	...	...	...
Cyclotella spp.	23	...	...	...
Melosira ambigua (Grun.) O. Muel.	57	7	155	4
Melosira distans var. *alpigena* Grun.	290	41	681	9
Melosira granulata Ralfs	...	...	66	1
Melosira italica subsp. *subartica* O. Müll.	...	...	...	...
Melosira varians Ag.	359	438	1 440	83
Microsiphona potamos Weber	23	5	...	2
Stephanodiscus astrea var. *minutula* (Kuetz.) Grun.	...	...	12	...
Stephanodiscus hantzschii Grun.	...	5	23	...
Stephanodiscus invisitatus Hohn and Hell.	11	14	101	6
Thalassiosira fluviatilis Hust.	...	...	...	...
Thalassiosira pseudonana Has. and Heim.	786	126	197	49
Pennate diatoms				
Achnanthes biasolettiana Kuetz.	...	...	...	...
Achnanthes lanceolata Breb.	...	...	12	5
Achnanthes minutissima Kuetz.	336	311	275	12
Amphipleura pellucida W. Sm.	...	...	...	...
Amphora montana Krass.	11	5	...	...
Amphora ovalis Kuetz.	...	...	...	...
Amphora submontana Hust.	...	...	...	...
Anomoeoneis exilis (Kuetz.) Cl.	...	...	...	...
Asterionella formosa Hass.	...	3	...	X
Bacillaria paradoxa Gmel.	34	18	12	6
Caloneis bacillum (Grun.) Meres.	111	14	23	1
Cocconeis placentula Ehr.	...	...	...	...
Cocconeis placentula var. *euglypta* (Ehr.) Cl.	...	3	50	20
Cocconeis placentula var. *lineata* (Ehr.) Cl.	...	...	...	...
Cymbella affinis Kuetz.	...	...	12	...
Cymbella delicatula Kuetz.	...	3	...	...
Cymbella prostrata (Berk.) Cl.	...	...	...	...
Cymbella sinuata Greg.	...	...	...	...
Cymbella tumida (Breb.) V. Heur.	96	5	...	...
Cymbella ventricosa Kuetz.	...	...	...	...
Cymbella spp.	...	...	...	...
Diatoma elongatum Ag.	...	...	...	X
Diatoma vulgare Bory	...	...	...	...
Eunotia spp.	...	...	...	...
Fragilaria capucina Desm.	...	...	43	...
Fragilaria construens (Ehr.) Grun.	57	...	...	...
Fragilaria crotonensis Kitt.	23	...	...	...
Fragilaria intermedia Grun.	...	...	...	...
Gomphoneis herculeana (Ehr.) Cl.	...	...	...	...
Gomphonema angustatum (Kuetz.) Rabh.	202	14	...	5
Gomphonema gracile Ehr.	...	...	...	...
Gomphonema lanceolatum Ehr.	...	...	...	X
Gomphonema olivaceum (Lyngb.) Kuetz.	...	3	...	...
Gomphonema parvulum Kuetz.	179	62	31	4
Gomphonema spp.	34	18	12	...
Gyrosigma kuetzingii (Grun.) Cl.	...	...	...	...
Gyrosigma scalproides (Rabh.) Cl.	...	...	...	X
Meridion circulare Ag.	...	...	...	...
Navicula cincta (Ehr.) Kuetz.	...	...	...	...
Navicula circumtexta Meist.	...	...	...	X
Navicula crytocephala Kuetz.	...	...	23	1
Navicula crytocephala var. *veneta* (Kuetz.) Grun.	...	...	...	...

Continued.

7 Oct.-14 Oct.	7 Oct.-21 Oct.	7 Oct.-5 Nov.	7 Oct.-2 Dec.	5 Nov.-11 Dec.	5 Nov.-18 Nov.	5 Nov.-2 Dec.
X	2	47	62	X[a]	X	8
2	16	125	342	X	1	100
...	...	...	103	...	X	5
...	...	...	...	...	X	...
...	...	10	31	...	...	...
2	9	10	404	X	X	47
...	16	57	62	...	X	21
...	...	...	135	...	X	8
...	...	...	...	...	X	...
44	316	1 049	777	X	16	113
...	...	16	31	...	...	...
...	...	...	...	X	...	...
...	...	10	...	X	X	...
...	...	21	73	...	X	...
...	...	...	...	...	...	...
X	5	224	125	X	X	...
...	...	...	...	...	...	...
...	...	...	10	X	6	47
...	5	10	62	X	X	150
...	...	...	10	...	...	...
...	2	21	41	...	...	5
...	...	...	...	X	...	...
...	...	...	...	X	...	...
...	...	10	...	...	...	...
...	...	10	62	X	X	...
...	7	21	41	X	X	5
...	...	...	41	X	...	5
...	...	...	...	...	...	...
...	...	...	103	X	X	32
X	...	...	...	...	...	...
...	...	10	...	...	...	...
...	...	...	...	X	...	...
...	...	...	...	X	...	...
...	...	...	...	...	...	...
3	12	...	10	...	X	...
X	...	...	...	X	X	...
...	...	10	...	...	...	...
...	...	...	...	...	...	...
...	...	...	...	...	...	...
...	...	16	...	...	...	...
...	...	42	...	...	...	...
...	...	...	...	X	...	...
...	...	...	385	X	X	...
...	...	...	...	...	...	...
X	...	...	...	...	...	...
46	126	145	1 388	X	15	222
...	...	...	...	...	...	...
...	...	...	73	X	...	5
...	...	...	...	X	X	...
46	245	574	1 347	X	5	1 530
33	94	303	104	...	...	...
...	...	...	...	...	...	...
...	...	...	...	...	...	...
...	...	...	...	...	...	...
...	...	10	...	...	...	...
...	...	...	...	...	...	...
...	5	16	93	X	X	5
...	...	...	...	...	...	...

TABLE 6—

	26 Aug.- 2 Sept.	26 Aug.- 9 Sept.	26 Aug.- 23 Sept.	26 Aug.- 5 Nov.
Navicula debilissima Grun.	...	...	...	...
Navicula dicephala Ehr.	...	...	...	...
Navicula frugalis Hust.	...	3	...	...
Navicula hungarica Grun.	...	...	...	...
Navicula meniscus Schum.	...	...	...	...
Navicula mutica Kuetz.	...	...	...	X
Navicula pupula Kuetz.	...	...	...	...
Navicula pygmaea Kuetz.	...	...	...	...
Navicula schroeteri var. *escambia* Patr.	...	...	...	...
Navicula seminulum var. *hustedtii* Patr.	11	3	...	4
Navicula tenera Hust.	...	...	...	...
Navicula tripunctata (O. Muel.) Bory	...	...	...	...
Navicula tripunctata var. *schizonemoides* (V.Heur.) Patr.	57	12	77	7
Navicula viridula Kuetz.	...	...	12	...
Navicula spp.	34	5	...	1
Nitzschia acicularis W. Sm.	...	3	...	...
Nitzschia amphibia Grun.	...	3	...	...
Nitzschia baccata Hust.	...	...	...	...
Nitzschia capitellata Hust.	...	...	...	...
Nitzschia clausii Hantz.	...	...	12	...
Nitzschia closterium (Ehr.) W. Sm.	...	...	...	...
Nitzschia dissipata (Kuetz.) Grun.	...	...	...	...
Nitzschia fasciculata Grun.	...	...	...	...
Nitzschia filiformis (W. Sm.) Hust.	23	9	31	3
Nitzschia fonticola Grun.	...	...	...	...
Nitzschia frustulum Kuetz.	23	...	...	1
Nitzschia gandersheimiensis Krass.	...	...	...	...
Nitzschia kuetzingiana Hilse	23	7	12	...
Nitzschia linearis W. Sm.	...	...	...	...
Nitzschia microcephala Grun.	11	3	...	1
Nitzschia palea (Kuetz.) W. Sm.	237	24	66	1
Nitzschia sublinearis Hust.	...	...	...	...
Nitzschia subrostrata Hust.	...	...	...	...
Nitzschia tryblionella Hantz.	...	5	43	...
Nitzschia spp.	145	...	31	5
Pinnularia braunii (Cl.) Grun.	...	...	...	X
Pinnularia spp.	...	...	...	...
Rhoicosphenia curvata (Kuetz.) Grun.	...	...	...	...
Stauroneis crucicula (Grun. *ex* Cl.) Ross	11	3	12	...
Surirella ovata Kuetz.	...	5	12	...
Synedra acus Kuetz.	...	3	...	X
Synedra amphicephala Kuetz.	...	...	...	...
Synedra nana Meist.	...	...	...	X
Synedra pulchella Kuetz.	...	...	...	X
Synedra rumpens Kuetz.	...	...	31	...
Synedra tabulata (Ag.) Kuetz.	69	41	54	1
Synedra tenera W. Sm.	...	...	...	...
Synedra ulna (Nitz.) Ehr.	76	41	89	X
Synedra ulna var. *oxyrhynchus* (Kuetz.) V.Heur.	...	...	...	...
Synedra vaucheriae Kuetz.	...	...	12	...
Synedra spp.	...	...	...	...
Unidentified pennate diatoms	96	12	12	1
Total (live + dead) diatoms	3 840	1 320	3 879	253

[a]X = less than 1 cell/mm^2.

Continued.

7 Oct.–14 Oct.	7 Oct.–21 Oct.	7 Oct.–5 Nov.	7 Oct.–2 Dec.	5 Nov.–11 Dec.	5 Nov.–18 Nov.	5 Nov.–2 Dec.
...	...	...	...	...	...	8
...	5	...	...	...	...	...
...	...	10	...	X	...	5
...	...	10	...	...	...	...
...	2	...	...	...	...	...
...	...	16	62	X	...	5
...	...	...	...	...	...	...
...	...	...	...	...	...	...
...	...	...	...	X	...	...
...	...	16	10	X	...	13
...	...	...	...	X	...	...
...	...	...	...	...	...	...
44	177	1 210	2 963	X	2	92
...	...	...	...	...	...	...
X	5	73	10	X	...	...
...	...	10	10	...	...	...
...	...	...	...	...	...	...
...	2	110	135	...	...	...
...	...	...	...	...	...	...
...	...	...	10	...	...	...
...	...	...	...	...	...	...
...	2	...	104	X	X	5
...	...	...	...	...	...	...
...	5	47	197	X	...	...
...	...	10	...	...	...	...
1	2	42	...	...	X	...
...	...	...	...	...	...	...
...	...	10	41	X	...	...
...	...	...	...	...	...	...
...	...	...	41	...	X	5
X	9	31	145	X	X	5
...	...	...	...	...	...	...
...	...	10	...	...	...	...
...	...	21	...	...	...	...
...	...	16	41	X	X	...
...	...	...	...	...	...	...
...	...	...	...	...	...	...
...	...	...	...	...	...	...
...	...	...	...	...	...	...
...	...	10	10	...	X	5
...	7	16	31	X	X	...
...	...	...	...	...	...	...
...	...	16	31	X	37	13
X	...	...	...	X	...	18
...	...	...	...	...	X	...
11	12	47	31	...	X	5
...	...	...	...	X	...	...
16	80	125	497	...	2	150
...	...	...	...	X	...	...
...	...	...	10	...	...	...
...	...	...	...	X	...	...
X	2	10	...	X	X	...
248	1 170	5 240	10 294	2	84	2 637

TABLE 7—*Species of diatoms in periphyton*

	5 June–12 June	5 June–17 June	5 June–1 July
Centric diatoms			
Cyclotella atomus Hust.	54	11	286
Cyclotella meneghiniana Kuetz.	36	22	154
Cyclotella pseudostelligera Hust.	36	169	38
Cyclotella woltereckii Hust.	...	107	...
Cyclotella sp. Kuetz.	...	...	...
Melosira crenulata (E.) Kg.	...	...	...
Melosira distans (E.) Kg.	...	...	38
Melosira granulata (E.) Ralfs	...	11	...
Melosira varians Ag.	...	...	...
Stephanodiscus hantzschii Grun.	72	17	38
Stephanodiscus invisitatus Hohn and Hell.	125	45	307
Thalassiosira fluviatilis Hust.	...	11	...
Thalassiosira sp.	...	...	17
Thalassiosira pseudonana Has. Heim.[b]	3 847	208	17
Pennate diatoms			
Achnanthes lanceolata Breb.	...	5	17
Achnanthes minutissima Kuetz.	...	...	...
Achnanthes sp. Bory	...	...	...
Amphiprora ovalis	...	...	...
Amphora montana Krass.	18	5	...
Amphora ovalis Kuetz.	72	11	210
Caloneis bacillum (Grun.) Meres.	...	...	...
Cocconeis pediculus Ehr.	...	...	...
Cocconeis placentula Ehr.	...	...	17
Cocconeis placentula var. *euglypta* (Ehr.) Cl.	18	...	...
Cymatopleura solea (Breb.) W. Sm.	...	...	...
Cymbella sinuata Greg.	...	...	...
Cymbella tumida (Breb.) V. Heur.	...	...	...
Cymbella ventricosa (Kuetz.)	18	...	...
Denticula thermalis Kuetz.	...	...	...
Diatoma vulgare Bory	18	...	...
Diploneis sp. Ehr.	...	...	...
Epithemia sorex Kuetz.	...	...	...
Fragilaria construens (Ehr.) Grun.	...	...	...
Fragilaria pinnata Ehr.	...	...	...
Frustulia vulgaris Thwaites	...	...	17
Gomphoneis herculeana (Ehr.) Cl.	...	...	...
Gomphonema angustatum (Kuetz.) Rabh.	36	5	...
Gomphonema augur Ehr.	...	...	...
Gomphonema gracile Ehr.	...	...	...
Gomphonema olivaceum (Lyngb.) Kuetz.	54	...	77
Gomphonema parvulum Kuetz.	...	5	56
Gomphonema sp. Ag.	90	...	...
Gyrosigma scalproides (Rabh.) Cl.	...	...	...
Navicula canalis	54	28	...
Navicula cincta (Ehr.) Kuetz.	...	...	...
Navicula contanta Grun.	...	...	...
Navicula cryptocephala Kuetz.	...	...	...
Navicula cryptocephala var. *intermedia* Grun.	...	...	...
Navicula cryptocephala var. *veneta* (Kuetz.) Grun.	143	5	286

from the Little Miami River, 1966 (cells/mm²).

5 June–28 July	1 July–9 July	1 July–15 July	1 July–28 July	4 Aug.–12 Aug.	4 Aug.–26 Aug.	4 Aug.–23 Sept.
127	63	636	92	199	74	24
1 797	62	989	830	2 278	219	94
48	10	...	22	33	2	13
...	...	...	...	...	...	...
...	...	85	...	...	...	...
...	...	...	11	46	2	...
...	3	...	11	23	...	...
12	5	...	...	...	...	...
...	...	...	...	...	...	...
...	...	85	...	10	...	X[a]
24	19	162	11	111	7	24
...	...	...	...	...	...	...
...	...	...	...	...	...	...
...	22	162	76	33	12	10
79	...	...	...	...	2	X
...	...	...	...	...	3	7
...	...	...	...	...	...	...
...	...	...	...	...	...	...
...	...	...	...	10	...	...
71	5	85	11	...	...	2
24	...	...	...	...	...	...
...	...	...	...	10	1	...
...	...	...	...	...	...	...
71	3	...	...	...	3	9
...	2	...	...	...	...	...
...	...	...	...	...	...	...
...	...	...	...	...	...	...
...	...	...	...	...	...	...
...	...	...	...	...	...	...
...	...	...	...	...	...	X
...	...	...	...	...	...	...
...	...	...	...	...	...	...
...	...	...	...	...	...	X
...	1	...	...	...	...	...
...	...	...	...	...	...	...
...	...	...	...	...	...	...
12	9	85	...	...	12	4
...	...	...	...	...	...	...
12	...	...	...	...	...	...
...	5	85	...	...	1	...
...	5	...	5	10	1	...
...	2	...	...	...	...	2
...	...	...	...	...	...	...
...	...	...	...	...	...	...
...	...	...	...	...	1	...
...	...	...	...	...	...	...
...	10	85	22	...	4	4
...	...	...	...	...	...	...
...	...	...	...	...	...	...

TABLE 7—

	5 June-12 June	5 June-17 June	5 June-1 July
Navicula debilissima Grun.	...	...	...
Navicula exigua (Gregory) O. Muller	...	...	...
Navicula gothlandica Grun.	...	...	...
Navicula hungarica Grun.	...	...	...
Navicula menisculus Schum.	...	5	133
Navicula menisculus var. *obtusa*	...	...	...
Navicula mutica Kuetz.	18	...	...
Navicula obtusa W. Sm.	...	...	...
Navicula popula Kuetz.	...	...	...
Navicula schroeteri var. *escambia* Patr.	...	...	...
Navicula seminulum var. *hustedtii* Patr.	...	5	...
Navicula tripunctata (O. Muel.) Bory	72	5	...
Navicula tripunctata var. *schizonemoides*	...	...	...
Navicula frugalis Hust.	...	28	...
Navicula viridula Kuetz.	18	5	...
Navicula sp.	18	17	534
Navicula pupula Kuetz.	...	11	...
Navicula schroeteri var. *escambia* Patr.	...	...	...
Navicula seminulum var. *hustedtii* Patr.	18	22	...
Navicula tripunctata (O. Muel.) Bory	...	...	...
Navicula tripunctata var. *schizonemoides*	...	11	...
Navicula frugalis Hust.	...	...	...
Navicula viridula Kuetz.	143	118	709
Navicula spp.	...	...	...
Nitzschia fonticola Grun.	...	...	...
Nitzschia frustulum Kuetz.	...	...	210
Nitzschia hungarica Grun.	...	22	...
Nitzschia kuetzingiana Hilse	...	...	...
Nitzschia lacuna Patr. and Freeze	...	...	...
Nitzschia linearis W. Sm.	...	...	...
Nitzschia microcephala Grun.	...	...	...
Nitzschia palea (Kuetz.) W. Sm.	54	39	56
Nitzschia punctata (W. Sm.) Grun.	...	...	...
Nitzschia sigma (Kuetz.) W. Sm.	...	...	...
Nitzschia stagnorum Rabh.	...	...	...
Nitzschia sublinearis Hust.	...	...	17
Nitzschia tryblionella Hust.	...	...	...
Nitzschia vermicularis (Kuetz.) Grun.	...	5	...
Nitzschia sp. Hassall	251	118	192
Rhoicosphenia curvata (Kuetz.) Grun.	143	17	...
Surirella angusta Kuetz.	...	...	...
Surirella angustata Kuetz.	...	...	...
Surirella ovalis Breb.	...	...	...
Surirella ovata Kuetz.	36	22	...
Synedra acus Kuetz.	...	...	...
Synedra nana Meist.	...	...	...
Synedra tabulata (Ag.) Kuetz.	...	...	...
Synedra ulna (Nitz.) Ehr.	...	...	...
Synedra vaucheriae Kuetz.	...	...	...
Unidentified pennate diatoms	36	28	56
Total (live + dead) diatoms	5 974	1 226	3 494

Continued.

5 June–28 July	1 July–9 July	1 July–15 July	1 July–28 July	4 Aug.–12 Aug.	4 Aug.–26 Aug.	4 Aug.–23 Sept.
...	...	...	...	...	...	...
...	...	...	...	...	...	...
...	...	...	...	...	...	...
...	...	...	...	...	...	...
...	...	85	...	...	...	...
...	...	...	...	...	...	...
...	...	...	...	...	...	...
...	...	...	...	...	...	...
...	1	...	...	...	...	...
...	...	...	...	...	...	...
833	1	...	152	...	36	76
21	...	...	...	...	...	3
...	...	...	...	...	1	...
...	1	...	65	10	95	9
...	...	...	...	23	...	2
36	4	169	5	10	3	X
...	...	...	...	...	...	...
174	1	...	38	10	26	8
...	...	...	...	...	...	...
...	...	...	...	...	...	...
...	...	...	...	...	...	...
...	...	...	...	...	...	...
301	1	...	...	...	...	...
...	...	...	...	...	...	...
...	1	...	...	...	...	...
103	49	332	159	...	42	13
...	7	247	...	...	...	...
...	...	...	16	...	...	...
...	...	...	...	...	...	...
...	...	...	...	...	...	...
...	6	...	87	...	2	...
24	236	565	224	430	10	7
...	...	...	...	...	...	...
...	...	...	...	...	...	X
...	...	...	...	...	...	...
...	...	...	...	...	...	...
...	...	...	...	...	...	...
...	...	...	...	...	...	...
...	1	3 157	32	...	...	X
...	13	...	5	...	...	2
...	...	...	...	...	...	2
...	...	...	...	...	...	...
...	...	...	...	...	...	X
...	...	...	...	...	...	X
...	...	...	...	...	...	...
...	...	...	...	...	...	...
...	...	...	...	...	...	...
...	1	...	...	...	...	...
...	...	...	...	...	...	...
95	...	...	11	23	1	6
3 967	550	7 062	1 808	3 255	561	328

TABLE 7—

	26 Aug.- 1 Sept.	26 Aug.- 9 Sept.	26 Aug.- 23 Sept.
Centric diatoms			
Cyclotella atomus Hust.	108	1 408	479
Cyclotella meneghiniana Kuetz.	340	2 499	1 553
Cyclotella pseudostelligera Hust.	8	71	144
Cyclotella woltereckí Hust.	...	...	...
Cyclotella sp. Kuetz.	6	...	...
Melosira crenulata (E.) Kg.	2	44	46
Melosira distans (E.) Kg.	...	...	...
Melosira granulata (E.) Ralfs	...	44	34
Melosira varians Ag.	...	...	...
Stephanodiscus hantzschii Grun.	...	...	23
Stephanodiscus invisitatus Hohn and Hell.	20	371	323
Thalassiosira fluviatilis Hust.	...	...	...
Thalassiosira sp.	...	...	...
Thalassiosira pseudonana Has. Heim.[b]	18	486	80
Pennate diatoms			
Achnanthes lanceolata Breb.	...	...	11
Achnanthes minutissima Kuetz.	...	...	80
Achnanthes sp. Bory	...	...	...
Amphiprora ovalis	...	...	...
Amphora montana Krass.	...	...	23
Amphora ovalis Kuetz.	...	...	11
Caloneis bacillum (Grun.) Meres.	...	...	...
Cocconeis pediculus Ehr.	...	...	...
Cocconeis placentula Ehr.	...	...	...
Cocconeis placentula var. *euglypta* (Ehr.) Cl.	...	...	...
Cymatopleura solea (Breb.) W. Sm.	...	...	...
Cymbella sinuata Greg.	...	...	...
Cymbella tumida (Breb.) V. Heur.	...	...	...
Cymbella ventricosa (Kuetz.)	...	...	...
Denticula thermalis Kuetz.	...	...	...
Diatoma vulgare Bory	...	...	...
Diploneis sp. Ehr.	...	...	...
Epithemia sorex Kuetz.	...	...	...
Fragilaria construens (Ehr.) Grun.	...	...	...
Fragilaria pinnata Ehr.	...	...	...
Frustulia vulgaris Thwaites	...	...	...
Gomphoneis herculeana (Ehr.) Cl.	...	...	...
Gomphonema angustatum (Kuetz.) Rabh.	...	...	...
Gomphonema augur Ehr.	...	...	...
Gomphonema gracile Ehr.	...	...	...
Gomphonema olivaceum (Lyngb.) Kuetz.	...	...	11
Gomphonema parvulum Kuetz.	...	...	...
Gomphonema sp. Ag.	...	...	...
Gyrosigma scalproides (Rabh.) Cl.	...	...	11
Navicula canalis	...	...	...
Navicula cincta (Ehr.) Kuetz.	...	...	...
Navicula contanta Grun.	...	...	...
Navicula cryptocephala Kuetz.	...	...	68
Navicula cryptocephala var. *intermedia* Grun.	...	...	...
Navicula cryptocephala var. *veneta* (Kuetz.) Grun.	...	...	...

Continued.

26 Aug.- 5 Nov.	7 Oct.- 14 Oct.	7 Oct.- 21 Oct.	7 Oct.- 5 Nov.	7 Oct.- 2 Dec.	5 Nov.- 12 Nov.	5 Nov.- 18 Nov.	5 Nov.- 2 Dec.
363	305	802	14	14	1	65	55
1 719	767	3149	79	30	2	61	411
334	129	534	44	20	2	102	110
...	...	...	...	...	...	...	...
...	...	59	3	...	...	...	...
65	...	114	11	...	...	...	16
80	...	...	...	2	...	...	...
...	...	57	...	2	...	...	6
80	291	1660	6	35	X	45	81
65	72	114	...	2	...	12	16
413	348	1450	32	14	1	102	159
...	...	...	...	...	...	12	6
...	...	...	...	...	...	...	...
334	1 044	6679	889	247	X	33	97
305	...	...	...	2	X	12	23
109	57	114	11	9	2	179	159
...	...	...	...	...	...	...	6
160	...	...	...	...	...	147	...
...	...	...	3	...	X	20	32
...	...	...	...	3	1	...	49
...	...	...	...	...	...	...	...
...	...	...	3	...	...	33	23
...	...	...	...	...	...	...	...
14	...	...	3	2	X	45	39
...	...	...	...	...	...	...	...
...	...	...	...	2	...	...	...
...	29	57	...	...	X	33	23
...	...	...	...	2	...	...	...
...	...	...	...	...	...	...	6
65	...	...	14	2	X	57	81
...	...	...	...	...	X	...	...
...	...	...	...	...	...	12	...
...	...	...	...	...	...	...	...
...	...	...	...	...	...	...	...
...	...	...	...	...	...	...	...
...	...	...	3	...	...	...	...
80	43	114	11	8	X	110	65
...	14	...	...	...	...	...	...
...	...	...	...	...	...	...	...
14	29	57	...	3	1	167	65
14	305	210	3	3	X	20	...
...	14	...	3	2	1	...	...
...	...	...	...	2	X	12	...
...	...	...	...	...	...	...	...
...	...	...	3	...	X	...	...
...	...	...	...	...	X	12	...
...	57	267	8	74	13	436	531
14	...	...	...	...	...	...	...
80	...	...	...	...	...	...	...

TABLE 7—

	26 Aug.- 1 Sept.	26 Aug.- 9 Sept.	26 Aug.- 23 Sept.
Navicula debilissima Grun.	...	...	34
Navicula exigua (Gregory) O. Muller	...	...	...
Navicula gothlandica Grun.	...	...	...
Navicula hungarica Grun.	...	...	...
Navicula menisculus Schum.	...	...	...
Navicula menisculus var. *obtusa*	...	...	...
Navicula mutica Kuetz.	...	...	...
Navicula obtusa W. Sm.	...	...	...
Navicula pupula Kuetz.	...	...	23
Navicula schroeteri var. *escambia* Patr.	...	...	...
Navicula seminulum var. *hustedtii* Patr.	...	...	23
Navicula tripunctata (O. Muel.) Bory	...	...	...
Navicula tripunctata var. *schizonemoides*	...	...	...
Navicula frugalis Hust.	...	...	91
Navicula viridula Kuetz.	...	...	...
Navicula spp.	...	...	156
Nitzschia acicularis W. Sm.	...	...	...
Nitzschia amphibia Grun.	...	...	23
Nitzschia apiculata (Gregory) Grun.	...	...	...
Nitzschia baccata Hust.	...	...	...
Nitzschia capitellata Hust.	...	...	...
Nitzschia denticula Grun.	...	...	...
Nitzschia dissipata (Kuetz.) Grun.	...	...	23
Nitzschia filiformis (W. Sm.) Hust.	...	...	...
Nitzschia fonticola Grun.	...	...	...
Nitzschia frustulum Kuetz.	4	...	57
Nitzschia hungarica Grun.	...	...	34
Nitzschia kuetzingiana Hilse	...	...	...
Nitzschia lacuna Patr. and Freeze	...	...	...
Nitzschia linearis W. Sm.	...	...	...
Nitzschia microcephala Grun.	...	...	...
Nitzschia palea (Kuetz.) W. Sm.	6	...	300
Nitzschia punctata (W. Sm.) Grun.	...	...	...
Nitzschia sigma (Kuetz.) W. Sm.	...	...	...
Nitzschia stagnorum Rabh.	...	...	...
Nitzschia sublinearis Hust.	...	...	...
Nitzschia tryblionella Hust.	...	...	...
Nitzschia vermicularis (Kuetz.) Grun.	...	...	...
Nitzschia sp. Hassall	...	...	...
Rhoicosphenia curvata (Kuetz.) Grun.	...	...	...
Surirella angusta Kuetz.	...	...	11
Surirella angustata Kuetz.	...	...	...
Surirella ovalis Breb.	...	...	...
Surirella ovata Kuetz.	...	...	23
Synedra acus Kuetz.	...	...	...
Synedra nana Meist.	6	...	11
Synedra tabulata (Ag.) Kuetz.	...	...	...
Synedra ulna (Nitz.) Ehr.	...	...	11
Synedra vaucheriae Kuetz.	...	...	...
Unidentified pennate diatoms	...	...	114
Total (live + dead) diatoms	517	5 457	3 798

[a]X = less than 1 cell/mm^2.
[b]May also include some *Cyclotella atomus*.

Continued.

26 Aug.- 5 Nov.	7 Oct.- 14 Oct.	7 Oct.- 21 Oct.	7 Oct.- 5 Nov.	7 Oct.- 2 Dec.	5 Nov.- 12 Nov.	5 Nov.- 18 Nov.	5 Nov.- 2 Dec.
94	...	...	...	2	...	20	...
...	...	...	...	...	X	...	...
...	...	...	...	...	X	12	...
...	29	...	...	...	X	12	6
...	...	...	3	2	...	20	16
44	...	...	...	...	...	...	...
...	...	...	...	2	X	...	32
14	...	...	...	...	...	...	...
...	14	...	...	...	X	...	6
109	...	...	...	...	...	...	...
798	14	57	...	...	1	57	6
...	...	...	...	...	X	57	...
...	...	...	...	2	...	...	23
373	29	114	...	8	1	57	22
...	...	...	...	2	...	20	16
...	276	172	22	22	1	237	49
14	...	57	...	...	...	12	16
94	...	...	...	2	X	...	...
...	...	...	...	...	...	...	...
...	...	...	...	...	X	...	...
...	...	57	...	...	...	...	...
...	...	...	...	...	...	12	...
65	14	...	11	56	3	167	201
...	...	...	...	2	X	...	6
...	...	...	...	2	...	...	6
558	86	57	6	14	3	102	23
80	14	57	14	59	X	90	201
...	...	57	...	2	X	...	...
...	...	...	...	...	...	...	39
...	...	...	...	6	...	20	39
29	...	267	92	9	2	78	175
348	505	2 519	58	59	3	481	16
44	...	...	3	2	X	...	6
...	...	...	...	3	...	...	6
...	...	...	...	2	...	...	...
...	...	...	...	...	...	...	...
...	...	...	3	2	...	...	16
...	...	...	...	...	...	...	...
...	29	...	...	2	X	78	194
29	14	...	...	2	X	...	65
...	...	...	11	2	X	45	71
29	...	...	...	...	...	...	...
...	...	...	...	...	...	...	...
...	29	210	6	3	X	167	136
...	...	...	...	...	...	...	6
...	...	...	...	...	...	...	...
...	...	...	...	...	X	...	...
...	29	...	3	...	...	20	...
...	...	...	...	...	...	...	16
...	172	420	...	2	X	57	6
7 252	4 767	19 084	1 380	752	50	4 079	3 239

Nitzschia dissipata (Kuetz.) Grun., *N. frustulum* Kuetz., and *N. palea* (Kuetz.) W. Sm.

Plankton—The phytoplankton in the Ohio River consisted largely of the centric diatoms, *Cyclotella meneghiniana* Kuetz., *Melosira ambigua* (Grun.) O. Mueller, *M. distans* (E.) Kg., and *Thalassiosira pseudonana* (Hust.) Has. and Heim. These species were also abundant in the periphyton. However, the relative abundance of the pennate diatoms in the plankton was much smaller than that in the periphyton (Table 8).

The phytoplankton of the Little Miami River was strongly dominated by the centric diatoms, but was also rich in coccoid green algae. The dominant diatoms were *C. atomus* Hust., *C. meneghiniana* Kuetz., *C. pseudostelligera* Hust., *Microsiphona potamos* Weber, *Stephanodiscus invisitatus* Grun., and *Thalassiosira pseudonana* (Hust.) Has. and Heim. (Table 9). The dominant coccoid green algae included species of *Actinastrum, Ankistrodesmus, Dictyosphaerium, Kirchneriella, Oocystis,* and *Scenedesmus* (Table 5). As in the Ohio River, the pennate diatoms contributed a much smaller fraction of the phytoplankton than of the periphyton.

Succession

Periphyton—There was little evidence of successional changes in the taxonomic composition of the periphyton within a given season. Successive increases in the abundance of blue-green and green filamentous algae were not observed in the 2, 4, and 8-week samples. The only consistent changes that were noted occurred in the pennate diatoms. In the June, July, and August series in the Ohio River, *Gomphonema parvulum* Kuetz. was dominant in the 1-week exposures, but was replaced by *Achnanthes minutissima* Kuetz. in the 2, 4, and 8-week samples (Fig. 4). This relationship changed in the September series, however, and the abundance of *A. minutissima* declined greatly in October and November.

In the June, July, and August series from the Little Miami River, the abundance of *Nitzschia palea* (Kuetz.) W. Sm. followed a pattern similar to that of *G. parvulum* in the Ohio River, but there was no counterpart to *A. minutissima.*

Plankton—Phytoplankton samples were not taken with sufficient frequency from the Ohio River to observe any seasonal changes in species composition. In the Little Miami River samples, however, distinct seasonal changes were observed in the species composition of centric diatoms. *Stephanodiscus invisitatus* Hohn and Hell. and *Thalassiosira pseudonana* (Hust.) Has. and Heim., were dominant in early summer, were replaced by *Cyclotella atomus* Hust. and *C. meneghiniana* Kuetz. during mid-summer, but assumed dominance again in early fall.

Biomass and Net Productivity Rates

The accumulation of organic matter on the slides, expressed as ash-free dry weight (AFW), followed a pattern similar to that of the total cell counts, frequently showing a decrease after 1 to 2 weeks, followed by a later recovery. The final weights, however, unlike the final cell counts, generally exceeded the 1 to 2-week values.

The AFW on the slides was usually less than 30 percent of the dry weight of the scrapings, and the values were similar to those found by Weber and Moore [*17*] for the seston in the Little Miami River.

There was no indication of a "lag phase" as reported by Kevern et al [*24*]. The maximum net productivity rates (NPR) were generally observed in the first or second week of exposure (Table 10, Fig. 5). The occurrence of high NPRs during the first 2 weeks of exposure was probably due primarily to colonization, whereas the later accumulation may have resulted largely from sedimentation.

Substrate Comparison

Paired glass and Plexiglas slides were collected from the Little Miami River after 2 weeks of exposure on five occasions during the period 26 Aug. to 5 Nov. (Table 11), to compare the AFW that accumulated on the two types of substrates. When the data from the four complete collections were subjected to an analysis of variance, the difference between the substrates was not statistically significant ($P = 0.05$). These samples were not examined for species composition.

Summary

During the summer months, maximum cell counts generally occurred after 1 or 2 weeks of exposure. As the water temperatures declined in the fall, however, the cell densities generally increased steadily with time throughout the exposure period. Successional patterns in the species composition of the periphyton were observed only in the relative abundances of *Gomphonema parvulum* and *Achnanthes minutissima* in samples from the Ohio River.

The periphyton community was distinct from the phytoplankton, and was generally dominated by pennate diatoms and filamentous green and blue-green algae. The phytoplankton, on the other hand, were dominated by centric diatoms and coccoid green algae. However, the dominant species of phytoplankton were also abundant in the periphyton. Maximum net productivity rates were usually observed during the first week of exposure, and declined rapidly thereafter. A lag phase in colonization was not observed. The accumulations of organic matter on glass and Plexiglas slides did not

TABLE 8—*Percent abundance of species of diatoms in plankton samples from the Ohio River, 1966.*

	8 June	6 July	10 Aug.	7 Sept.	5 Oct.	9 Nov.
Centric diatoms						
Coscinodiscus rothii (E.) Grun.	...	...	...	...	...	X[a]
Cyclotella atomus Hust.	X	1	7	X	5	4
Cyclotella meneghiniana Kg.	34	12	15	7	17	9
Cyclotella pseudostelligera Hust.	2	1	...	3	1	1
Melosira ambigua (Grun.) O. Mueller	5	13	15	...	19	36
Melosira distans (E.) Kg.	15	16	27	10	11	4
Melosira granulata (E.) Ralfs	2	...	2	...	13	11
Microsiphona potamos Weber	2	...	2	2	X	X
Stephanodiscus hantzschii Grun.	2	X	1	...	1	1
Stephanodiscus invisitatus (Hohn and Hell.)[b]	10	...	2	X	2	3
Stephanodiscus spp.	22	3	...	...	...	...
Thalassiosira pseudonana (H.) Has. and Heim.	...	11	21	70	8	14
Pennate diatoms						
Achnanthes minutissima Kuetz.	X	X	...	X	1	X
Amphora ovalis Kuetz.	...	...	...	X	...	...
Asterionella formosa Hass.	2	1	...	X	3	5
Bacillaria paradoxa Gmelin	...	...	...	X	...	...
Diatoma vulgare Bory	...	...	...	...	X	...
Fragilaria construens (E.) Grun.	...	17	...	...	...	...
Fragilaria crotonensis Kitt.	...	12	...	...	...	6
Meridion circulare (Grev.) Ag.	...	X	...	...	...	...
Navicula cryptocephala Kuetz.	...	...	...	...	3	1
Navicula hungarica Grun.	...	...	...	...	X	...

Navicula subrotundata Hust.	...	...	...	...	...	X
Navicula tenera Hust.	...	...	...	...	X	...
Navicula frugalis Hust.	...	...	...	...	X	...
Navicula viridula Kuetz.	...	...	...	...	X	...
Navicula spp.	...	...	...	...	...	X
Nitzschia acicularis W. Sm.	...	3	...	...	X	X
Nitzschia amphibia Grun.	...	...	...	...	...	X
Nitzschia dissipata Kuetz.) Grun.	...	...	...	...	X	...
Nitzschia fonticola Grun.	...	...	...	...	X	X
Nitzschia gracilis Hantz.	...	...	...	...	...	X
Nitzschia kuetzingiana Hilse	...	...	...	3	1	X
Nitzschia microcephala Grun.	...	...	...	...	...	X
Nitzschia palea (Kuetz.) W. Sm.	...	...	...	...	3	X
Nitzschia subrostrata Hust.	...	...	...	...	...	X
Nitzschia tryblionella Hust.	...	...	X	...	...	...
Nitzschia spp.	X	3	2	...	...	X
Pinnularia braunii (Grun.) Cl.	...	...	...	...	X	...
Surirella ovata Kuetz.	...	...	...	...	X	...
Synedra acus Kuetz.	...	2	2	...	1	X
Synedra nana Meist.	...	...	...	...	...	X
Synedra tabulata (Ag.) Kuetz.	...	...	...	...	X	...
Synedra ulna (Nitz.) Ehr.	...	2	1	...	2	X
Other unidentified pennates	...	...	...	...	X	X

[a] X = less than 1 percent.

[b] *Stephanodiscus invisitatus* may include some *S. tenuis* Hust. and *S. subtilis*. (van Goor) Cl. Eul. *Thalassiosira pseudonana* may include some *Cyclotella atomus.*

TABLE 9—*Percent abundances of species of diatoms in*

	1 June	8 June	15 June	22 June	29 June
Centric diatoms					
Coscinodiscus rothii (E.) Grun.	...	...	...	...	...
Cyclotella atomus Hust.	2	3	6	73	72
Cyclotella meneghiniana Kuetz.	X	X	2	8	7
Cyclotella pseudostelligera Hust.	2	29	5	...	5
Cyclotella stelligera Cl. and Grun.	...	...	...	...	...
Cyclotella woltereckii Hust.	...	...	...	...	...
Melosira ambigua (Brun.) O. Mueller	...	...	...	...	...
Melosira distans var. *alpigena* Grun.	...	...	...	X	...
Melosira granulata (E.) Ralfs	...	...	...	...	...
Melosira varians Ag.	...	...	...	...	...
Microsiphona potamos Weber	...	...	...	...	...
Stephanodiscus hantzschii Grun.	X	5	...	...	...
Stephanodiscus invisitatus Hohn and Hell	9	22	25	13	13
Thalassiosira fluviatilus Hust.	...	...	...	...	...
Thalassiosira pseudonana (H.) Has. and Heim.	67	26	52	5	...
Pennate diatoms					
Achnanthes brevipes Ag.	...	...	...	...	...
Achnanthes exigua Grun.	...	...	...	...	...
Achnanthes lanceolata Breb.	...	...	...	...	...
Achnanthes minutissima Kuetz.	...	...	...	...	...
Amphora montanam Krass.	...	...	...	...	...
Amphora ovalis Kuetz.	...	...	...	...	...
Amphora veneta Kuetz.	...	...	...	...	...
Asterionella formosa Hass.	...	...	...	...	...
Caloneis bacillum (Grun.) Meres.	...	...	...	...	...
Cocconeis pediculus Ehr.	...	...	...	...	...
Cocconeis placentula Ehr.	...	...	...	X	X
Cylindriotheca gracilis (Breb.) Grun.	...	...	...	...	...
Cymbella sinuata Greg.	...	...	...	...	...
Cymbella tumida (Breb.) V. Heur.	...	...	...	...	...
Cymbella ventricosa Kuetz.	...	...	...	...	...
Diatoma vulgare Bory	...	X	...	...	...
Fragilaria pinnata Ehr.	...	...	...	...	...
Gomphonema angustatum (Kuetz.) Rab.	...	...	...	...	...
Gomphonema montanum Schum.	...	...	...	...	...
Gomphonema olivaceum (Lyngb.) Kuetz.	X	1	X	...	X
Gomphonema parvulum Kuetz.	...	...	...	...	X
Gyrosigma scalproides (Rab.) Cl.	...	...	...	...	...
Meridion circulare Ag.	...	...	...	...	...
Navicula canalis Patr.	X	...	...	...	...
Navicula cari Ehr.	...	...	...	...	...
Navicula contenta Grun.	...	...	...	...	...
Navicula cryptocephala var. *veneta* K. Gr.	2	1	...	...	X
Navicula debilissima Grun.	...	...	...	...	...
Navicula frugalis Hust.	...	...	...	...	...
Navicula hungarica Grun.	...	...	X	...	...
Navicula menisculus Schum.	...	...	...	...	...
Navicula mutica Kuetz.	...	...	...	...	...
Navicula pupula Kuetz.	...	...	...	...	...
Navicula seminulum var. *hustedtii* Patr.	...	...	...	...	...
Navicula tripunctata (O. Muel.) Bory	...	1	...	...	...
Navicula spp.	...	1	X	...	...
Nitzschia acicularis W. Sm.	8	...	X	...	...
Nitzschia amphibia Grun.	...	...	...	...	...
Nitzschia baccata Hust.	...	...	...	...	...
Nitzschia capitellata Hust.	...	...	...	...	...
Nitzschia clausii Hantz.	...	...	...	...	...
Nitzschia closterium (Ehr.) W. Sm.	...	...	...	...	...

plankton samples from the Little Miami River, Ohio, 1966.

6 July	13 July	20 July	27 July	3 Aug.	10 Aug.	17 Aug.	24 Aug.	31 Aug.
...	...	...	...	...	...	...	X[a]	...
50	30	28	8	8	9	18	35	27
31	36	49	87	54	81	55	49	62
7	9	7	2	9	3	...	5	6
...	...	...	...	...	...	...	X	...
...	...	...	...	...	...	...	...	...
...	...	...	...	...	...	X	...	...
...	1	...	...	...	...	...	...	...
...	...	...	X	...	X	...	...	X
...	...	...	...	...	...	...	...	...
5	10	...	...	23	2	...	5	3
...	...	...	...	...	...	6	X	X
3	2	4	...	1	...	6	X	1
...	...	...	...	...	...	...	...	...
...	2	1	...	X	...	1	3	...
...	...	...	...	...	...	...	...	...
...	...	...	...	...	...	...	...	...
...	...	...	...	...	...	...	...	...
...	...	...	...	...	...	...	...	...
...	...	...	...	...	...	...	...	...
...	...	...	...	...	...	...	...	...
...	...	...	...	...	...	...	...	...
...	...	...	...	...	...	...	...	...
...	...	...	...	...	...	...	...	...
X	...	...	...	...	...	...	...	...
...	...	X	...	...	...	...	...	...
...	...	...	...	...	...	...	...	...
...	...	...	...	...	...	...	...	...
...	...	...	...	...	...	...	...	...
...	...	...	...	...	...	...	...	...
...	...	...	...	...	...	...	...	...
...	...	...	...	...	...	...	...	...
...	...	...	...	...	...	...	X	...
...	...	...	...	...	...	...	...	...
...	...	1	...	...	...	...	...	...
...	...	X	...	...	...	...	...	...
...	...	...	...	...	...	...	...	...
...	...	...	...	...	...	X	...	...
X	...	...	...	...	...	X	...	...
...	...	...	...	...	...	...	...	...
...	...	...	...	...	...	...	...	...
X	X	X	X	...	...	X	...	...
...	...	...	...	...	...	...	...	...
...	...	...	...	...	...	...	...	...
...	...	...	...	...	...	...	...	...
...	X	X	...	...	...	5	...	X
...	...	...	...	...	...	...	...	...
...	...	...	...	...	...	...	...	...
...	...	...	...	...	...	...	...	...
...	...	1	...	...	...	X	...	...
...	...	X	...	...	X	X	X	...
...	...	...	X	...	...	X	...	...
...	...	...	...	...	...	X	...	...
...	...	...	...	...	...	...	...	...
...	...	...	...	...	...	X	...	...
...	...	...	...	...	...	...	...	...
...	...	...	...	...	...	...	...	...

TABLE 9—

	1 June	8 June	15 June	22 June	29 June
Nitzschia denticula Grun.	...	...	...	...	...
Nitzschia dissipata (Kuetz.) Grun.	4	4	X	...	...
Nitzschia fonticola Grun.	...	...	...	...	...
Nitzschia frustulum Kuetz.	...	X	...	...	X
Nitzschia gracilis Hantz.	...	...	...	...	...
Nitzschia hosatica Hust.	...	...	...	...	...
Nitzschia hungarica Grun.	...	...	X	...	...
Nitzschia kuetzingiana Hilse	...	...	...	...	...
Nitzschia microcephala Grun.	...	...	...	...	...
Nitzschia palea (Kuetz.) W. Sm.	4	1	X	...	...
Nitzschia paleacea Grun.	...	...	...	...	...
Nitzschia punctata (W. Sm.) Grun.	...	...	...	...	...
Nitzschia tryblionella Hantz.	...	...	...	...	...
Nitzschia spp.	...	...	X	X	...
Rhoicosphenia curvata (Kuetz.) Grun.	...	X	...	...	...
Surirella angustata Kuetz.	...	...	...	...	...
Surirella ovata Kuetz.	...	...	X	...	...
Surirella ovalis Breb.	...	...	...	...	...
Synedra acus A.-V. H.	...	1	...	...	...
Synedra nana Meist.	X	...	...	...	...
Synedra ulna (Nitz.) Ehr.	...	...	...	...	...
Synedra vaucheriae Kuetz.	...	...	...	...	...

	7 Sept.	14 Sept.	21 Sept.	28 Sept.
Centric diatoms				
Coscinodiscus rothii (E.) Grun.	...	...	...	...
Cyclotella atomus Hust.	15	13	8	17
Cyclotella meneghiniana Kuetz.	65	52	46	38
Cyclotella pseudostelligera Hust.	4	3	9	3
Cyclotella stelligera Cl. and Grun.	...	...	...	...
Cyclotella woltereckii Hust.	...	...	...	X
Melosira ambigua (Brun.) O. Mueller	...	...	...	...
Melosira distans var. *alpigena* Grun.	...	...	...	...
Melosira granulata (E.) Ralfs	3	X	...	...
Melosira varians Ag.	...	...	...	...
Microsiphona potamos Weber	6	5	3	X
Stephanodiscus hantzschii Grun.	...	...	1	1
Stephanodiscus invisitatus Hohn and Hell	4	22	10	12
Thalassiosira fluviatilus Hust.	...	...	...	...
Thalassiosira pseudonana (H.) Has. and Heim.	2	3	9	15
Pennate diatoms				
Achnanthes brevipes Ag.	...	...	...	...
Achnanthes exigua Grun.	...	...	...	...
Achnanthes lanceolata Breb.	...	...	...	...
Achnanthes minutissima Kuetz.	...	...	5	...
Amphora montanam Krass.	...	...	...	X
Amphora ovalis Kuetz.	...	...	...	...
Amphora veneta Kuetz.	...	...	...	...
Asterionella formosa Hass.	...	...	...	...
Caloneis bacillum (Grun.) Meres.	...	...	...	...
Cocconeis pediculus Ehr.	...	...	...	...
Cocconeis placentula Ehr.	...	...	...	...
Cylindriotheca gracilis (Breb.) Grun.	...	...	...	...
Cymbella sinuata Greg.	...	...	...	...
Cymbella tumida (Breb.) V. Heur.	...	...	...	...
Cymbella ventricosa Kuetz.	...	...	...	...

Continued.

6 July	13 July	20 July	27 July	3 Aug.	10 Aug.	17 Aug.	24 Aug.	31 Aug.
...	...	...	...	...	...	...	...	...
...	X	1	...	...	...	...	...	...
...	...	...	...	...	...	...	...	...
...	...	1	X	X	...	...	X	...
...	...	...	...	...	...	X	...	...
...	...	...	...	...	1	...	...	...
...	...	X	...	...	...	...	...	...
...	...	...	...	...	...	...	...	...
...	...	...	...	...	...	...	...	...
...	2	X	2	...	X	1	1	X
...	...	...	...	...	...	X	...	...
...	...	...	...	...	...	...	...	...
...	...	...	...	...	...	...	...	...
X	1	1	...	...	X	1	...	...
...	X	...	...	...	...	...	...	...
...	...	...	...	...	...	...	...	...
...	...	...	...	...	...	1	...	...
...	...	...	...	...	...	...	...	...
...	...	...	...	...	...	...	...	...
...	...	...	...	...	...	...	...	...
...	...	...	...	...	...	X	...	...
...	...	...	...	...	...	...	...	...

5 Oct.	12 Oct.	19 Oct.	26 Oct.	2 Nov.	9 Nov.	16 Nov.	23 Nov.	30 Nov.
...	...	...	...	...	...	...	...	...
10	7	9	4	4	4	4	2	2
22	14	15	13	11	6	12	12	13
1	...	2	1	2	2	3	2	6
...	...	...	...	...	...	...	...	...
...	...	...	...	...	...	...	...	...
...	...	...	...	...	...	...	...	...
...	...	2	1	1	...	...	5	1
...	...	...	...	...	...	...	...	X
...	...	...	...	X	5	...	...	X
4	2	6	14	X	...	...	2	2
X	X	4	4	...	X	1	2	5
31	34	19	40	45	14	11	46	20
...	...	...	...	...	...	...	X	X
22	38	36	15	15	6	7	6	4
...	...	...	...	...	X	...	...	...
...	...	...	...	...	...	...	X	...
...	...	...	...	...	X	X	1	X
X	...	...	X	2	X	4	1	3
...	...	...	...	...	...	...	...	...
X	...	X	...	X	X	2	X	X
...	...	...	...	...	X	...	...	...
...	...	...	...	...	...	2	X	4
...	...	...	...	...	1	...	...	...
...	...	...	...	...	1	...	...	X
...	...	...	...	...	...	...	...	X
...	...	...	...	...	...	X	...	...
...	X	...	...	...	...	...	...	...
...	...	...	...	X	...	...	...	...
...	...	...	...	...	...	X	...	X

TABLE 9—

	7 Sept.	14 Sept.	21 Sept.	28 Sept.
Diatoma vulgare Bory	...	...	...	...
Fragilaria pinnata Ehr.	...	...	...	...
Gomphonema angustatum (Kuetz.) Rab.	...	...	...	...
Gomphonema montanum Schum.	...	...	...	...
Gomphonema olivaceum (Lyngb.) Kuetz.	...	...	...	...
Gomphonema parvulum Kuetz.	...	...	...	X
Gyrosigma scalproides (Rab.) Cl.	...	...	...	...
Meridion circulare Ag.	...	...	...	...
Navicula canalis Patr.	...	...	...	...
Navicula cari Ehr.	...	...	...	...
Navicula contenta Grun.	...	...	...	...
Navicula cryptocephala var. *veneta* K. Gr.	...	X	X	X
Navicula debilissima Grun.	...	...	X	...
Navicula frugalis Hust.	...	...	...	...
Navicula hungarica Grun.	...	...	...	...
Navicula menisculus Schum.	...	...	...	...
Navicula mutica Kuetz.	...	...	...	...
Navicula pupula Kuetz.	...	...	...	...
Navicula seminulum var. *hustedtii* Patr.	...	...	...	...
Navicula tripunctata (O. Muel.) Bory	...	...	...	...
Navicula spp.	X	...	X	X
Nitzschia acicularis W. Sm.	...	X	...	...
Nitzschia amphibia Grun.	...	...	X	...
Nitzschia baccata Hust.	...	...	...	...
Nitzschia capitellata Hust.	...	...	...	...
Nitzschia clausii Hantz.	...	...	...	...
Nitzschia closterium (Ehr.) W. Sm.	...	...	...	...
Nitzschia denticula Grun.	...	...	...	...
Nitzschia dissipata (Kuetz.) Grun.	...	...	...	X
Nitzschia fonticola Grun.	...	...	X	...
Nitzschia frustulum Kuetz.	...	X	X	X
Nitzschia gracilis Hantz.	...	...	...	...
Nitzschia hosatica Hust.	...	...	...	...
Nitzschia hungarica Grun.	...	...	...	X
Nitzschia kuetzingiana Hilse	...	...	X	X
Nitzschia microcephala Grun.	...	...	...	...
Nitzschia palea (Kuetz.) W. Sm.	2	1	X	X
Nitzschia paleacea Grun.	...	...	...	...
Nitzschia punctata (W. Sm.) Grun.	...	...	...	...
Nitzschia tryblionella Hantz.	...	...	...	...
Nitzschia spp.	...	...	...	...
Rhoicosphenia curvata (Kuetz.) Grun.	...	...	...	...
Surirella angustata Kuetz.	...	...	X	...
Surirella ovata Kuetz.	...	...	...	...
Surirella ovalis Breb.	...	...	...	...
Synedra acus A.-V. H.	...	...	...	...
Synedra nana Meist.	...	...	...	...
Synedra ulna (Nitz.) Ehr.	...	...	...	...
Synedra vaucheriae Kuetz.	...	...	...	...

[a] X = Less than 1 percent.

Continued.

5 Oct.	12 Oct.	19 Oct.	26 Oct.	2 Nov.	9 Nov.	16 Nov.	23 Nov.	30 Nov.
X	...	...	X	10	18	1	X	2
...	...	...	...	...	...	...	...	X
...	...	...	...	...	...	...	...	...
...	...	...	...	...	...	...	...	X
...	...	...	2	X	1	...	X	X
X	...	...	...	...	1	...	X	X
...	...	...	...	...	1	...	...	...
...	...	...	...	...	...	...	...	...
...	...	...	...	...	...	...	...	...
...	...	...	...	X	...	...	...	...
...	...	...	...	...	...	...	X	...
X	X	...	...	X	7	5	3	10
...	...	...	...	...	...	X	...	...
...	...	X	...	X	...	X	...	X
...	...	...	...	...	...	...	...	...
...	...	...	...	...	...	...	X	...
X	...	...	...	...	X	X	...	...
...	...	...	...	...	X	...	...	X
...	...	...	...	...	...	X	...	X
...	...	...	...	...	X	...	...	X
...	...	X	...	X	...	3	X	...
...	...	X	...	...	...	1	X	4
...	...	...	...	...	...	...	X	X
...	...	...	...	...	...	...	...	X
...	...	X	...	...	...	1	...	...
...	...	...	...	...	...	...	...	X
...	...	...	...	X	...	...	X	...
...	...	...	...	...	...	...	...	X
X	...	...	...	...	3	5	2	3
...	...	...	...	...	X	...	X	X
X	1	...	X	X	...	1	1	X
...	...	...	...	...	...	...	...	...
...	...	...	...	...	...	...	...	...
...	...	...	...	...	...	...	X	X
...	X	X	X	X	...	1	X	X
...	...	...	...	...	...	...	...	X
X	1	2	1	X	2	11	4	5
...	...	...	...	...	...	...	...	...
...	...	...	...	...	X	...	...	...
...	...	...	...	...	X	...	...	...
...	...	...	...	...	...	...	...	...
1	...	...	...	X	4	1	X	1
...	...	...	...	...	X	...	...	...
...	...	...	...	...	2	...	X	1
...	...	...	...	...	...	X	...	...
...	...	...	...	...	...	...	...	...
...	...	...	...	...	...	X	1	...
...	...	...	X	...	1	...	X	X
...	...	...	...	...	...	...	...	X

TABLE 10—*Ash-free dry weights and dry weights of periphyton, 1966*

Little Miami River[b]					Ohio River				
Date	Weeks	AFW[a]/DW,[b] %	SC,[c] g/m^2	NPR,[d] $mg/m^2/d$	Date	Weeks	AFW/DW, %	SC, g/m^2	NPR, $mg/m^2/d$
5 June–12 June	1	15.9	9.72	1390	5 June–12 June	1	26.8	3.08	440
3 June–17 June	2	25.2	4.77	340	5 June–17 June	2	44.8	2.41	170
3 June– 1July	4	28.7	7.37	280	5 June–1 July	4	22.2	3.16	120
3 June–28 July	8	28.1	10.61	200	5 June–28 July	8	11.5	11.75	220
1 July– 9 July	1	...	...	...	1 July– 9 July	1	24.6	5.92	740
1 July–15 July	2	12.9	7.24	520	1 July–15 July	2	...	...	...
1 July–28 July	4	26.3	6.27	230	1 July–28 July	4	11.8	16.95	630
1 July–26 Aug.	8	...	...	...	1 July–26 Aug.	8	31.6	14.08	250
4 Aug.–12 Aug.	1	...	...	...	28 July– 4 Aug.	1	18.3	7.29	1040
4 Aug.–26 Aug.	3	...	...	...	28 July–12 Aug.	2	19.6	6.90	460
4 Aug.–23 Sept.	7	...	...	...	28 July–26 Aug.	4	24.0	10.13	350
					28 July–23 Sept.	8	19.8	16.32	290
26 Aug.– 2 Sept.	1	20.5	8.16	1170	26 Aug.– 2 Sept.	1	19.4	2.16	310
26 Aug.– 9 Sept.	2	15.3	12.63	900	26 Aug.– 9 Sept.	2	23.7	3.68	260
26 Aug.–23 Sept.	4	14.7	11.63	420	26 Aug.–23 Sept.	4	26.5	11.39	410
26 Aug.– 5 Nov.	10	23.3	21.88	310	26 Aug.– 5 Nov.	10	30.0	9.93	140
7 Oct.–14 Oct.	1	20.7	3.16	450	7 Oct.–14 Oct.	1	43.3	0.56	80
7 Oct.–21 Oct.	2	15.7	9.49	680	7 Oct.–21 Oct.	2	26.6	3.10	220
7 Oct.– 5 Nov.	4	19.8	6.07	210	7 Oct.– 5 Nov.	4	...	...	...
7 Oct.– 2 Dec.	8	12.3	9.24	160	7 Oct.– 2 Dec.	8	18.8	9.74	170
5 Nov.–12 Nov.	1	16.2	0.76	110	5 Nov.–12 Nov.	1	...	...	...
5 Nov.–18 Nov.	2	13.1	3.54	270	5 Nov.–18 Nov.	2	40.6	0.55	40
5 Nov.– 2 Dec.	4	...	...	...	5 Nov.– 2 Dec.	4	20.4	1.45	50

[a] AFW = ash-free dry weight.
[b] DW = dry weight.
[c] SC = standing crop.
[d] NPR = net production rate of organic matter.

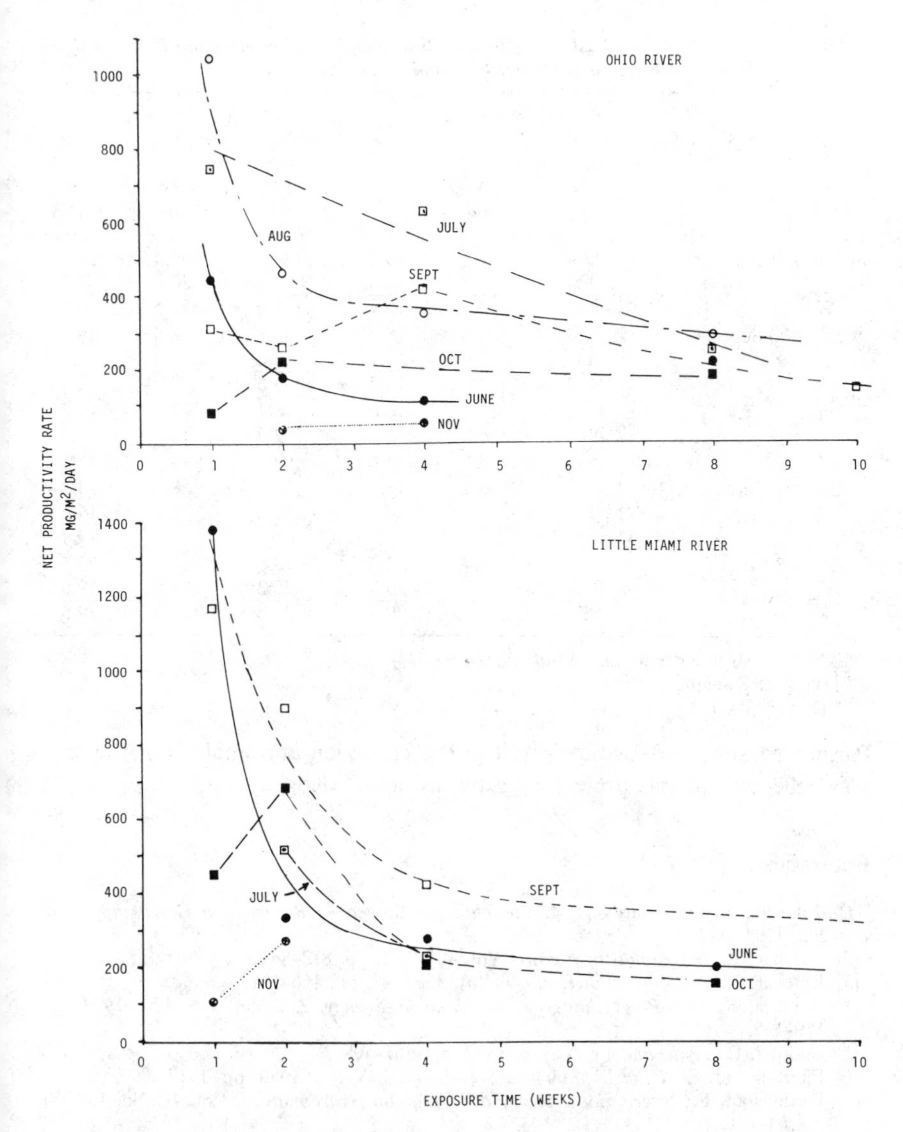

FIG. 5—*Net periphyton productivity in the Ohio and Little Miami Rivers, 1966.*

differ significantly during five exposure periods between 26 Aug. and 5 Nov. The taxonomic composition of the samples was not examined, however, and may not have been similar.

Acknowledgments

The field and laboratory assistance of Gretchen Cooper, Donald Moore,

TABLE 11—*Dry weights and ash-free dry weights of scrapings from glass and Plexiglas slides exposed in the Little Miami River, Ohio, 1966.*

	Glass Substrate			Plexiglas Substrate		
Date	Dry Weight, mg	AFW,[a] mg	AFW, % DW[b]	Dry Weight, mg	AFW, mg	AFW, % DW
26 Aug.- 9 Sept.	125.4	31.9	25.4	128.4	36.4	28.3
	168.0	44.3	26.4	94.8	30.6	32.3
$\overline{X}$	146.7	38.1	25.9	111.6	33.5	30.3
9 Sept.-23 Sept.	62.7	9.6	15.3	43.9	8.5	19.4
	56.1	9.6	17.1	53.6	9.3	17.4
$\overline{X}$	59.4	9.6	16.2	48.8	8.9	18.4
23 Sept.- 7 Oct.	63.6	10.4	16.4	66.7	9.5	14.2
	45.0	6.9	15.3	72.9	12.7	17.4
$\overline{X}$	54.3	8.6	15.8	69.8	11.1	15.8
7 Oct.-21 Oct.	50.6	6.2	12.2	58.8	8.4	14.3
	ND[c]	ND	...	48.6	7.1	14.6
$\overline{X}$				53.7	7.8	14.4
21 Oct.- 5 Nov.	16.6	3.7	22.3	18.5	4.2	22.7
	20.5	4.0	19.5	25.2	4.8	19.0
$\overline{X}$	18.6	3.8	20.9	21.8	4.5	20.8

[a] AFW = ash-free dry weight of organic matter.
[b] DW = dry weight.
[c] ND = no data.

Paralee Mason, and Lydia Correll in the collection and analysis of the samples collected for this project is gratefully acknowledged.

References

[*1*] Kolkwitz, R. and Marsson, M., *Berichte der Deutschen Botanischen Gesellschaft*, Vol. 26a, 1908, p. 118.
[*2*] Butcher, R. W., *Annals of Botany*, Vol. 46, 1932, pp. 813-861.
[*3*] Butcher, R. W., *Journal of Ecology*, Vol. 35, 1947, pp. 186-191.
[*4*] Butcher, R. W., *Proceedings of the Linnaeus Society, London*, Vol. 170, 1959, pp. 159-165.
[*5*] Blum, J. L., *Hydrobiology*, Vol. 9, 1957, pp. 361-408.
[*6*] Fjerdingstad, E., *Folia Limnologica Scandinovica*, Vol. 5, 1950, pp. 1-123.
[*7*] Fjeringstad, E., *International Revue der Gesamlin Hydrobiologie*, Vol. 49, No. 1, 1964, pp. 63-131.
[*8*] Fjeringstad, E., *Hydrobiology*, Vol. 50, No. 4, 1965, pp. 475-604.
[*9*] Sladeckova, A. and Sladecek, V., *Technol. Vod.*, Vol. 7, No. 1, 1963, pp. 507-561.
[*10*] Castenholz, R. W., *Limnology and Oceanography*, Vol. 5, 1960, pp. 1-28.
[*11*] Cushing, C. E., *Hydrobiology*, Vol. 29, 1967, pp. 125-139.
[*12*] Hohn, M. and Hellerman, J., *Transactions of the American Microscopical Society*, Vol. 92, 1963, pp. 250-329.
[*13*] Weber, C. I., "Methods of Collection and Analysis of Plankton and Periphyton samples in the Water Pollution Surveillance System," *Application and Development Report*, No. 19, Water Pollution Surveillance System, U.S. Public Health Service, Department of Health, Education and Welfare, Cincinnati, Ohio, 1966.
[*14*] Weber, C. I. and Raschke, R. L., "Use of a floating periphyton Sampler for Water Pollution Surveillance," *Application and Development Report*, No. 20, Water Pollution

Serveillance System, U.S. Public Health Service, Department of Health, Education and Welfare, Cincinnati, Ohio, 1966.

[*15*] U.S. Public Health Service: "The National Water Quality Network," U.S. Department of Health, Education, and Welfare, Division of Water Supply and Pollution Control, Basic Data Branch, Water Quality Section, Cincinnati, Ohio, 1962.

[*16*] U.S. Public Health Service: "Annual National Water Quality Network. Compilation of Data," U.S. Department of Health, Education, and Welfare, Division of Pollution Surveillance, Cincinnati, Ohio, 1958-1963.

[*17*] Weber, C. I. and Moore, D. R. *Limnology and Oceanography,* Vol. 12, 1967, pp. 311-318.

[*18*] Williams, L. G. and Scott, C., *Limnology and Oceanography* Vol. 7, 1962, pp. 365-379.

[*19*] Williams, L. G., Kopp, J. F., and Tarzwell, C. M., *Journal of the American Water Works Association,* Vol. 58, 1966, pp. 333-346.

[*20*] Patrick, R., Hohn, M. H., and Wallace, J. H., *Notulae Naturae*, Vol. 259, 1954, pp. 1-12.

[*21*] Weber, C. I., *Transactions of the American Microscopical Society,* Vol. 87, No. 1, 1968, pp. 80-81.

[*22*] Albin, R. W., "Colonization by Periphyton Algae on an Artificial Substrate Compared to a Natural Substrate. Master's thesis. University of South Dakota, Vermillion, S. Dak., 1965.

[*23*] Starrett, W. C. and Patrick, R., *Proceedings of the Academy of Natural Sciences of Philadelphia,* Vol. 104, 1952, pp. 219-243.

[*24*] Kevern, N. R., Wilhm, J. L., and Van Dyne, G. M., *Limnology and Oceanography,* Vol. 11, No. 4, 1966, pp. 499-502.

H. D. Putnam,[1] *J. H. Sullivan, Jr.,*[1] *B. C. Pruitt,*[1] *J. C. Nichols,*[1]
M. A. Keirn,[1] *and D. R. Swift*[1]

Impact of Trinitrotoluene Wastewaters on Aquatic Biota in Lake Chickamauga, Tennessee

---

REFERENCE: Putnam, H. D., Sullivan, J. H., Jr., Pruitt, B. C., Nichols, J. C., Keirn, M. A., and Swift, D. R., **"Impact of Trinitrotoluene Wastewaters on Aquatic Biota in Lake Chickamauga, Tennessee,"** *Ecological Assessments of Effluent Impacts on Communities of Indigenous Aquatic Organisms, ASTM STP 730,* J. M. Bates and C. I. Weber, Eds., American Society for Testing and Materials, 1981, pp. 220-242.

ABSTRACT: Field surveys were conducted in Lake Chickamauga, Tenn., to determine the effects from wastewater discharge from the Volunteer Army Ammunition Plan. Trinitrotoluene (TNT) is the principal munitions component manufactured at this facility, and process wastes contain a complex mixture of compounds associated with the production of this explosive. Effects from the waste stream via selected components were established in periphyton and the macroinvertebrate communities, utilizing the Pinkham-Pearson Biotic Similarity Analysis. Less success was achieved in assessing impact by employing the Shannon-Weaver diversity theory. The utility of standard artificial substrates in periphyton studies was realized by comparing effects between community structures on natural substrates and glass slides. The results of the investigation show that effects on fixed biological communities from wastewater components of TNT manufacture can be detected at concentrations in the microgram per litre range.

KEY WORDS: trinitrotoluene, Lake Chickamauga (Tenn.), periphyton, macroinvertebrates, munitions chemistry, ecology, effluents, aquatic organisms

Biological Assessment of Ecological Stress

Pollution biology has traditionally dealt with the populations of organisms responding to man-made stress. In Britain, where much of the early work in water pollution abatement was done, corrective measures for sanitary wastes were instituted that would maintain dissolved oxygen levels in the Thames River to protect fish. These and other events in the latter part of the nine-

[1]Aquatic biologist, environmental engineer, aquatic biologist, environmental engineer, aquatic biologist, and phycologist, respectively, Water and Air Research, Inc., Gainesville, Fla. 32602.

teenth century led to our current understanding of ecology as a practical science in the pollution field. Biologists working in this area have strived toward establishing cause and effect relationships but have been misled many times by making naive judgments about extremely complex environmental interactions.

The field of aquatic biology is sufficiently developed to recognize that these systems are composed of distinct units or compartments. Organisms of various complexity are found therein and together form the biotic portion of the ecosystem. Populations of organisms in an aquatic system are usually characterized as inhabitants of the water column or colonizers of a substrate. Thus, we can refer to components as plankton, periphyton, macrophyton, or benthic invertebrates. All of these are distinct and, for example, are addressed in separate sections of *Standard Methods for the Examination of Water and Wastewater* (1976) [*1*].[2]

Although each of the aforementioned compartments is useful to the biologist, certain of these are of particular importance in assessing effects from wastewaters. Organisms that colonize and remain fixed in or on surfaces respond to continually changing environmental conditions. Changes or variability may be a reflection of natural events due to season or to a stress from the exposure to wastewater. The importance to the biologist in examining fixed populations is that these organisms respond to the environment and are integrators of environmental events. As such, they can be used as monitors of dynamic conditions.

Field Study Characteristics

Using a protocol developed by Cairns and Dickson [*2*], the U.S. Army Medical Research and Development Command conducted extensive field studies at selected ammunition plants as part of an overall effort to establish no-effect levels of munitions compounds on receiving waters. As a part of this overall program, surveys were completed at the Volunteer Army Ammunition Plant (VAAP), which is located near Chattanooga, Tenn. This trinitrotoluene (TNT) manufacturing facility discharges liquid wastes into Waconda Bay. The embayment is a part of the Lake Chickamauga Reservoir on the Tennessee River.

In the surveys, sampling was carried out in Waconda Bay, as shown in Fig. 1, and in an adjacent unnamed reference bay unaffected by munitions plant effluent. Both chemical and biological analyses were conducted. Water and sediment samples were analyzed for major ions, nutrients, and munitions residues. The biological components selected were periphyton and benthic macroinvertebrates.

The relative sizes of Waconda Bay and the reference bay are 95 and 21

[2]The italic numbers in brackets refer to the list of references appended to this paper.

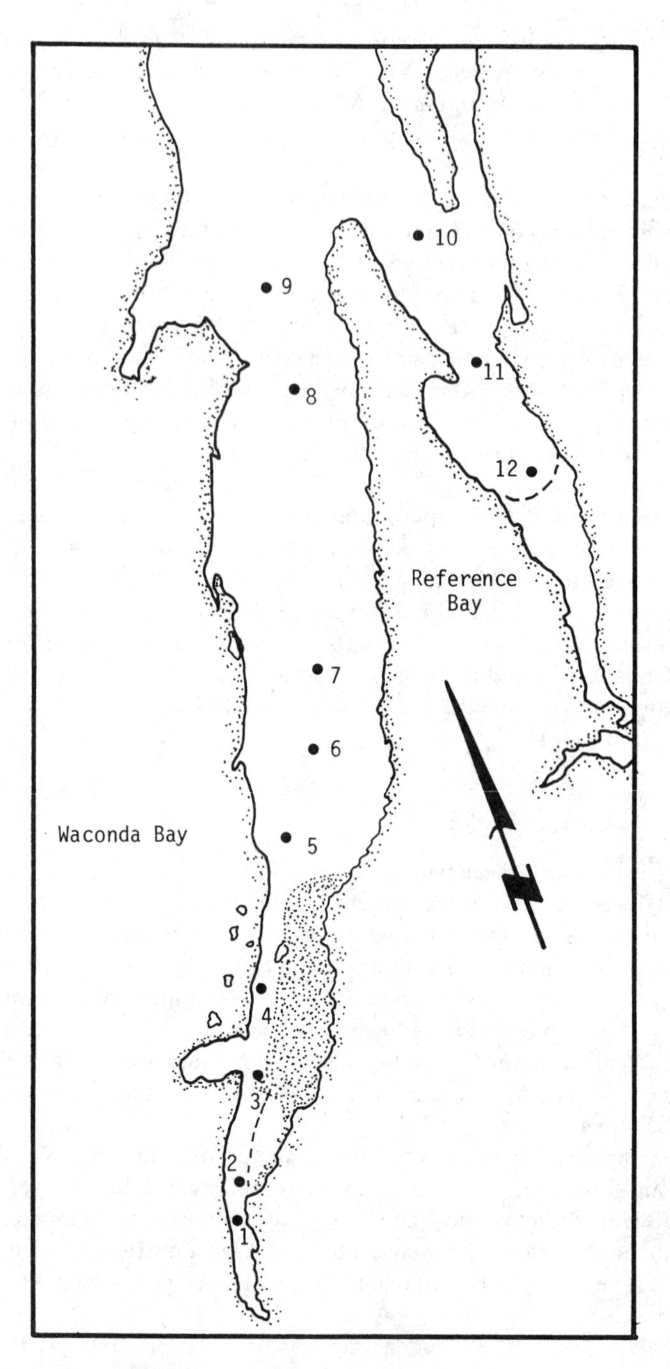

FIG. 1—*Sampling sites in Lake Chickamauga. Scale of distance: 1 cm = 160 m.*

hectares, respectively (Fig. 2). The mean depth is about 3 m for both bays. Storm-water runoff from residential areas and VAAP enters Waconda Bay. The watershed of the reference bay is largely residential.

Flushing of the bays is caused by (1) runoff, (2) stage fluctuations, and, in the case of Waconda Bay, (3) effluent from VAAP. Present estimates for the flushing of Waconda Bay are about 20 percent per month.

Periodic discharges of TNT and nitrobodies historically have exceeded the 0.3 mg/litre National Pollutant Discharge Elimination System (NPDES) permit limits. Analyses conducted for monitoring purposes in 1974 showed TNT levels ranging above 1 mg/litre in the upper bay.

Methods Employed for Water Quality

The methods employed for collecting, preserving, and analyzing the routine water quality parameters followed accepted methods in *Standard Methods for the Examination of Water and Wastewater* [*3*] or U.S. Environmental Protection Agency (EPA) [*4*] procedures. *Chemistry Laboratory Manual, Bottom Sediments* [*5*] was the source of the routine methods utilized for collection, preservation, and analysis of the sediment samples. Where existing methods, particularly for trace metals and munitions, were insufficient to provide the desired levels of detection, alternative analytical procedures were employed after their accuracy and precision had been verified.

In addition to *in situ* measurements of dissolved oxygen, temperature, pH, and specific conductance, the water quality parameters monitored are listed in Table 1. The sediments were characterized by analyzing the parameters listed in Table 2.

Biological characterization was made by sampling natural and artificial substrates. The periphyton and macroinvertebrate communities were selected as being the most representative to reflect the impact of TNT wastewaters from the VAAP.

Standard glass microscope slides were placed in an eight-unit cartridge and suspended in place 2.5 cm (1 in.) below the water surface. The artificial sampling unit was manufactured by Design Alliance, Inc., and was so designed that multiple units could be placed at a single station. Individual slides were examined for the structure of the community and the biomass of the periphyton assemblage.

For estimates of plant pigments, the following procedure was employed. While in the field, periphyton slides were retrieved and placed in 50 ml of a 90 percent acetone (by volume), 10 percent saturated magnesium carbonate solution and immediately stored in the dark on dry ice. Prior to analysis, chlorophyll was extracted for 24 h in the dark at 4°C. To facilitate extraction, the slides were scraped and the acetone suspension ground for 30 s at 500 rpm in a Potter-type tissue homogenizer.

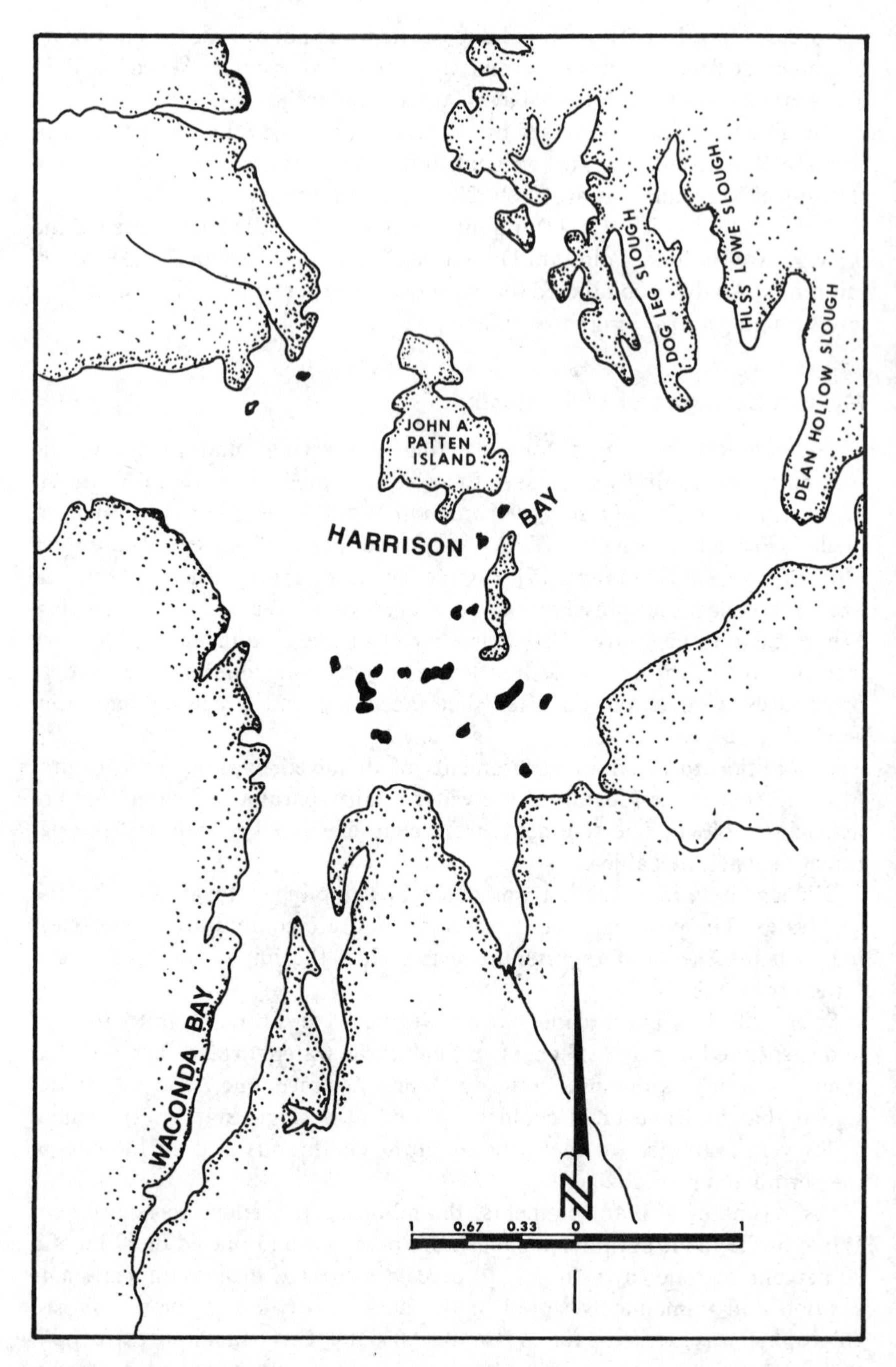

FIG. 2—*Overview of study area showing relationship of Waconda Bay to western Lake Chickamauga. Scale of distance: 1 cm = 0.44 km.*

TABLE 1—*Water quality parameters monitored.*

Major ions	Oxygen demand
Total alkalinity	Chemical oxygen demand
Chloride	Total organic carbon
Total hardness	Trace metals
Sulfate	Cadmium
Total dissolved solids	Copper
Suspended materials	Chromium (VI)
Suspended solids	Iron
Total solids	Lead
Biological organisms	Mercury
Periphyton	Nickel
Macroinvertebrates	Zinc
Plant nutrients	Munitions compounds
Ammonia nitrogen	2,4-Dinitrotoluene (2,4-DNT)
Total Kjeldahl nitrogen	2,6-Dinitrotoluene (2,6-DNT)
Nitrite nitrogen	α-Trinitrotoluene (2,4,6-TNT)
Nitrate nitrogen	1,3-Dinitrobenzene (1,3-DNB)
Total phosphorus	1,3,5-Trinitrobenzene (1,3,5-TNB)

TABLE 2—*Parameters characterizing sediments.*

Nutrients	Lead
Chemical oxygen demand	Manganese
Total Kjeldahl nitrogen	Mercury
Nitrate nitrogen	Nickel
Nitrite nitrogen	Zinc
Total phosphorus	Munitions compounds
Total solids	2,4-Dinitrotoluene (2,4-DNT)
Total volatile solids	2,6-Dinitrotoluene (2,6-DNT)
Trace metals	α-Trinitrotoluene (2,4,6-TNT)
Cadmium	1,3-Dinitrobenzene (1,3-DNB)
Copper	1,3,5-Trinitrobenzene (1,3,5-TNB)
Iron	

Following extraction, chlorophyll *a*, corrected for phaeophytin, was determined fluorometrically after the methods of Yentsch and Menzel [*6*], Holm-Hansen et al [*7*], Lorenzen [*8*], and Moss [*9*], using a Turner Design Model 10 fluorometer. The chlorophyll *a* reference solution was a purified spinach chlorophyll standard (Product No. C5753, Sigma Chemicals, St. Louis, Mo.) calibrated by spectrophotometric chlorophyll analysis.

For biomass determinations, accumulated material was scraped from the slide into a graduated cylinder and then resuspended in a total volume of 50 ml distilled water. A portion of the suspension was filtered on a tared, fired-glass filter (Gelman, GFA), and the ash-free dry weight was determined [*1*] and converted to grams of organic matter per square metre.

A minimum of five replicate slides per station was examined for diatom community structure and cell density estimates. Periphyton growth was scraped from glass slides into tall labeled beakers using razor blades and a rubber policeman. The samples were oxidized with 20 ml of 50 percent hydrogen peroxide and approximately 50 mg of potassium dichromate. The solution was cooled, allowed to settle for 24 h, decanted, and brought to an appropriate volume for counting. Following preparation of a permanent mount in Hyrax medium, the cell densities were estimated by performing field counts at six-slide coordinates randomly selected on the coverslip. A total of ten microscope fields were examined at each of the six selected coordinates for a total of 60 fields. Each field represented an area of 0.0182 mm^2. The diatom populations were determined by the following formula

$$\text{Cells/mm}^2 = \frac{\text{diatom counts} \times \dfrac{\text{total area of coverslip } (324 \text{ mm}^2)}{\text{total area examined } (1.092 \text{ mm}^2)} \times \dfrac{\text{original volume of periphyton suspension } (50 \text{ ml})}{\text{volume of sample dried on coverslip } (0.4 \text{ ml})} \times \text{dilution factor}}{\text{original surface area of slide } (3871 \text{ mm}^2)}$$

Hester–Dendy plates were employed for macroinvertebrate colonization. Incubation was completed using square eight-plate units. The Hester–Dendy artificial substrates were suspended approximately 0.5 to 1.0 m below the water surface. Three Hester–Dendy units were collected at each location to minimize natural variability. At the time of collection, the individual plates were stored in cotton drawstring bags and immersed in 10 percent formalin. The bags containing the plates were transported in bulk using Roper pails.

Five replicates were obtained of lake sediments for infauna using a petite Ponar dredge. In the field, dredge samples of the natural substrate were washed in a bucket sieve (U.S. Standard No. 30 mesh) and bottled. Rose bengal dye was added to facilitate laboratory sorting. The samples were preserved in 10 percent formalin. The natural substrate samples were rewashed in the laboratory and sorted in a white enamel pan partially filled with water. After being sorted, the organisms were placed in vials containing 95 percent ethanol. Chironomid larvae were mounted on standard-sized glass slides in polyvinyllactophenol. Identifications were made to the lowest practical taxonomic level and verified according to W. C. Beck [*10*] of Florida A & M University. The key taxonomic references were Edmondson [*11*] and Pennak [*12*].

Water Exchange Characteristics

Prior to locating the stations along Waconda Bay, dye studies were completed to estimate flushing. Water movement was defined using rhodamine B, the level of munitions residues, and selected *in situ* measurements such as conductivity. Results showed that the dispersion of effluent into the bay was most influenced by reservoir stage changes. A rising stage had the effect of holding the effluent in the upper part of the bay, while a draw-down period moved the effluent rapidly downbay.

Chemical Characteristics of the Waconda Bay Environment

The impact of TNT wastewaters can be manifested by biostimulation of primary producers, as well as a toxic response in aquatic plants and animals. Eutrophication of bay water occurs from both reduced and oxidized forms of nitrogen, which is available for the growth of green plants. Substances potentially toxic, in the form of nitrobodies, TNT, and heavy metals, are discharged into the aquatic environment in varying concentrations. These were of primary concern in examination of lake water and sediments. During the study, significant gradients and enrichment of the upper end of Waconda Bay by oxidized nitrogen forms occurred. Figure 3 illustrates this effect. Reduced nitrogen showed a pattern similar to that of nitrate. The trend can be seen in Fig. 4.

In March, total nitrogen in the channel averaged 14.2 mg/litre, with approximately 79 percent occurring as nitrate. Concentrations downbay showed a gradual reduction. The nitrogen levels in the medium zone averaged 5.6 mg/litre, with nitrate occurring again in the same proportion. The outer bay stations reflected a mean nitrogen content of 4 mg/litre, with nitrate concen-

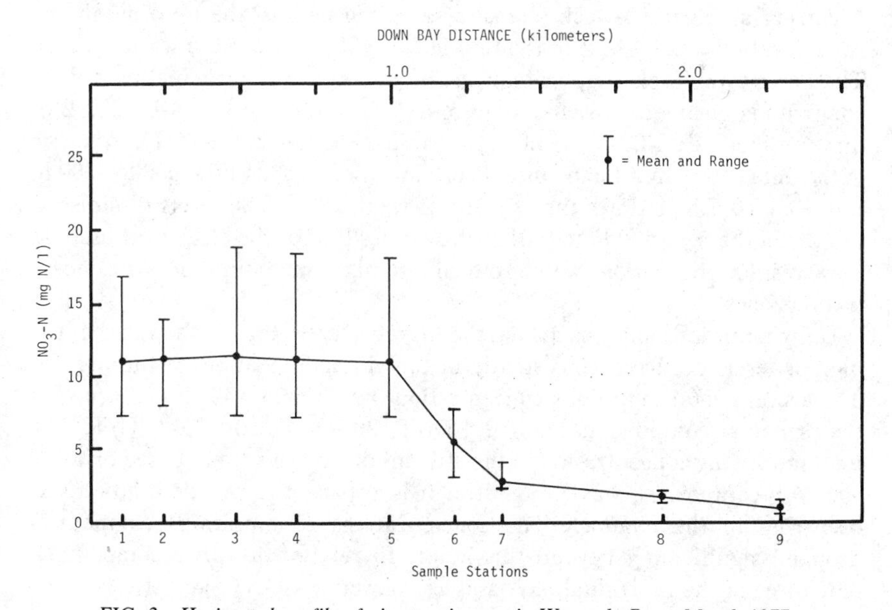

FIG. 3—*Horizontal profile of nitrate-nitrogen in Waconda Bay—March 1977.*

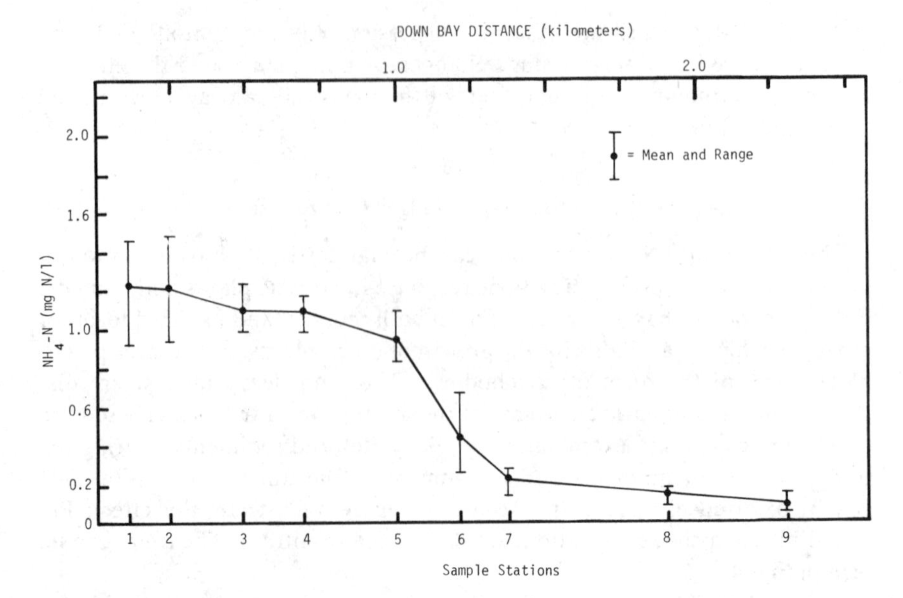

FIG. 4—*Horizontal profile of ammonia nitrogen in Waconda Bay—March 1977.*

tration as the most significant form. Other constituents, such as chloride, total hardness, and sulfate, followed a similar pattern of decreasing concentrations downbay.

During an earlier 1-week intensive sampling period, the total munitions concentration was highest in the first 900 to 1000 m from the waste outfall. The mean total levels were as high as 397 μg/litre. The principal munitions compounds near the discharge were 2,4-dinitrotoluene (2,4-DNT), 2,6-dinitrotoluene (2,6-DNT), and 2,4,6-trinitrotoluene (2,4,6-TNT). Analysis of the data shows that the pollutant concentration in the initial 900 to 1000 m was about 0.5 to 1 times the effluent concentration. The levels diminished rapidly in the next 500 m to 0.01 to 0.05 times that of the effluent. Based on these results, the stations were grouped into high, medium, and low concentration zones.

Daily munitions analysis during the March survey (Fig. 5) showed that the highest levels occurred consistently in the channel (Stations 1 through 5). The average total munitions concentration in this area was 78 μg/litre with the principal components being 2,4-DNT, 2,6-DNT, and 2,4,6-TNT. The total munitions concentration in the effluent during this time, based on average 24-h composites, was 89 μg/litre. It is apparent that little dilution was occurring in the channel. The concentrations of munitions compounds dropped significantly beyond this point. In relation to other compounds, only three of the individual bay samples showed 2,4-DNT, and five samples

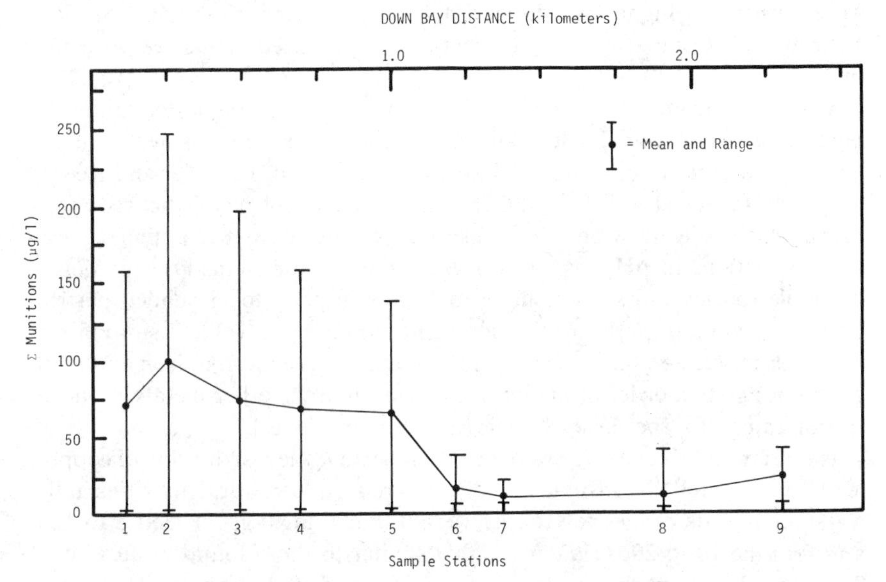

FIG. 5—*Horizontal profile of total munitions in Waconda Bay—March 1977.*

contained 1,3-dinitrobenzene (1,3-DNB). Only 1,3,5-trinitrobenzene (1,3,5-TNB) showed increased levels downbay. It appears that this compound is a breakdown product of 2,4,6-TNT. No 1,3,5-TNB was detected in the plant effluent.

Trinitrotoluene is subject to microbial action, and at least some of the products observed in the bay are the result of this biological mechanism. The breakdown occurs stepwise, giving rise to dinitrated amines and hydroxylamines such as 4-amino-2,6-dinitrotoluene. Resistance of aromatic rings to attack is measured by the increased number of nitroradicals [*13*], and the toxicity of intermediate breakdown products may be as great as or greater than that of the parent compound. Weitzel et al [*14*] were able to characterize a major TNT breakdown in sediment at the Iowa Army Ammunition Plant as a monohydroxylaminodinitrotoluene that exhibited comparable toxic properties with TNT and, further, found concentrations of this daughter in sediments equal to or greater than corresponding TNT concentrations.

The lake sediments were analyzed for munitions compounds. The results show that 1,3,5-TNB and 2,4,6-TNT predominate over 1,3-DNB, 2,4-DNT, and 2,6-DNT. The first two compounds were found in sediments at all stations and in all samples except one replicate. For both compounds the concentrations at Stations 6 through 9 were higher than those at Stations 1 through 5—263 and 179 μg/kg for 1,3,5-TNB and 148 and 119 μg/kg for 2,4,6-TNT. For the remaining compounds (2,4-DNT, 1,3-DNB, and 2,6-DNT), the levels were consistently higher in sediments at stations closer to the outfall (Stations 1 through 5). Comparative values for the water and sediments are presented in Table 3.

The results from metals analysis showed none to be in the water column in potentially toxic levels. Little is known about the fate of metals in sediments, although several in-depth studies have been conducted. Iskandar and Keeney [*15*] and Holmes et al [*16*] found that redistribution of metals between sediments and overlying waters in lakes and estuaries occurs seasonally, paralleling variations in pH, dissolved oxygen, temperature, etc. Precipitation of insoluble species and scavenging of metals by sorption to suspended particulates are two main pathways for sediment enrichment. Table 4 summarizes the levels in selected bay sediments for our study. As expected, iron and manganese ranged an order of magnitude higher than the other metals. Concentration ranges for the various species are given in Table 4.

Iskandar and Keeney [*15*] were able to show that the distribution of copper, lead, and zinc in Wisconsin lake sediments related to cultural activities in the watersheds. The ranges for copper were 0 to 400 mg/kg, for lead 5 to 160, and for zinc 10 to 200, ranges roughly similar to those found in our study. The upper bound of the ranges for iron, manganese, chromium(VI), copper,

TABLE 3—*Mean values for munitions in water and sediment—March 1977.*

Station	Mean Values, μg/litre					
	1,3-DNB	1,3,5-TNB	2,4-DNT	2,6-DNT	2,4,6-TNT	Total
			Water—Daily Analyses March 1 Through 5			
1	4.1	2.9	<0.10	17.9	48.0	73.0
2	<0.25	<0.75	22.1	38.7	40.3	102.0
3	<0.25	<0.75	<0.10	32.0	43.7	76.7
4	0.4	<0.75	17.8	37.2	13.3	69.5
5	0.5	17.1	<0.10	25.7	26.2	69.7
6	<0.25	1.2	<0.10	12.1	1.8	≤15.6
7	<0.25	1.6	<0.10	7.6	0.2	≤9.8
8	<0.25	10.6	<0.10	1.3	0.1	≤12.4
9	<0.25	20.6	1.3	2.3	0.3	24.8
24 h composite effect	<0.25	<0.75	28.3	15.5	44.5	89.4
			Sediment[a]			
1	≤13.0	73.0	≤3.4	10.0	91.0	14%
2	≤12.0	≤179.0	≤7.9	17.0	119.0	12%
3	≤14.0	76.0	<2.5	17.0	119.0	17%
4	11.0	135.0	<2.5	≤4.3	86.0	32%
5	14.0	80.0	<2.5	11.0	70.0	27%
6	<6.3	221.0	<2.5	<1.3	138.0	23%
7	<6.3	250.0	<2.5	<1.3	105.0	22%
8	<6.3	304.0	<2.5	<1.3	142.0	9%
9	<6.3	263.0	<2.5	<1.3	148.0	35%

[a] Values are the average of three sediment replicates taken 3 March 1977 and reported on a wet weight basis.

TABLE 4—*Concentration ranges for various metals.*

Element	Concentration Range, mg/kg dry weight
Iron	1400 to 140 000
Manganese	30 to 10 000
Cadmium	<1 to 3.5
Chromium (VI)	23 to 740
Copper	<1 to 130
Mercury	<0.1 to 1.90
Nickel	2 to 110
Lead	5 to 860
Zinc	4 to 690

lead, and zinc was found near the discharge point. Our data show that occasional release of manganese, iron, chromium, and copper at levels above permit specifications may occur. Overall, the data agree with levels found in an earlier survey by Wapora, Inc. [*17*].

Impact of TNT Residues on Periphyton

Glass microscope slides were incubated for 30 days for the purpose of periphyton colonization. Following retrieval of these artificial substrates, biomass estimates were made of the attached growth, and community structure was determined via diatom composition.

In excess of 100 species representing 23 genera were recorded during the study. *Achnanthes minutissima* was the dominant organism. Other common species were *Fragilaria capucina, Cymbella affinis, Melosira ambigua, Diatoma tenue, Fragilaria vaucheriae, Melosira varians, Synedra rumpens,* and *Gomphonema parvulum.*

During the winter, the zone of influence extends further downbay, and both the population and the biomass correlate inversely with the concentration of VAAP waste. Table 5 and Fig. 6 show these trends for chlorophyll and organic biomass. Figure 7 represents a clustering of stations based on the Pinkham-Pearson biotic similarity concept for periphyton and selected effluent constituents. The data show a similar clustering of stations and suggest a close relationship between water chemistry and the periphyton population.

The diatom population, as presented in Table 6, ranged from 27/mm^2 at Station 2 to 37 267 at Station 11. The numbers of species per station ranged from 22 to 52 in Waconda Bay and from 76 to 83 in the reference bay. To simplify data presentation and eliminate rarely occurring species, a culling routine was completed, and all species whose numbers did not constitute at least 1 percent of the total population at any one station were eliminated. This procedure resulted in a reduction in the number of species from 124 to 30. However, the maximum drop in population density at any one station was

TABLE 5—*Chlorophyll* a, *biomass, and autotrophic index of periphyton colonizing artificial substrates—March 1977.*

Station	Chlorophyll *a*, mg/m²		Organic Biomass, gm/m²		Autotrophic Index
	Mean	Range	Mean	Range	
1	0.02	0.01 to 0.06	1.1	0.61 to 2.2	55 000
2	0.02	0.01 to 0.04	0.53	0.27 to 1.0	26 500
3	0.08	0.02 to 0.1	0.34	0.02 to 1.3	4 250
4	0.2	0.1 to 0.4	0.10	0.02 to 0.26	500
5	0.5	0.2 to 0.8	0.50	0.22 to 1.0	1 000
6	14.0	7 to 19	2.3	1.2 to 2.8	164
7	41.0	29 to 52	3.7	2.6 to 4.4	90
8	23.0	13 to 34	4.8	2.9 to 6.0	209
9	17.0	5 to 35	3.4	0.57 to 6.1	200
10	...	...	...	...	...
11	28.0	16 to 44	6.2	4.7 to 8.2	221
12	14.0	11 to 18	4.7	3.6 to 6.8	336

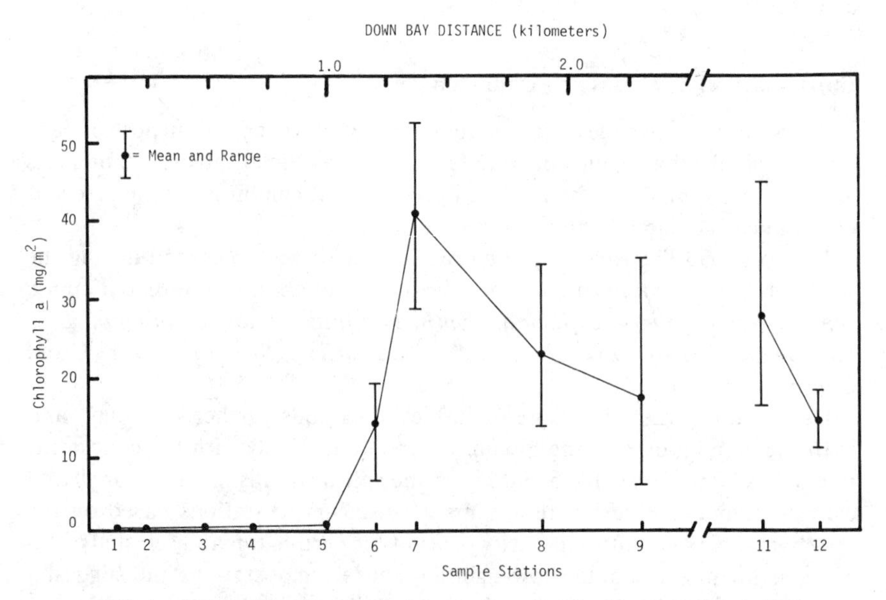

FIG. 6—*Chlorophyll* a *levels of periphyton on artificial substrates—March 1977.*

7 percent. Table 6 shows the differences in the density and number of species before and after culling. The organism density is significantly different at Stations 1 through 5 in comparison with Stations 6 through 12.

The dominant winter species in Waconda Bay are *Achnanthes minutissima, Cymbella prostrata, Diatoma tenue* var. *elongatum, Fragilaria vaucheriae, Melosira varians,* and *Synedra rumpens.* These species reached population

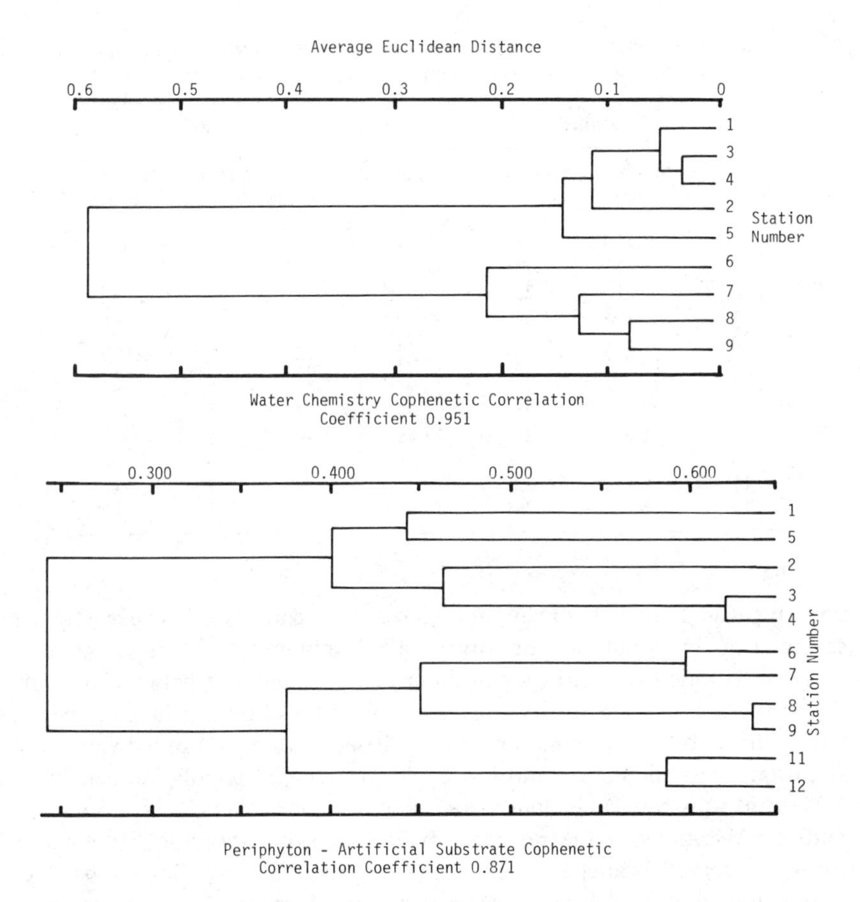

FIG. 7—*Comparative phenograms of chemical and periphyton data in Waconda Bay—March 1977.*

levels of at least 1000 individuals/mm^2 at one or more stations. The diatom populations were reduced in the impact area from Stations 1 to 5, although the actual number of species was approximately the same as that downbay.

Considering the dominant species, the ratio of population density for the dominant organisms between Stations 6 and 5 where the bay widens is as follows: *Achnanthes minutissima,* 48; *Diatoma tenue,* 25; *Fragilaria vaucheriae,* 14; *Melosira varians,* 9; *Synedra rumpens,* 51. Shannon-Weaver diversity values ranged from 2.2 in the immediate area of the outfall to 0.6 at Station 6. The latter station represents a zone beyond the high-stress area.

Theoretically, the higher the diversity, the "healthier" the community. However, examination of the results shows that diversity in the area of maximum stress, Stations 1 through 5, is higher than at the downbay Sta-

TABLE 6—*Comparison of population size and numbers of species in culled and unculled data at 1 percent periphyton—artificial substrate.*

Station	Culled		Unculled		Percent Difference	
	Number/ mm^2	Number of species	Number/ mm^2	Number of species	Number/ mm^2	Number of species
1	135	24	141	39	4	38
2	27	22	27	22	0	0
3	76	25	78	46	3	46
4	165	25	169	46	2	46
5	727	25	745	52	2	52
6	14 721	18	14 916	37	1	51
7	18 714	21	18 941	40	1	48
8	27 021	25	27 785	52	3	52
9	21 970	22	22 458	39	2	44
10	...	...	...	...	...	...
11	35 158	22	37 267	76	6	71
12	18 354	24	19 696	83	7	71

tions 6 through 9. This presumably anomalous condition is caused by the severe drop in population density of all dominant species at Stations 1 through 5, which results in a condition of better balance between the surviving species. Since diversity considers only relative proportions of species and ignores absolute population density, it seems to be of limited value in a situation where the main input is seen as reduction in population density.

Natural substrate collections were made at nine stations in Waconda Bay and three stations in the reference bay. Generally, scrapings were made from submerged branches, rocks, and styrofoam floats. Three substrates were sampled at each location except Stations 4 and 8. Counts were made to approximate 500 diatom valves. Comparisons among stations can be made only relative to numbers of individual species colonizing these substrates. The overall total number of species ranged from 11 to 77, with no marked differences related to station locations. In fact, the two extreme values both occurred at Station 3. The data are culled to eliminate species that do not comprise at least 1 percent of the total of at least one station. The five most commonly occurring species were (1) *Achnanthes minutissima*, (2) *Fragilaria capucina*, (3) *Fragilaria vaucheriae*, (4) *Stephanodiscus invisitatus*, and (5) *Synedra rumpens*. In considering the distribution of individual species, it is interesting to note that *Achnanthes minutissima* occurred as commonly in the impact zone as in other areas of Waconda Bay and the reference bay. This suggests that *Achnanthes* is not as susceptible to munitions waste as might be expected.

Pinkham-Pearson comparisons among the various samples are shown in Fig. 8. The clustering of stations using natural substrates results in a grouping based primarily on substrate type rather than location in relation to the

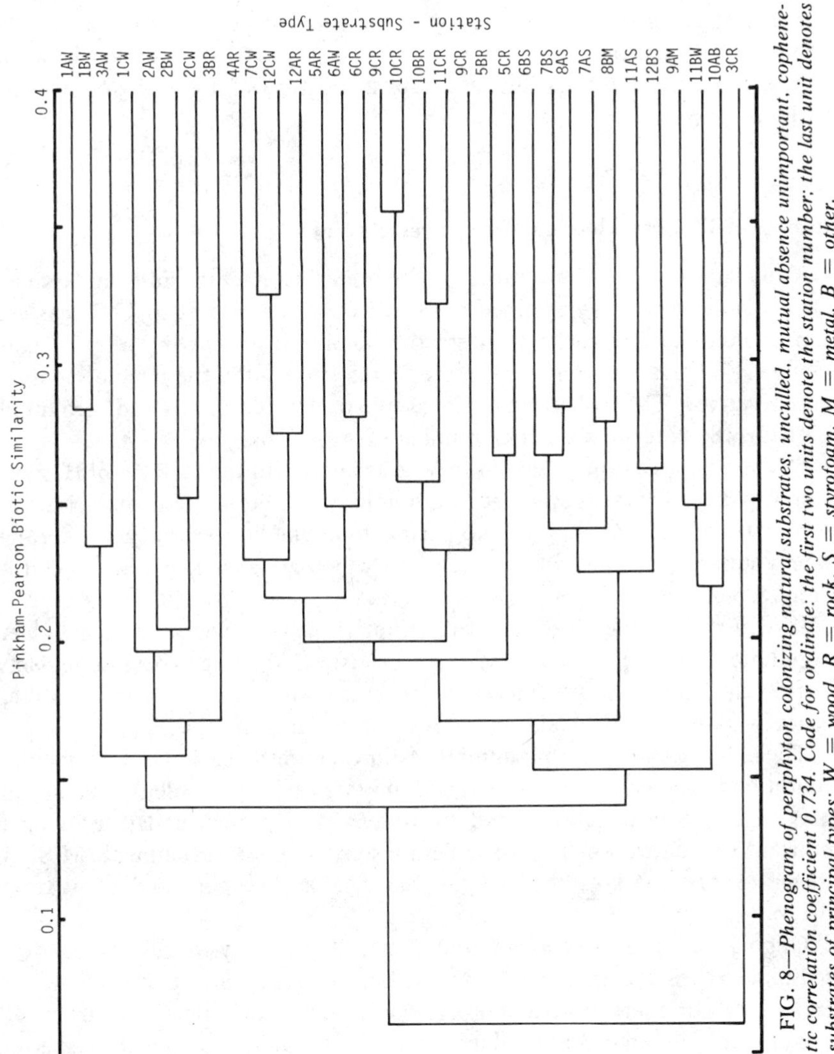

FIG. 8—*Phenogram of periphyton colonizing natural substrates, unculled, mutual absence unimportant, cophenetic correlation coefficient 0.734. Code for ordinate: the first two units denote the station number; the last unit denotes substrates of principal types: W = wood, R = rock, S = styrofoam, M = metal, B = other.*

waste discharge. For example, the first station cluster, with one exception, has a common denominator of wood scrapings as a substrate type. The second, beginning with 4AR, includes predominantly rock scrapings. Styrofoam substrates principally compose the next group and include samples from Stations 6 through 8, 11, and 12. Treatment of the data by the Pinkham-Pearson analysis is especially useful in this case to point out the substrate dependence of periphyton diatoms. The phenographic display in this instance further demonstrates the utility of uniform substrates for periphyton investigations.

Impact of TNT Residues on Macroinvertebrates

The macroinvertebrate community from lake sediments and from Hester-Dendy substrates was examined to determine its responses to TNT wastes. Chironomids and oligochaetes were the dominant groups in Lake Chickamauga. Midge larvae comprised more than 50 percent of the population during the survey. They colonized artificial substrates abundantly and accounted for nearly 80 percent of the total number of organisms.

Of the 54 taxa enumerated, 44 were associated with the surfaces of Hester-Dendy plates. This suggests that chironomids preferentially colonize artificial substrates of this type to graze on periphyton. The lake sediments generally were composed of clay, silt, and detritus, which are also conducive to chironomid colonization.

In contrast to the chironomid distribution, oligochaetes reached higher populations in sediments. This is to be expected, since these organisms burrow into the substrate and recycle organic materials. This kind of habitat is not available on artificial substrates as used in this study.

These two groups exhibit similarities in their responses to environmental conditions. The chironomids, which dominate, are adaptable to changes in pH, oxygen concentrations, and turbidity. They prefer moderate to high nutrient concentrations and will tolerate some organic enrichment. Most of them are scavengers—filtering algae, bacteria, and suspended detritus from the water.

Oligochaetes are commonly found in standing or slow-moving waters over sediments enriched in organic matter. The presence of aquatic plants increases the numbers of taxa and organisms. Plants provide shelter from the current and predators and produce detritus as food. Most oligochaete species are able to tolerate or even thrive in low concentrations of dissolved oxygen. Many can survive anaerobic conditions for extended periods of time.

The results show that three midge forms were the most abundant in sediments. These were *Procladius*, *Coelotanypus*, and *Chironomus*. *Procladius* was the dominant form at Stations 1 through 5 and may represent a form more tolerant to the TNT waste. Small numbers of *Hexagenia* were also found in zones where TNT waste was elevated, indicating at least some tolerance to these substances.

Chironomids dominated artificial substrates at Stations 7 through 9 in Waconda Bay and Stations 10 and 11 in the reference bay. *Cricotopus* accounted for about two-thirds of the Chironomidae. Several species of *Cricotopus* were involved; however, at present there are no effective means for separation of taxa. Four species of *Dicrotendipes* and two species of *Glyptotendipes* comprised most of the remaining chironomids. These six feed upon planktonic materials. *Cricotopus* was present throughout Waconda Bay and was abundant in the reference bay. *Dicrotendipes* and *Glyptotendipes* were relatively common in the reference bay and were frequently found at Stations 6 through 9. Two other genera, *Ablabesmyia* and *Tanytarsus*, were frequently found in the reference bay but rarely found in Waconda Bay.

Of the nonchironomid fauna, only two of the nine found on artificial substrates were present in significant numbers. The larvae of *Caenis*, a small herbivorous mayfly, reached its greatest population at Station 6, where it formed 61 percent of the population. Its densities declined both upbay and downbay. It was not collected at Stations 1, 2, 9, and 10. Oligochaetes occurred sporadically in Waconda Bay, but were the dominant taxa on reference bay artificial substrates. The larvae of *Argia*, a predaceous damselfly, were found only at Stations 1 and 2. This genus has previously been reported only from unpolluted situations [*10*] in Florida, but Beck (personal communication)[3] believes this may not hold true for some non-Florida species.

The highest densities and numbers of species occurred in the reference bay stations, which suggests some inhibition in all the Waconda Bay samples. Considering the total number of species per station, the stations seem to group in the following manner. Stations 1 through 5 averaged 3.3 species; Stations 6 through 9 averaged 9.5 species; the reference bay stations averaged 13.5 species.

The macroinvertebrate data from lake sediments, as analyzed by cluster analysis, is presented in Fig. 9. The phenogram is representative of two options in the analysis procedure. Organisms that were present at 4 percent or less of the total population at any station were culled and no longer considered in the analysis. Using the Pinkham-Pearson biotic similarity procedure, the option of mutual absence important was employed. The second procedure shows the clustering pattern based on all organisms that were collected and where mutual absence was considered unimportant. These two routines were run to examine the impact on station clustering of organisms occurring on an infrequent or rare basis. As the figure shows, the clustering is relatively similar. In both cases, Stations 1 and 2 are highly unrelated to the remaining stations. Stations 3, 4, and 5 are somewhat similar. No con-

[3]Beck, W. F., Jr., Florida A. & M. University, Tallahassee, Fla., 1976.

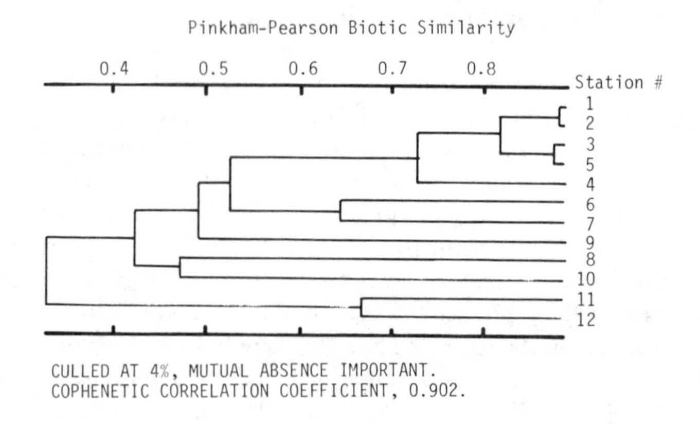

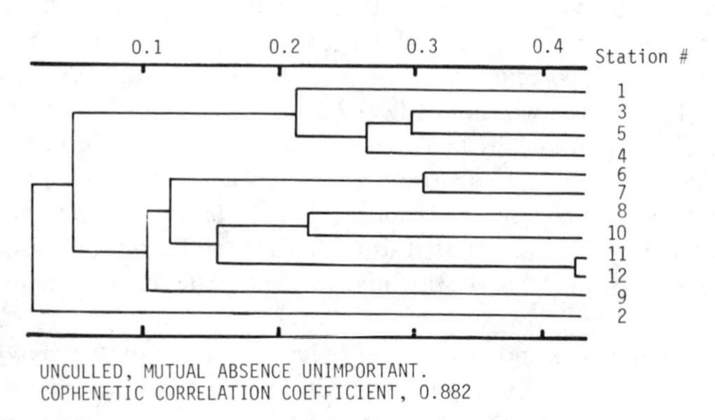

FIG. 9—*Phenogram showing station similarity of macroinvertebrates in sediments of study area.*

sistent pattern exists for the remaining stations, indicating that Stations 6 through 9 in Waconda Bay are not distinguishable from Stations 10 through 12 in the reference bay.

Cluster analysis using the Pinkham-Pearson approach was performed on data from the Hester-Dendy units (Fig. 10). The unculled data were analyzed on a mutual absence unimportant basis, whereas when the data were culled, mutual absence was considered important. In the culled data, Stations 1 through 5 group separately, as do Stations 11 and 12. In the unculled data, Stations 1, 3, 4, and 5 are grouped together, Station 2 is unlike any other station, and Stations 11 and 12 are highly similar. Thus, in both cases, Stations 1 through 5 seem to separate from the remaining group.

Discussion and Conclusions

The study at VAAP offered a unique opportunity to observe the response

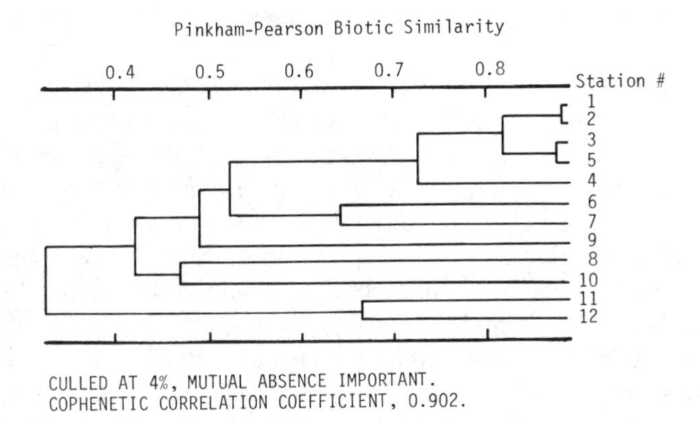

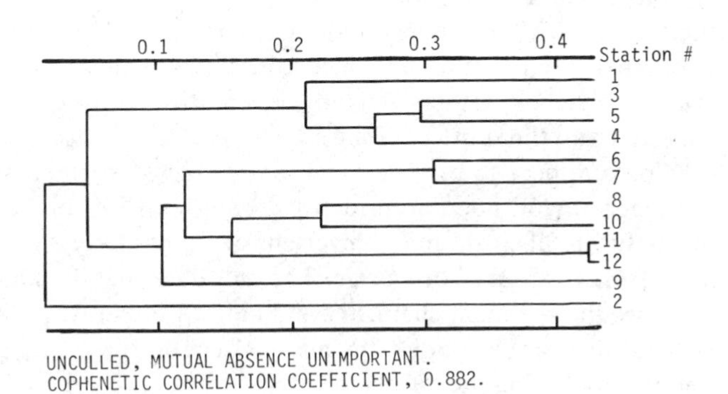

FIG. 10—*Phenogram showing station similarity of macroinvertebrates colonizing Hester-Dendy samplers.*

of the Waconda Bay ecosystem to TNT waste input. The chemical data showed that the discharge affected water quality in the upper portion of Waconda Bay by increasing the concentration of TNT, 2,4-DNT, 2,6-DNT, total dissolved solids, hardness, chlorides, ammonia, organic nitrogen, nitrates, and nitrites. Chemically, this effect was noticeable downbay at least 1 km from the bayhead. Samples taken in two reference bays verified that the elevated concentrations observed in Waconda Bay were due to the VAAP wastewaters.

Biological response by the periphyton and macroinvertebrate communities was observed in the same area of Waconda Bay where the chemical characteristics were altered. At the bayhead, toxic effects were noted in both communities. However, downbay, biological stimulation of periphyton was clearly evident. This pattern of toxicity followed by biological stimulation suggests that one or more components in the waste stream are at toxic concentrations in the bayhead water. As this material moves downbay, it either is diluted or degrades. When this occurs, nutrients become the controlling factor, resulting in biological stimulation of green plants.

In relating environmental effects to munitions concentrations in a field study such as this, one should understand the several assumptions and limitations inherent in such studies. First, the concentrations at any site vary with time. The munitions levels in sediments are presumed to be "normal" or "average," and the biologic communities are assumed to be in equilibrium with these concentrations.

The artificial substrates are exposed to bay water that is assumed to have a more variable concentration of munitions compounds than sediments. Ideally, daily samples of bay water collected during the incubation of substrates could have provided an estimate of the exposure level. Logistics and cost factors prevented this option. As an alternative, measurements were made of munitions in the effluent during the incubation period. Using the effluent concentrations, along with an understanding of how the effluent mixes with bay waters, one can predict the bay concentrations. On this basis, certain conclusions can be reached. Others can only be tentative because of the complexity of the waste and the subtlety of the effects.

Overall, the data appear to be very consistent. The water chemistry, artificial substrate periphyton population density, chlorophyll *a*, biomass, and artificial and natural substrate macroinvertebrate community structure all indicate effects from the VAAP discharge. The only data that do not support this conclusion are the sediment chemistry and natural substrate periphyton. Since the effluent does appear to be having an effect on these, the problem thus becomes one of relating the effects to specific causes.

In the natural substrate during winter, no effect could be discerned in number of species, density, or community structure for both macroinvertebrates and periphyton at Stations 6 through 9 in comparison with reference bay Stations 10 through 12. During the sampling period, the total concentration of the five munitions compounds averaged less than 25 μg/litre at Stations 6 through 9. Hence, it appears that for total munitions concentrations of less than 25 μg/litre, no effects would be expected.

At the opposite end of this range, effluent munitions concentrations during the incubation period for the artificial substrate samples averaged about 575 μg/litre and reached a peak of 2370 μg/litre. The concentrations at Stations 1 through 5 should not have exceeded these values. At this level, definite effects were noted in periphyton population density and community structure. It cannot be determined whether the peak or average concentration was most influential in the result.

Between these extremes, effects were subtle and less pronounced. In the 5-day period prior to the incubation of the artificial substrates, munitions concentrations in the upper channel, Stations 1 through 5, were averaging about 80 μg/litre. Results of prior studies suggest that the long-term average munitions concentration in this area might actually be somewhat higher.

Even so, the only observed effect on macroinvertebrate infauna was a slight shift in community structure. Substrate variation in periphyton samples may have masked slight shifts in community structure in that compartment also.

During the incubation of the artificial substrate samples, it is estimated that the concentration of munitions compounds at Stations 6 through 9 could have gone up to 50 to 100 μg/litre. For macroinvertebrates, the density, number of species, and community structure at Stations 6 through 9 were different from those at stations in the reference bay. For periphyton, the numbers of species seemed to be slightly lower at Stations 6 through 9 than at the reference bay stations. These differences were not as substantial as those observed between the upbay and reference bay areas. However, the differences were of such magnitude as to eliminate Stations 6 through 9 from the definite no-effect category.

In summary, the work reported here did not indicate environmental effects on periphyton or macroinvertebrates from a complex TNT manufacturing effluent at a total munitions concentration of 25 μg/litre or less. Definite effects were noted during a period when the total munitions concentration averaged 500 to 600 μg/litre, with a peak of over 2000 μg/litre. The data further suggest the tentative conclusion that minimal effects are associated with concentrations in the range of 50 to 100 μg/litre. Again, it should be noted that the concentrations of nonmunitions compounds varied as a function of the munition levels. These changes probably were factors also in the biological response observed during the study.

Acknowledgments

This research was supported by the U.S. Army Medical Research and Development Command, Ft. Detrick, Frederick, Md., and by the Ecological Research Office, U.S. Army Armament Research and Development Command, Aberdeen Proving Ground, Md., under Contract No. DAMD-17-75-C-5049.

Appreciation is also extended to the Ecological Research Office at Edgewood Arsenal for assistance in the completion of the field study.

The authors are especially grateful to J. Gareth Pearson and Edward Bender for helpful criticism and encouragement throughout the entire investigation. The assistance of Keith Furman, Jeanne Dorsey, and Deborah Nickelson in the preparation of the manuscript is also acknowledged.

References

[1] *Standard Methods for the Examination of Water and Wastewater*, 14th ed., American Public Health Association, Washington, D.C., 1976.

[2] Cairns, J., Jr., and Dickson, K. L., "Protocol for Aquatic Field Studies of the Toxicity of Environmental Contaminants," Center for Environmental Studies, Virginia Polytechnic Institute and State University, Blacksburg, Va., 1976, unpublished manuscript.

[3] *Standard Methods for the Examination of Water and Wastewater,* 13th ed., American Public Health Association, Washington, D.C., 1971.
[4] *Manual of Methods for Chemical Analysis of Water and Wastes,* Office of Technical Transactions, U.S. Environmental Protection Agency, Washington, D.C., 1974.
[5] *Chemistry Laboratory Manual, Bottom Sediments,* compiled by the Great Lakes Region, Committee on Analytical Methods, U.S. Environmental Protection Agency, Lake Michigan Basin Office, December 1969.
[6] Yentsch, C. S. and Menzel, D. W., *Deep Sea Research,* Vol. 10, pp. 221-231.
[7] Holm-Hansen, O., Lorenzen, C. J., Holmes, R. W., and Strickland, J. D. H., *Journal Conseil Permanent International Exploration de la Mer,* Vol. 30, 1965, pp. 3-15.
[8] Lorenzen, C. J., *Deep Sea Research,* Vol. 13, 1967, pp. 223-266.
[9] Moss, B., *Journal of Limnology and Oceanography,* Vol. 12, No. 2, 1968, pp. 335-340.
[10] Beck, W. M., Jr., *Quarterly Journal,* Florida Academy of Science, Vol. 17, No. 4, 1954, pp. 221-227.
[11] Ward, H. B. and Whipple, G. C. in *Freshwater Biology,* W. R. Edmondson, Ed., Wiley, New York, 1959.
[12] Pennak, R. W., *Freshwater Invertebrates of the United States,* Ronald Press, New York, 1953.
[13] McCormick, N. G., *Microbial Breakdown of Munitions Wastes,* American Society for Metals Meeting, Chicago, Ill., 1974.
[14] Weitzel, R. L., Simon, P. B., Jerger, D. E., and Schenk, J. E., "Report No. 1, Evaluation of Effects of Munitions Wastes on Aquatic Life—Iowa Army Ammunition Plant," Interim Report to U.S. Army Medical Research and Development Command, *Environ. Contr. Tech. Corp.,* Ann Arbor, Mich., 1975.
[15] Iskandar, I. K. and Keeney, D. R., *Environmental Science and Technology,* Vol. 8, No. 2, 1974, pp. 165-170.
[16] Holmes, C. W., Slade, E. A., and McLerran, C. J., *Environmental Science and Technology,* Vol. 8, No. 3, pp. 255-259.
[17] *Evaluation of Effects of Volunteer Army Ammunition Plant Wastes on Aquatic Life in Waconda Bay, Chickamauga Lake, Chattanooga, Tennessee,* Wapora, Inc., Charleston, Ill., 1977.

R. K. Guthrie,[1] *D. S. Cherry,*[2] *E. M. Davis,*[1] *and H. E. Murray*[1]

Establishment of Biotic Communities Within a Newly Constructed Ash Settling Basin and Its Drainage System

---

REFERENCE: Guthrie, R. K., Cherry, D. S., Davis, E. M., and Murray, H. E., **"Establishment of Biotic Communities Within a Newly Constructed Ash Settling Basin and Its Drainage System,"** *Ecological Assessments of Effluent Impacts on Communities of Indigenous Aquatic Organisms, ASTM STP 730,* J. M. Bates and C. I. Weber, Eds., American Society for Testing and Materials, 1981, pp. 243-254.

ABSTRACT: A newly constructed coal-ash settling basin and its drainage system at the 400D area power plant of the Savannah River Project, in Aiken, S.C., were studied from April 1977 to December 1978. The basin was filled in August 1977 by a flow of 7500 litres/min that consisted of Savannah River water and coal ash. The effluent was released to an excavated canal approximately 300 m in length, which drained into an older stream (300 m long) and then into a 3-km^2 swamp area that had received effluent from a previous ash settling basin. Between March and August 1977, the stream and swamp received only runoff from rainfall. During this time, when no basin effluent was released to the stream and swamp, the water quality characteristics were altered, and aquatic biota were reduced throughout the system.

The water quality characteristics were tested in the new ash settling system approximately every two months between April 1977 and December 1978. Samples were collected to determine the aerobic bacterial, phytoplankton, zooplankton, benthic invertebrate, and vertebrate populations at five stations in the basin system and one in the Savannah River at the pump station.

During the period of study, there were no major changes in water quality characteristics within the new basin. Seasonal peaks occurred in the bacterial populations in warmer months, and the composition of those populations within the basin system was determined to be different from that in the Savannah River. Sparse planktonic populations within the basin increased in numbers and diversity as the basin stabilized. Benthic invertebrates were first collected in the new basin in December 1977 and then increased in density and diversity at each subsequent sampling. Tadpoles were first collected in the basin in April 1978, and mosquito fish were first collected in May 1978.

[1]Professor of microbiology and ecology, associate professor of environmental sciences, and research associate in environmental sciences, respectively, School of Public Health, University of Texas, Houston, Texas 77025.

[2]Assistant professor of biology, Center for Environmental Studies, Department of Biology, Virginia Polytechnic Institute and State University, Blacksburg, Va. 24061.

Data obtained indicate that an efficient settling basin for coal ash has provided a suitable habitat for varied aquatic biotic communities, although some characteristics of the effluent have affected the normal balance and stability of communities within this environment. Such findings are important in that the suitability of receiving water habitats to aquatic life should be little affected by the release of effluent from such a basin.

KEY WORDS: aquatic communities, coal ash, water quality, bacteria, plankton, invertebrates, ecology, effluents, aquatic organisms

Indigenous aquatic biotic populations were found in the coal-ash settling basin, stream, and swamp system at the 400D area electric generating plant of the Savannah River Project, in Aiken, S.C., between 1973 and 1976 [*1,2*][3] in previous studies. In 1976, when the ash basin in use had completely filled, a new basin system was constructed adjacent to the original one. This system, which consisted of two U-shaped primary settling basins and one secondary basin, emptied into the same swamp drainage area. Changes in the settling efficiency of the basins can have significant effects upon biota in the renewing system because of turbidity and the settling action of the ash [*3,4*]. Likewise, populations of phytoplankton, zooplankton, and aquatic insects can be altered in the settling basin, depending upon their ages and morphologies and the amount of ash accumulated.

The purpose of this study was to compare the biotic populations in a new settling basin system with those present in an older basin system that had been used for at least 25 years. It is important that effluent from an ash settling basin be suitable for some reuse purposes, and the suitability can be partially assessed by study of the aquatic biota existing within the system.

Methods and Materials

The new settling basin system consisted of two primary settling basins (capacity of 25 000 m^3 each), which emptied into a secondary basin of approximately 410 000 m^3 (Fig. 1). This settling system differed from the original one in that the older system was generally an elongated settling basin that consisted of two equal sections. When one section was filled, effluent was diverted into the other half while the filled section was excavated. Samples wre collected at five sites from the pump station located on the Savannah River (Site 5) where water was brought into the basin system (Fig. 1). The sampling stations within the basin drainage system provided samples from the discharge from the original basins (Site 1), the new basin effluent (Site 2), the first swamp station 300 m from the ash effluent of the original basin outfall (Site 3), and the farthest site (1200 m) from Site 3 within the swamp proper (Site 4).

The old basin system, sampled from November 1973 through August 1978, except for periods of complete filling, and immediately after fly ash addition,

[3]The italic numbers in brackets refer to the list of references appended to this paper.

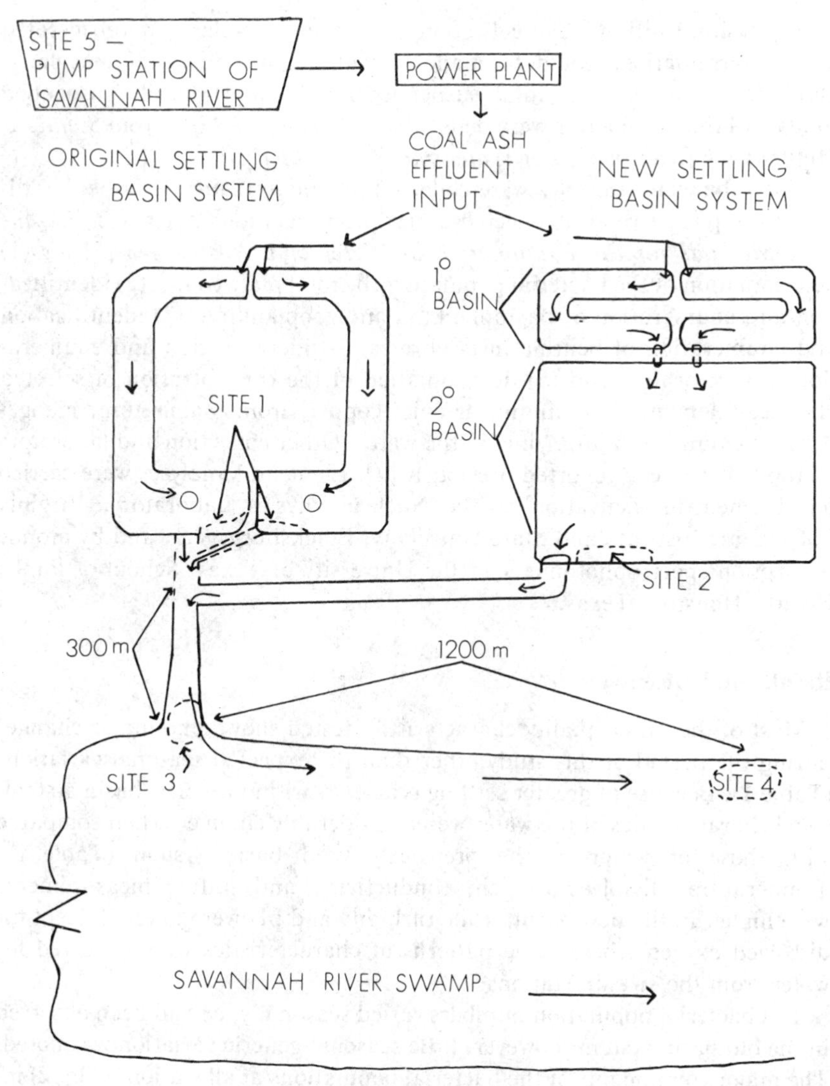

FIG. 1—*Sampling sites in the original and new settling basins and in the swamp drainage system.*

maintained some populations of organisms from bacteria through invertebrates within the basin. Some of the data from previous sampling periods are presented in order that comparisons may be made between the populations of the old and new systems.

The temperature and dissolved oxygen were measured in the field at approximately 2-month intervals between April 1977 and August 1978. Other water quality parameters, including pH, turbidity, and specific conductance, were measured within 2 h of collection. Additional samples were refrigerated until determinations could be made for sulfate, nitrate, orthophosphate, alkalinity, total carbon, total suspended solids, and volatile suspended solids. All the parameters were determined by using methods from *Standard Methods for Examination of Water and Wastewater* [5].

When the water samples were collected for the tests just described, additional samples were also collected according to methods described in *Standard Methods for the Examination of Water and Wastewater* [5] for (1) determination of the bacterial populations by total plate count, (2) identification and enumeration of phytoplankton and zooplankton, (3) identification and enumeration of benthic invertebrates, (4) identification and enumeration of vertebrates, and (5) determination of the concentration of selected chemical elements (aluminum, arsenic, copper, iron, magnesium, manganese, selenium, titanium, zinc) in the water. Other collection and laboratory methods have been reported previously [1]. Elemental analyses were carried out by neutron activation at the Nuclear Physics Laboratory, Virginia Polytechnic Institute and State University, Blacksburg, Va., and by atomic absorption spectrophotometry at the University of Texas, School of Public Health, Houston, Texas.

Results and Discussion

Most of the water quality characteristics tested showed no major changes during the period of this study other than the expected seasonal variations (Table 1). Because of greater settling efficiency within the new basin system, some characteristics of the water were considerably changed when compared with those of water in the previously used basin system (Table 2). Temperature, dissolved oxygen, conductivity, and sulfate measurements were higher in the new basin, while turbidity and pH were lower. Except for dissolved oxygen, these same patterns of characteristics were observed for water from the swamp stations.

The bacterial population numbers varied seasonally, as had been observed in the old basin system; however, little seasonal generic variation was noted. The major components of the bacterial populations at all stations (Fig. 2) included members of the genera *Bacillus*, *Flavobacterium*, *Pseudomonas*, *Achromobacter*, *Alcaligenes*, *Brevibacterium*, and *Enterobacter*. The Savannah River populations had greater numbers of *Enterobacter* species and fewer *Brevibacterium* and *Flavobacterium* species. Ninety-two percent of the isolates from the basin system were nitrate reducers. These percentages reflect one result of the water quality characteristics that were altered in the basin effluent (Table 1), in comparison with the Savannah River.

The increase in plankton numbers and diversity was most readily seen in the secondary basin (Figs. 3 and 4). These population numbers reached levels comparable to those of the old basin system only after the new basin

TABLE 1—*Mean water quality parameters in the new basin system.*

Sampling Station and Time Period	Temperature, deg C	Dissolved Oxygen, ppm	Turbidity, JTU[a]	pH	Specific Conductance, μmhos/cm	Sulfate, ppm
Station 1—Primary basin effluent						
Feb. 1977 to Aug. 1977 basin filling	27.5 (26.1 to 30.4)	8.0 (6.7 to 8.6)	17 (15 to 19)	6.1 (6.0 to 6.3)	200 (15 to 300)	72 (50 to 94)
Aug. 1977 to Aug. 1978 basin releasing effluent	26.6 (22.5 to 31.5)	6.5 (5.8 to 7.8)	51 (13 to 70)	6.0 (4.7 to 7.0)	280 (225 to 400)	77 (59 to 94)
Station 2—Secondary basin effluent						
Feb. 1977 to Aug. 1977 basin filling	27.7 (26.1 to 30.8)	8.1 (7.0 to 8.6)	17 (14 to 20)	5.9 (5.8 to 6.0)	118 (55 to 150)	60 (55 to 69)
Aug. 1977 to Aug. 1978 basin releasing effluent	24.6 (14.0 to 32.0)	7.3 (5.4 to 8.4)	32 (19 to 47)	6.6 (6.2 to 7.2)	218 (160 to 260)	67 (61 to 75)
Station 3—First swamp station						
Feb. 1977 to Aug. 1977 basin filling	23.8 (21.8 to 25.8)	6.8 (4.0 to 8.2)	14 (12 to 16)	5.5 (5.4 to 5.8)	350 (300 to 450)	112 (100 to 122)
Aug. 1977 to Aug. 1978 basin releasing effluent	24.4 (15.0 to 32.2)	7.4 (6.6 to 8.0)	27 (17 to 37)	6.7 (6.3 to 7.5)	225 (140 to 280)	62 (48 to 75)
Station 4—Second swamp station						
Feb. 1977 to Aug. 1977 basin filling	26.4 (25.0 to 29.0)	6.7 (6.5 to 7.0)	9 (6 to 10)	5.9 (5.8 to 5.9)	320 (300 to 350)	94 (69 to 112)
Aug. 1977 to Aug. 1978 basin releasing effluent	23.0 (16.0 to 28.0)	7.6 (6.9 to 8.2)	27 (18 to 40)	6.9 (6.3 to 7.4)	180 (95 to 220)	63 (51 to 80)
Station 5—Savannah River sluice water pump station						
Feb. 1977 to Aug. 1977 basin filling	25.5 (22.2 to 28.8)	6.6 (4.4 to 7.8)	10 (8 to 11)	6.9 (6.3 to 7.8)	125 (90 to 160)	10 (9.5 to 11)
Aug. 1977 to Aug. 1978 basin releasing effluent	24.9 (15.7 to 28.8)	6.4 (4.4 to 8.4)	28 (11 to 48)	7.4 (6.8 to 7.8)	110 (50 to 160)	9 (8 to 11)

[a]JTU = Jackson turbidity units.

TABLE 2—*Mean water quality parameters in the old and new basin systems.*

Station	Temperature, deg C	Dissolved Oxygen, ppm	Turbidity, JTU[a]	pH	Specific Conductance, μmhos/cm	Sulfates, ppm
Station 1						
Old settling basin	12.8 (12.2 to 33.0)	7.1 (6.1 to 9.5)	98 (40 to 180)	7.3 (7.1 to 7.5)	93 (40 to 180)	17 (9.3 to 28)
New settling basin	14.9 (14.0 to 32.0)	7.7 (5.4 to 8.6)	27 (14 to 47)	6.3 (5.8 to 7.2)	175 (55 to 260)	64 (50 to 75)
Station 3—First swamp station						
Old system	13.5 (12.0 to 33.0)	7.6 (5.9 to 10.4)	96 (13 to 250)	7.4 (7.1 to 8.0)	82 (30 to 200)	19 (11.3 to 33)
New system	13.1 (15.0 to 32.2)	7.1 (4.0 to 8.2)	22 (12 to 37)	6.1 (5.4 to 7.1)	277 (140 to 450)	87 (48 to 122)
Station 4—Second swamp station						
Old system	12.0 (10.0 to 30.0)	8.0 (6.8 to 10.5)	96 (25 to 185)	7.4 (7.2 to 7.6)	96 (25 to 185)	16 (14 to 21)
New system	13.3 (16.0 to 29.0)	7.2 (6.5 to 8.2)	19 (6.40)	6.3 (5.8 to 7.4)	237 (95 to 350)	81 (51 to 112)

[a]JTU = Jackson turbidity units.

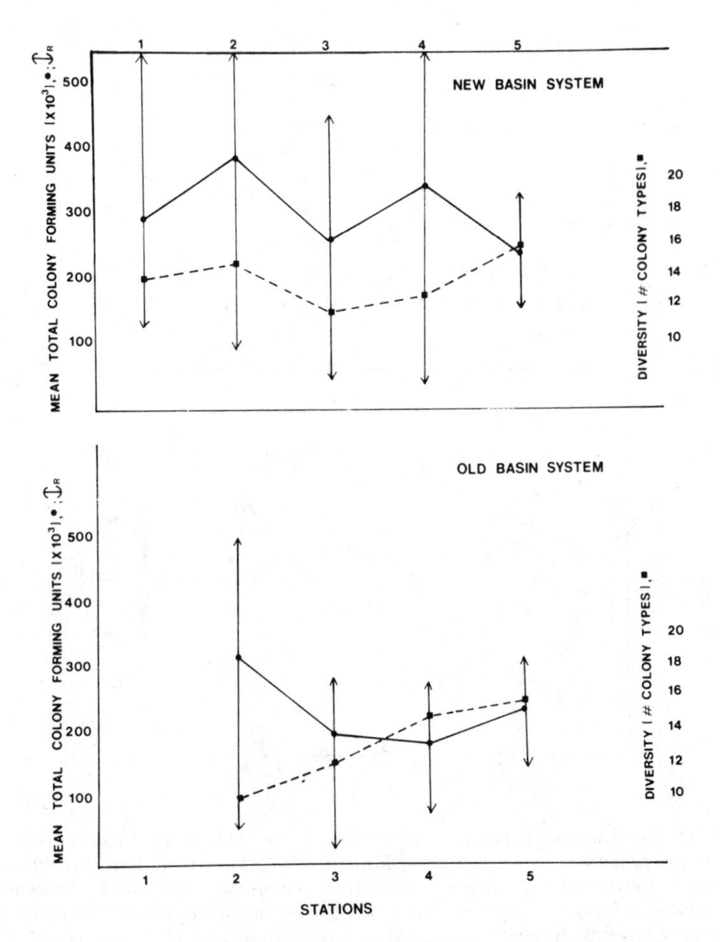

FIG. 2—*Bacterial populations at stations in the old and new basin systems: Station 1—old basin effluent; Station 2—new secondary basin effluent; Station 3—first swamp station; Station 4—most distant swamp station; Station 5—Savannah River pump station. All values are reported as means for the sampling period.*

had been in full operation for at least 6 months (Table 3). No zooplankters were observed in the new basin system before September 1977, when the basin had completely filled and effluent was being released. Undoubtedly, a major reason for the absence of these organisms at the swamp stations was higher concentrations of several elements (arsenic, magnesium, manganese, zinc), probably due to leaching from the previously sedimented ash deposits (Table 4).

Few benthic invertebrates were found in the new basin until the end of 1977, when the basin had been releasing effluent for at least 4 months. During the latter part of 1977, however, dipterans and odonates rapidly populated the area (Table 5). By December 1978, the densities of

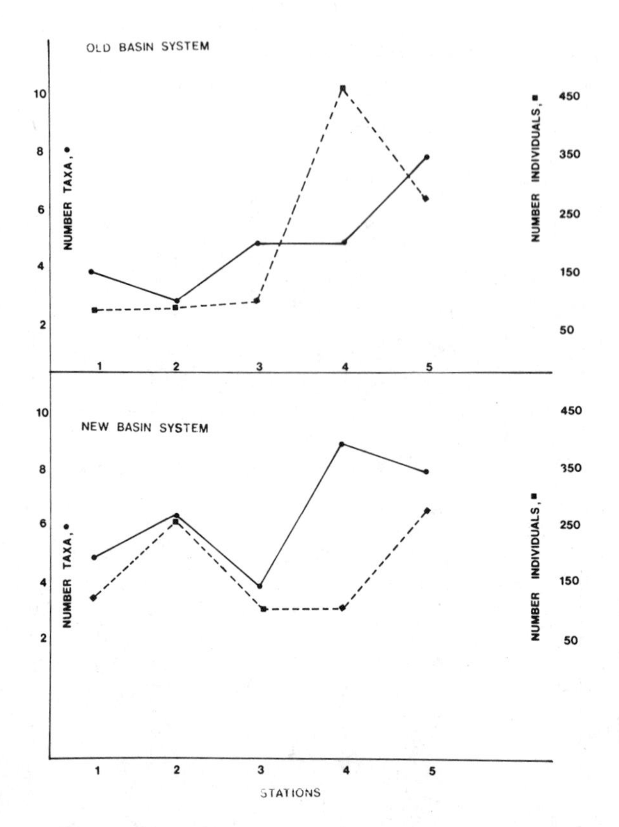

FIG. 3—*Phytoplankton populations at stations in the old and new basin systems: Station 1—old basin effluent, new primary basin effluent; Station 2—new secondary basin effluent; Station 3—first swamp station; Station 4—most distant swamp station; Station 5—Savannah River pump station. All values are reported as means for the sampling period. The number of individuals per taxon is reported.*

chironomids and the odonates *Enallagma* and *Libellula* spp. were greater in the basin sediments than they were in the original basin prior to excavation in 1974. Aquatic conditions in the basin and swamp drainage system were most severe after excavation and the addition of fly ash in the original basin, because of the low pH, although the densities of most taxa of biota were frequently higher in the swamp station. Crayfish and snails were rarely found in either settling basin throughout the study. The sparsity of invertebrates in the new basin before the latter part of 1977 was probably due to a combination of leached elemental concentrations in the standing water and the slowly rising pH (5.8 to 6.2) [*4*].

Mosquito fish (*Gambusia affinis*) were not found in the old ash settling basin. Although some fish may have been entrained through the plant in sluice water, it was apparent that they were unable to reproduce and maintain populations in that system. Mosquito fish were observed in small

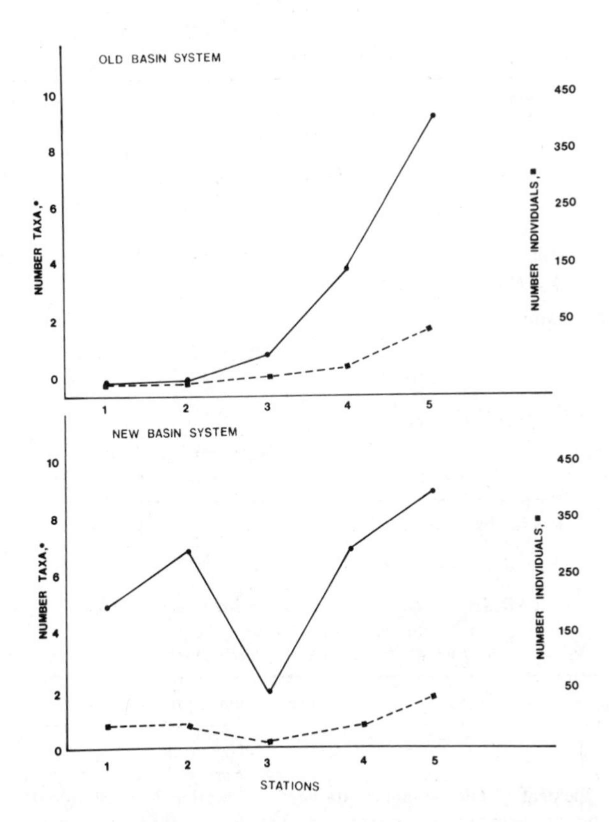

FIG. 4—*Zooplankton populations at stations in the old and new basin systems: Station 1—old basin effluent, new primary basin effluent; Station 2—new secondary basin effluent; Station 3—first swamp station; Station 4—most distant swamp station; Station 5—Savannah River pump station. All values are reported as means for the sampling period. The number of individuals per taxon is reported.*

numbers in the new secondary basin in May 1978 after 6 months of full operation but were not observed after that time within the basin. The low numbers of mosquito fish in the lower part of the drainage system in the early part of the study period (Table 5) were no doubt due to the restrictions on movement of the fish when the water flow had diminished through the swamp portion of the drainage system. The fish numbers increased by 1978, when a higher flow was added from the operation of the new basin system. Frogs were observed in large numbers in both the old and new basin systems throughout the study. Tadpoles were not observed in the new basin system until April 1978; however, this absence was probably due to the time of the breeding season rather than the condition of the basin operation, since they were most numerous in the most heavily polluted primary basin water in April 1978.

It is apparent in the results presented here that effluent from sluiced heavy

TABLE 3—*Plankton populations in the old and new drainage systems.*

Station	Phytoplankton		Zooplankton	
	Number of Taxa	Number of Organisms	Number of Taxa	Number of Organisms
Station 1				
Old settling basin	22	82	20	15
New secondary settling basin	5	89	3[a]	11
Station 3—First swamp station				
Old system	15	75	15	12
New system	5	84	2[a]	8
Station 4—Second swamp station				
Old system	35	20	30	32
New system	9	300	7[a]	19

[a]No zooplankton were observed before September 1977.

TABLE 4—*Mean elemental concentrations in water in the old and new basin systems (no water flow in the new basin system).*

Element	Concentration, ppm			
		New System		
	Old System	Station 1	Station 3	Station 4
Al	13.0	16.7	7.1	4.7
As	0.06	0.23	3.3	0.1
Cu	0.39	0.63	0.61	2.5
Fe	16.9	14.2	17.2	9.2
Mg	4.1	5.8	9.4	2.8
Mn	0.07	0.11	0.18	0.13
Se	0.10	0.09	0.06	0.06
Ti	0.90	1.2	0.69	0.56
Zn	0.40	2.0	2.2	2.0

and fly ash from coal combustion is immediately usable as a substrate for the growth of aquatic bacteria and some algae. These microbial populations are able to increase as a secondary sedimentation basin stabilizes, and, as this occurs, populations of zooplankton, benthic invertebrates, and some vertebrates, successively, become established within the system. These results agree with those of Cherry and Guthrie [*6*], Gutenmann et al [*7*], Patrick and Loutit [*8*], and Chu et al [*9*] that effluents may be suitable for some reuse purposes; however, there remains the possibility that certain potentially toxic elements in the coal ash may permit elemental biomagni-

TABLE 5—*The density of macroinvertebrate and fish populations in the original basin (from November 1973) before and after ash excavation, with fly ash addition, and during operation of the new basin during early 1977 (January through August) through 1978.*

		Density (m^2)					
					New Basin Operation		
Taxon	Sampling Station	Before Excavation	After Excavation	After Fly Ash Addition	January through August 1977	September through December 1977	1978
Chironomidae	basin	0.8	0	0	2.3	12.1	11.5
	swamp	11.9	4.2	8.9	7.7	8.2	6.8
Coenagrionidae	basin	0.4	0.1	0.1	0.2	0.2	0.9
(*Enallagma* sp.)	swamp	1.0	0	<0.1	0.1	0.2	0.3
Libellulidae	basin	0.3	0	0	0	0.1	1.0
(*Libellula* spp.)	swamp	1.1	0	2.3	1.4	1.6	1.7
Astacidae	basin	0.2	0	0.1	0	0	0
(*Procambarus* spp.)	swamp	3.2	0.1	0	0.3	0.4	2.0
Gastropoda	basin	0	0	0	0	0	0 to 5.4[a]
(*Physa* and *Lymnaea* spp.)	swamp	9.4	0.2	1.9	4.7	8.4	9.6
Poeciliidae	basin	0	0	0	0	0	0
(*Gambusia affinis*)	swamp	10.4	0.4	1.4	1.1	1.5	8.3

[a]Empty shells.

fication by some organisms and that different biotic communities in receiving waters may be affected by these elements.

Acknowledgment

This work was done in the Houston and Blacksburg laboratories and at the Savannah River Project, Aiken, S.C., with partial support from Oak Ridge Associated Universities travel contracts and from E. I. du Pont de Nemours Co. Facility Use Agreement No. 28.

References

[*1*] Guthrie, R. K., Cherry, D. S., and Rogers, J. H., Jr., *Proceedings*, Seventh Mid-Atlantic Industrial Waste Conference, Philadelphia, Pa., 1975, pp. 17–43.
[*2*] Guthrie, R. K. and Cherry, D. S., *Water Resources Bulletin*, Vol. 12, No. 5, 1976, pp. 889–902.
[*3*] Cherry, D. S., Guthrie, R. K., Rodgers, J. H., Jr., Cairns, J., Jr., and Dickson, K. L., *Transactions of the American Fisheries Society*, Vol. 105, No. 6, 1976, pp. 686–694.
[*4*] Cherry, D. S., Guthrie, R. K., Larrick, S. R., and Sherberger, F. F., *Hydrobiologia*, Vol. 62, 1979, pp. 257–267.
[*5*] *Standard Methods for the Examination of Water and Wastewater*, 14th ed., American Public Health Association, Washington, D.C., 1976.
[*6*] Cherry, D. S. and Guthrie, R. K., *Water Resources Bulletin*, Vol. 13, No. 6, 1977, pp. 1227–1236.
[*7*] Gutenmann, W. H., Bache, C. A., Youngs, W. D., and Lisk, D. J., *Science*, Vol. 191, 1976, pp. 966–967.
[*8*] Patrick, F. M. and Loutit, M., *Water Research*, Vol. 10, 1976, pp. 333–335.
[*9*] Chu, T.-Y. J., Ruane, R. J., and Krenkel, P. A., *Journal of the Water Pollution Control Federation*, Vol. 50, No. 11, 1978, pp. 2474–2508.

W. P. Kovalak[1]

Assessment and Prediction of Impacts of Effluents on Communities of Benthic Stream Macroinvertebrates

REFERENCE: Kovalak, W. P., **"Assessment and Prediction of Impacts of Effluents on Communities of Benthic Stream Macroinvertebrates,"** *Ecological Assessments of Effluent Impacts on Communities of Indigenous Aquatic Organisms, ASTM STP 730,* J. M. Bates and C. I. Weber, Eds., American Society for Testing and Materials, 1981, pp. 255–263.

ABSTRACT: Diversity indexes, particularly those based on information theory, have been widely promoted as the best method for assessing the impact of effluents on macroinvertebrate communities. Despite their superiority, neither the Shannon-Weaver index (SWI) nor the Brillouin index (BI) is universally applicable. The SWI is preferred where effluents are selective (for example, organics) and BI preferred where effluents are nonselective (for example, heavy metals). Despite their simplicity, it should be remembered that diversity indexes supplement, not replace, community analyses based on species richness and population densities. Covariance analyses and more tightly defined stratified sampling programs are discussed as ways of improving the accuracy of density estimates.

Impact prediction depends on models predicting changes in species richness and population densities. A mechanistic model predicting changes in species richness as a function of oxygen supply and demand is offered as a step in this direction. Despite its accuracy in undisturbed streams, the applicability of the model to organic and thermal pollution awaits testing. Implications of the model for impact prediction are discussed.

KEY WORDS: water pollution, water quality, impact assessment, impact prediction, diversity, species richness, streams, ecology, effluents, aquatic organisms

Impacts of effluents on macroinvertebrate communities are usually assessed by changes in community structure. The most commonly used measures of community structure include the presence and absence and (or) abundance of critical species (indicator organisms), species richness (number of species), relative or absolute population densities, and diversity (a numerical expression of community composition based on the numbers and

[1]Assistant professor of biology, University of Michigan—Dearborn, Dearborn, Mich. 48128.

kinds of individuals in a community or collection). Although any or all of these parameters may be applicable in certain situations, most have easily demonstrated weaknesses that make them inappropriate in a large number of cases, and some require training in biology to interpret them properly. This has led to recent attempts to provide a limited repertoire of methods for community analysis that are sufficiently robust to be applicable to a wide variety of circumstances and that are readily interpretable by those using them.

Diversity indexes based on information theory have been widely promoted as the best measures of community structure [*1-4*].[2] Recent reviews include discussions of the usefulness of these indices in relation to others [*5*], criteria for determining which index should be used under certain circumstances [*6-8*], and the validity of the diversity concept [*9*]. My purpose is not to review these papers, but rather to discuss a few key points from them as well as some often overlooked considerations that directly affect the applicability of these techniques. A focal point of this discussion will be an assessment of the assumption that diversity indexes based on information theory provide an adequate solution to the problem of collation and interpretation of cumbersome species lists, which are the basis of environmental impact assessment.

Whereas new methods for impact assessment have developed rapidly, the development of models for impact prediction has lagged behind. Models developed to describe specific situations usually lack the realism and generality [*10*] necessary to make them applicable to a wide variety of circumstances because they are based on a description of events rather than on the processes that control the events.

A preliminary model predicting changes in the species richness of benthic stream communities in relation to oxygen supply and demand in undisturbed streams is presented as an example of a process-oriented model that has sufficient realism and generality to be used to predict changes in species richness resulting from oxygen-demanding effluents. The model is used to identify some critical areas for future research in impact prediction.

Impact Assessment

Most of the techniques used for impact assessment assume that effluents cause selective elimination of sensitive (intolerant) species and enhancement of insensitive (tolerant) species. This situation is exemplified by conditions below sewage outfalls, where low oxygen concentrations eliminate intolerant forms (for example, stone flies and mayflies), which are replaced by large populations of tolerant forms (for example, Tubificidae and Chironomidae). Under these conditions, any of the proposed measures of community structure will document alteration of the macroinvertebrate community.

[2]The italic numbers in brackets refer to the list of references appended to this paper.

Recently, the Shannon-Weaver index (SWI)

$$H' = - \sum_{i=1}^{n} p_i \log_2 p_i$$

where p_i is the proportion of the ith species in a collection [7], has been most widely used. Three reasons account for its popularity. First, SWI is recommended where samples from a community rather than the complete community are being analyzed [7]. Second, for purposes of uniformity, it has been recommended by the U.S. Environmental Protection Agency (EPA) [4]. Third, and most important, it is relatively insensitive to sample size. As the sample size increases (either as a function of the area of quadrats used or of the cumulative number of samples analyzed), initially the diversity values increase rapidly but then level off to some asymptotic value, which remains essentially unchanged with further increases in sample size. In benthic sampling, asymptotic diversity values are normally reached when the cumulative area sampled is 0.5 to 0.6 m^2 (5 to 6 ft^2). This stability is attributable to the fact that SWI is based on the proportion each species contributes to the total population rather than on the absolute abundances of species. A number of samples [with a cumulative area of 0.5 to 0.6 m^2 (5 to 6 ft^2)] are necessary to establish the proportional composition of the community, because most macroinvertebrate species exhibit a contagious (clumped) distribution. In contrast, other measures of diversity, such as Margalef's index (MI) [11], are much more sensitive to changes in sample size. The response of MI to sample size varies with the community diversity, making it virtually impossible to compare polluted (low diversity) and unpolluted (high diversity) situations using the same sample size at both stations.

Whereas most diversity indexes provide an adequate measure of change in macroinvertebrate communities where effluents are selective, most are of limited value where effluents are nonselective. This situation is typified by conditions below discharges of heavy metals or other toxicants that nonselectively reduce population densities of all species. Under these conditions, many diversity indexes, including SWI, exhibit the highest diversity where the conditions are worst. This is attributable to the fact that SWI depends on two components—species richness and equitability (distribution of individuals among species). Changes in equitability have a greater effect on SWI than do changes in species richness. Normal (healthy) communities typically are composed of a few numerically dominant species and a larger number of rare species. The dominant species, which most strongly influence equitability, control SWI. The most important effect of nonselective effluents, apart from reducing population densities and species richness, is to increase the equitability of surviving species. Since SWI is determined primarily by equitability, SWI tends to be greatest where effects of the effluent are most pronounced. The overall insensitivity of SWI to change in species richness also accounts for the similarity of diversity values achieved, whether indentification of the macroinvertebrates is at the family, genus, or species level.

Since nonselective effluents increase equitability and decrease population densities, an index that gives greater emphasis to lowered densities would be a more sensitive measure of impact. One index based on information theory that is sensitive to changes in density is the Brillouin index (BI)

$$H = \frac{1}{N} \log \frac{N!}{\Pi N_i!}$$

where N_i is the number of the ith species, and N is the total number of individuals in a community [7]. In the case of nonselective effluents, BI exhibits the lowest values where the impact is greatest.

Although it has been argued that BI is superior to and more appropriate than SWI in impact assessment [8], justifications given for its use are debatable. Diversity measured with BI is appropriate where all the species and individuals in a community have been collected [7]. Since benthic sampling normally represents subsampling from a larger community, H' (SWI), which is an estimator of H, is more appropriate than H, which assumes that all species and individuals have been collected.

Use of BI depends on whether or not the investigator is willing to assume that the fauna in his samples constitutes the community [8]. It has been argued that it is impossible to define the limits of benthic communities in running waters because, conceptually, they are infinite owing to drift and seasonal changes in composition [8]. If one cannot define the limits of community, in theory one cannot randomly sample it, and, therefore, by default it is easier to define the samples as the community.

Assumptions about what is being sampled aside, it should be stated that although BI is sensitive to changes induced by nonselective effluents, it is too strongly influenced by sample size to be used where effluents are selective. Where effluents are selective, it is no more reliable than MI.

From the foregoing discussion it should be clear that no single index of diversity is appropriate for use under all circumstances. However, indexes based on information theory (SWI and BI) appear to possess fewer liabilities than other measures proposed to date. As stated by Peet [5] the characteristics of diversity indexes and their responses to changes in community structure should be understood as the basis for their use. It also bears restating that diversity indexes are interpretative tools, which aid in community analysis but do not necessarily supersede other measures of community structure because how diversity relates to community structure is not clear [9]. Other standard measures, such as species composition and population densities, must still be considered the basis of impact assessment.

The collection of statistically reliable estimates of population densities, which are essential to impact assessment, has created problems because of the highly contagious (clumped) distributions of most species. Although this problem has been recognized for some time [*12*], it has not been dealt with in a meaningful fashion. Rather, the variability in density estimates has been dismissed as a problem beyond the cost-effectiveness of most assessment programs.

It should be obvious that if organisms are clumped, either they are gregarious by nature or some extrinsic factor or factors are controlling their distribution. Egglishaw [*13,14*] showed that in moderate-size streams, about 60 percent of the variability in faunal numbers from sample to sample was correlated with differences in the detrital standing crop. Although it has not been conclusively demonstrated that this relationship is causal, the detrital standing crop could be used as a covariate to increase the accuracy of density estimates. It should not be inferred that detritus is a universal covariate applicable under most circumstances. Rather, the main point is that we have not explored the possibility of using covariates to improve the accuracy of density estimates, despite the large number of factors that have been identified as potentially important determinants of macroinvertebrate distribution and abundance [*15*].

The use of covariance analysis presumes a linear relationship between the dependent variable and the covariate. It may happen that few, if any, useful linear relationships exist. Alternatively, where definable relationships are nonlinear (for example, relationships between faunal abundance and substrate type or current velocity), a portion of the relationship where the response is constant or at least predictable could be used as a basis for stratified sampling. Although this approach has been advocated for some time [*4*], rarely is it incorporated into field programs. Where it has been used, often the strata identified have been too broad (for example, riffles and pools) to improve the accuracy of density estimates effectively. The strata can be neither too broad nor too narrow. Identification of the sampling strata imposes additional cost, but this should be more than repaid by the fewer number of samples necessary to achieve statistically reasonable density estimates.

Impact Prediction

Comparatively little has been done to produce predictive models of sufficient realism and generality to be applicable to a wide variety of circumstances. The absence of good predictive models seems to be related to two considerations. First, despite the wealth of descriptive data on undisturbed systems [*15*] we have little understanding of the mechanisms by which various factors control the distribution and abundance of macroinvertebrates. Second, even where it should be possible to obtain good pre- and postoperation data on the impact of specific effluents, we have failed to do so because of a lack of support by industry. Consequently, the best we can hope to achieve at the present time are very simple deterministic models fitted to sparse descriptive data for specific situations. Unfortunately, even though these models may achieve some precision based on the data used to generate them, there is no guarantee they have any predictive value.

In making predictions about the impacts of effluents on macroinvertebrates, the one thing about which we can be certain is that anything added to a receiving body is apt to have some impact on the indigenous fauna. Our experience with the variability of faunas in adjacent, unpolluted rivers should be sufficient evidence of the tremendous impact that small differences in environmental conditions can have on macroinvertebrate communities. The problem is then not one of predicting whether or not there will be a change but, rather, the magnitude of that change. From our earlier discussion, it should be clear that whether or not we are interested in changes in diversity, we are fundamentally interested in predicting changes in species richness and population densities.

As an initial step in this direction, in the following section I present a simple model that may be used to predict changes in species richness resulting from changes in oxygen supply. This model is of interest for two reasons. First, many kinds of effluents (for example, organic and thermal) affect oxygen supply. Second, this model presumes to be mechanistic because it is based on principles derived from unrelated studies. It should be emphasized that originally the model was designed to account for changes in species richness along undisturbed stream courses, not to predict effluent impacts; its application here is an extension of its original purpose.

Because oxygen concentrations in turbulent waters remain near saturation, it has been tacitly assumed that oxygen is not an important limiting factor in fast-flowing streams [*15,16*]. However, most stream insect larvae depend on passive diffusion of oxygen through the body surface or gills or both to satisfy oxygen requirements [*15*]. Therefore, oxygen supply to insects depends not only on the oxygen content of water, but also on the rate of water renewal at respiratory surfaces [*17*]. In streams water renewal is primarily a function of current, turbulence, or both. Current, therefore, increases oxygen richness.

Because of spatial variations in current velocity, the oxygen supply to insects can vary widely on both a macro- and microscale. Madsen [*17*] measured the oxygen diffusion rate ($\approx$oxygen richness) in several microhabitats and showed that the oxygen supply at the tops of stones was 19 times greater than at the bottoms of stones and was four times greater than in the interstitial spaces of gravel. The lower oxygen supplies under stones and in gravel, which were due primarily to differences in current velocity, are significant because most stream insects are photonegative and show a preference for these cryptic microhabitats.

On a macroscale, most insect larvae are found over a wide range of current velocities but show preferences for specific currents [*15*]. The preferred current varies with temperature and in relation to body size. As temperature increases, insects move into faster currents, presumably because at higher temperatures the oxygen content of water (at saturation) is lower and the oxygen demand of insects is higher. As they grow, insects also move into faster currents, presumably because the larger individuals, which have a lower surface area:volume ratio [*15*], have greater difficulty satisfying oxygen demands.

Similarly, on a microscale, most insect larvae are found on all faces of stones but show preferences for specific faces [*15*]. These preferences presumably reflect current preferences, which Hynes [*15*] suggested are related to the respiratory requirements of the species. Kovalak [*18*] tested Hynes' hypothesis by assessing changes in positioning of *Glossosoma nigrior* larvae on bricks in relation to current velocity, water temperature, and time of day. He reported that at higher temperatures ($>19°C$) and lower current velocities (<70 cm/s) larvae aggregated on the upstream faces of bricks, where oxygen richness was greatest because of greater turbulence. Selection of the upstream face was reduced at currents greater than 70 cm/s by the formation of a standing wave, which reduced turbulence and, therefore, oxygen richness. At higher currents the insects moved to the top and sides of the bricks where oxygen richness was greater because of the current.

If the relationship between oxygen supply and demand determines the macro- and microdistribution of benthic invertebrates, then ultimately it controls the species richness and population densities within communities. To test this hypothesis, the relationship between species richness and the quotient of oxygen supply and demand along an altitudinal gradient was examined. The quotient of oxygen supply and demand is given by the following expression

$$I_o = CO/M$$

where I_o is the oxygen index, C is the mean current velocity in metres per second, O is the oxygen concentration in milligrams per litre, and M is an index of the metabolic rate of the insects. M is determined by assuming that the metabolic rate of insects at 0°C is 1.0 and that the respiratory quotient (Q_{10}) is 2.0.

I_o was determined for a series of stations along Cement Creek in Colorado and compared with observed species richness [*19*] (Fig. 1). There was a linear relationship between I_o and species richness given by the regression equation $Y = -11.3 + 7.6X$, where Y is the species richness, and X is I_o. Species richness at an individual station (S_2) also could be predicted from the relationship

$$S_2 = S_1 (I_{o2}/I_{o1})^2$$

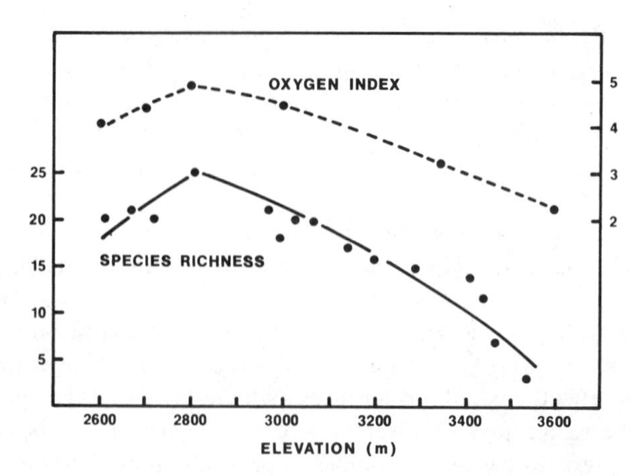

FIG. 1—*Changes in species richness and oxygen index (see text for explanation) along an altitudinal gradient, Cement Creek, Colorado. Data from Allan* [19].

where I_{o2} is the oxygen index for that station, and S_1 and I_{o1} are values from a reference station. Of the two relationships, the latter is the more interesting because, if it is a fixed relationship that applies to most riverine systems, predictions could be made from a minimal amount of information. The regression equation requires data from several stations to establish the relationship. Moreover, the regression equation is likely to vary from river to river because of natural differences in species richness among rivers. If the second relationship holds despite changes in the regression equation, then predictions could be made from as little as one estimate of the oxygen index and species richness at a single station.

That oxygen is an important determinant of species richness in undisturbed streams should not be surprising, as it may be inferred from earlier studies [*20*], which showed that stream faunas vary along altitudinal gradients primarily as a function of temperature and current velocity. Water temperature and current velocity are the two key factors determining oxygen supply and demand. Moreover, it has been demonstrated that alpine insects have a greater gill area to body weight ratio than do lowland species, presumably to compensate for lower oxygen supplies at higher altitudes [*21*].

The importance of this model in predicting faunal changes resulting from organic or thermal effluents awaits serious testing. At present, there are no published studies that provide sufficient data to estimate I_o. Therefore, the applicability of the model can only be inferred. For example, it has long been known that untreated or partially treated sewage effluents cause a reduction in species richness. Some of the earliest and best data documenting this phenomenon were presented by Gaufin and Tarzwell [*22*]. In the study, where oxygen concentrations and presumably I_o were lowest, species richness was lowest. Unfortunately, it was impossible to test the accuracy of the model using these data.

It also seems probable that the model could be used to predict some effects of thermal effluents that affect both oxygen supply and oxygen demand. The applicability of the model to other types of effluents (for example, heavy metals) should also be investigated. Recently, it was demonstrated that sublethal concentrations of copper increase the oxygen consumption rates in stone flies [*23*].

This model, coupled with models of effluent dispersal or oxygen depletion curves, should provide realistic predictions of changes in species richness. However, considerably more information will be necessary to predict which species will be among the survivors and likely changes in population densities.

The foregoing discussion assumes a single, essentially pure effluent entering the recipient body. Unfortunately, in practice, new effluents probably will be introduced in streams and rivers already burdened with other effluents. At the least, it complicates environmental prediction since each effluent would have to be modelled separately and then joined into a larger master model. The most important problem to be encountered in this procedure will be the modeling of synergistic and antagonistic interactions.

References

[*1*] Wilhm, J. L. and Dorris, T. C., *Bioscience*, Vol. 18, 1968, pp. 477-481.
[*2*] Wilhm, J. L., *Annual Review of Entomology*, Vol. 17, 1972, pp. 223-252.
[*3*] Cairns, J., Jr., and Dickson, K. L., *Journal of the Water Pollution Control Federation*, Vol. 43, 1971, pp. 755-772.
[*4*] Weber, C. I., *Biological Field and Laboratory Methods for Measuring the Quality of Surface Waters and Effluents*, EPA-670/4-73-001, U.S. Environmental Protection Agency, 1973.
[*5*] Peet, R. K., *Annual Review of Ecology and Systematics*, Vol. 5, 1974, pp. 285-307.
[*6*] Pielou, E. C., *Journal of Theoretical Biology*, Vol. 13, 1966, pp. 131-144.
[*7*] Pielou, E. C., *Ecological Diversity*, Wiley Interscience, New York, 1975.
[*8*] Kaesler, R. L. and Herricks, E. E., *Water Resources Bulletin*, Vol. 12, 1976, pp. 125-135.
[*9*] Hurlbert, S. H., *Ecology*, Vol. 52, 1971, pp. 577-586.
[*10*] Odum, E. P., *Fundamentals of Ecology*, Saunders, Philadelphia, 1971.
[*11*] Margalef, D. R., *General Systems*, Vol. 3, 1957, pp. 36-71.
[*12*] Needham, P. R. and Usinger, R. L., *Hilgardia*, Vol. 24, 1956, pp. 383-409.
[*13*] Egglishaw, H. J., *Journal of Animal Ecology*, Vol. 33, 1964, pp. 463-476.
[*14*] Egglishaw, H. J., *Journal of Animal Ecology*, Vol. 38, 1969, pp. 19-33.
[*15*] Hynes, H. B. N., *The Ecology of Running Waters*, University of Toronto Press, Toronto, 1970.
[*16*] Macan, T. T., *Freshwater Ecology*, Wiley, New York, 1963.
[*17*] Madsen, B. L., *Oikos*, Vol. 19, 1968, pp. 304-310.
[*18*] Kovalak, W. P., *Canadian Journal of Zoology*, Vol. 54, 1976, pp. 1585-1594.
[*19*] Allan, J. D., *Ecology*, Vol. 56, 1975, pp. 1040-1053.
[*20*] Illies, J., *Schweizerische Zeitschrift für Hydrologie*, Vol. 24, 1962, pp. 433-435.
[*21*] Dodds, G. S. and Hisaw, F. L., *Ecology*, Vol. 5, 1924, pp. 262-271.
[*22*] Gaufin, A. R. and Tarzwell, C. M., *Sewage and Industrial Wastes*, Vol. 28, 1956, pp. 906-924.
[*23*] Kapoor, N. N. and Griffiths, W., *Zoological Journal of the Linnean Society*, Vol. 59, 1976, pp. 209-215.

M. G. Tesmer[1] *and D. R. Wefring*[1]

Annual Macroinvertebrate Sampling—A Low-Cost Tool for Ecological Assessment of Effluent Impact

REFERENCE: Tesmer, M. G. and Wefring, D. R., **"Annual Macroinvertebrate Sampling—A Low-Cost Tool for Ecological Assessment of Effluent Impact,"** *Ecological Assessments of Effluent Impacts on Communities of Indigenous Aquatic Organisms, ASTM STP 730*, J. M. Bates and C. I. Weber, Eds., American Society for Testing and Materials, 1981, pp. 264–279.

ABSTRACT: Annual macroinvertebrate sampling is viewed as a low-cost tool for effluent impact assessment. The specific objectives for such studies are discussed, and the necessary restrictions required to provide meaningful results are presented. An example study is provided for demonstrating the data analysis and interpretation. The effluent discharge from the Champion International Corp. pulp and paper mill into the Tennessee River at Courtland, Ala., was monitored annually between 1970 and 1978. The study revealed that annual sampling effectively identified the causes of major fluctuations in the macroinvertebrate community structure. The effluent discharge studied was found to have a negligible effect upon the macroinvertebrate community, while the major fluctuations were attributed to natural variations in physical and hydrological conditions. The net impact of the effluent discharge upon the receiving stream was considered to be minimal over the entire study period.

KEY WORDS: macroinvertebrates, sampling, effluents, data processing, rivers, discharge, ecology, aquatic organisms

The vast majority of organisms inhabiting the aquatic environment have been examined for their ability to reflect water quality conditions. An excellent account of the early history of pollution biology is presented by Mackenthun and Ingram [*1*].[2] Most early investigators and many contemporary investigators have been concerned with finding the ultimate indicator or most realistic index of water pollution. Some investigators have turned

[1] Research fellow and research fellow, respectively, The Institute of Paper Chemistry, Appleton, Wis.

[2] The italic numbers in brackets refer to the list of references appended to this paper.

toward the concept of "saturation sampling" in an attempt to evaluate as many aspects of the aquatic ecosystem as possible.

Any monitoring program, no matter what degree of sampling intensity is applied, requires a set of specific objectives and guidelines. Mount [*2*] outlined several important objectives in biotic monitoring and bioassay. These were (1) the establishment of baseline conditions, (2) the measurement of incurred damage, and (3) the comparison of impacts from various polluters. In addition, a monitoring program should be continued at some interval to display possible temporal changes as well. This objective is exemplified by Patrick [*3*], who showed, after 25 years of monitoring a river system, that man's effects upon that system can be evaluated over time. All too often, a single monumental effort is applied with no provision for future checks.

While no organism or group of organisms has proven to be a universal indicator of all pollution, the alternate approach, that of complete ecosystem evaluation, is typically cost-restrictive. As with most problems, a compromise can be reached, provided that a defined set of restrictions is adhered to. First, the selection of a representative group of organisms can be made for monitoring purposes. Second, the selection should be based upon the particular type of effluent to be monitored and the type of adversity expected to develop. Third, the group of organisms selected should be easily sampled at a relatively low cost per sample. Fourth and finally, the conclusions drawn from the data should not exceed the obvious.

For the majority of wastes entering our waterways, the resident macroinvertebrate fauna have proven sensitive to the ensuing environmental insults. Macroinvertebrates meet the necessary restrictions and allow the attainment of the objectives just stated. Much of the early interest in the macroinvertebrate community stems from their relative ease of capture, their predominantly sessile nature, and the existence of a median annual life cycle. Today's interest in the macroinvertebrates also stems from these attributes plus one more very important aspect. Our increased knowledge of aquatic ecosystems over the recent past has demonstrated that, like any group of organisms, the macroinvertebrates are not independent. They are subject to external stresses, such as waste inputs applied to the ecosystem, as well as to internal stresses, such as competition with other community members. It is this ability of theirs to react and interact with the surrounding environment that allows us to use the macroinvertebrates as a valuable monitoring tool.

Study Design

The basic concept of annual macroinvertebrate sampling is that perturbations occurring within an aquatic ecosystem can be monitored and related to a causative agent. Significant variation in macroinvertebrate community structure, between polluted and unpolluted waters, results from the wide range in environmental requirements [*4*] under which the various species can exist. The predominantly sessile nature of macroinvertebrates, combined with their behavioral patterns under stressed conditions, allows the accumulation of information concerning the applied stress. When sampled on an annual basis, these major differences in macroinvertebrate community structure between polluted and unpolluted areas become visible. The macroinvertebrate community inhabiting a specific aquatic ecosystem is, therefore, a product of its environment. It can be used as such to evaluate the existing conditions, as well as to evaluate the degree of perturbation produced by a specific waste source. The annual macroinvertebrate sampling program is not a definitive assessment of the effluent impact. It cannot fully address all of the specific interactions that may occur between an effluent and the receiving environment. Specific problems, such as the accumulation of a pollutant (heavy metals, recalcitrant organics) or the tainting of fish flesh, are problems that demand special attention. It must be stressed that the annual macroinvertebrate sampling program is a tool designed to provide information concerning the net impact of an effluent source upon the general health and well-being of the biota of the receiving waters.

Annual macroinvertebrate sampling programs are typically designed to monitor the effects of a single point-source discharge or a group of discharges into a specific reach of receiving water. The degree of sophistication for any study is highly variable and usually dependent upon cost restrictions. A basic quantitative macroinvertebrate study for a single effluent source should range from $8000 to $15 000 (1978 dollars). The major cause for variability stems from the type of receiving water habitat (lentic versus lotic) and the boundaries of the zone of impact. A comprehensive impact assessment of this same effluent source, including (1) water chemistry profiling, (2) periphyton, zooplankton, and phytoplankton community dynamics, (3) basic fisheries inventory, and (4) the macroinvertebrates, would cost approximately $75 000 to $125 000. Annual macroinvertebrate sampling is not a replacement for such an in-depth impact assessment. However, under cost restrictions, the macroinvertebrate sampling program offers a higher return (in information obtained) per research dollar spent than equal expenditures on other community components, provided the macroinvertebrates are sensitive to the type of waste being discharged. Again, this is in light of the second restriction stated above.

The fundamental differences in ecological structure that exist between receiving waters are beyond the scope of this paper. Weber [*5*] provides a good introductory discussion on macroinvertebrate sampling strategy and field techniques. The following study is an example of a typical annual macroinvertebrate sampling program. It serves to highlight the basic steps involved in the examination and evaluation of such data.

The macroinvertebrate community inhabiting a portion of the Tennessee River in the vicinity of Courtland, Ala., has been studied annually since

1970. A map of the study area is presented in Fig. 1. The study was initiated for the purpose of evaluating the impact of a pulp and paper mill effluent upon representative fauna (the macroinvertebrates) of the indigenous biological community inhabiting the receiving waters. Champion Papers Division began operation of their Courtland facility in 1971. The mill produces bleached kraft pulp [6] for the on-site manufacture of fine papers. Waste treatment consists of primary clarification and secondary biological treatment via an aerated oxidation ponding system.

Data Evaluation

The first objective, that of establishing baseline conditions, is composed of two important factors. First, it is necessary to define the physical characteristics of the receiving waters and identify the sources of natural variation that may influence the macroinvertebrate data set. Second, it is desirable to gather macroinvertebrate information prior to any new effluent inputs. The remaining study objectives—incurred damage, comparability, and temporal changes—are intricately related. The macroinvertebrate community, when disturbed, will typically respond in one of three ways. If the damage to a receiving stream is acute, the classic zonation of degradation and recovery will be noted. If the damage is slight, trends in the annual data will be evident only after several years. The third, and most often neglected, response is one of no general impact. The natural ability of the receiving stream to accept the waste source has not been exceeded.

It is important that the third typical response be clearly understood. In

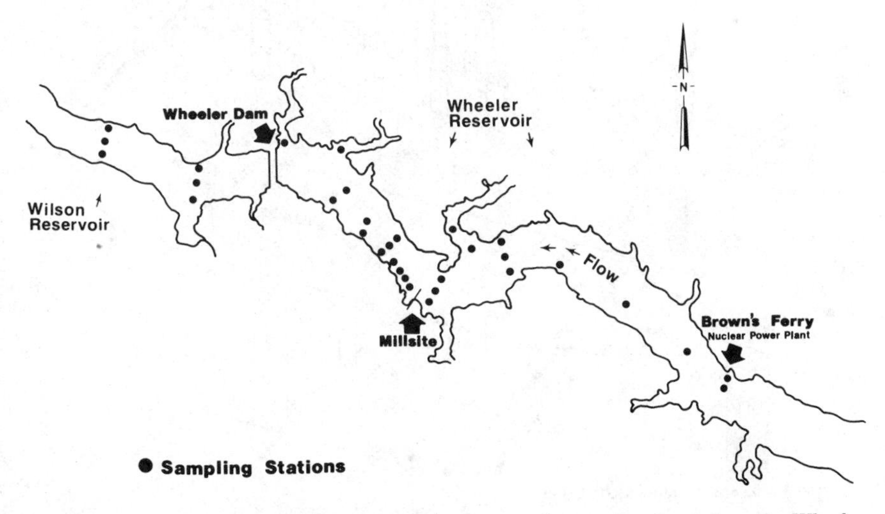

FIG. 1—*A map of the study area displaying the sampling stations located on the Wheeler and Wilson Reservoirs of the Tennessee River near Courtland, Ala. The solid circles indicate the sampling stations.*

any monitoring program the absence of any measurable response to an effluent input is not necessarily a failure of the applied monitoring system. When sufficient data have been gathered, the absence of a response can be real and indicate that there was no measurable effect upon the monitoring agent. Fortunately, the scientific literature supporting the use of macroinvertebrates for monitoring purposes is well documented [7-10]. The study reported here belongs to this third category of responses. Studies of this nature are by far the most difficult to assess. They often require additional effort to ensure confidence in the end result.

The Tennessee River study area consists of the lower portion of the Wheeler Reservoir and the upper portion of the Wilson Reservoir. The study area is predominantly lentic. Sediment cores were periodically collected to identify potential areas of differing habitat. Little annual fluctuation in sediment composition was noted. An isometric projection of the 1978 core data is presented in Fig. 2. The sediments for the most part are relatively uniform. They consist predominantly of soft clay, low in organic matter (1.0 to 10 percent by dry weight). Within the uppermost reaches of the study area, a higher percentage of sand has been noted. This particular area is subject to the transition from a more lotic environment to the lentic environment existing in the pool area. During periods of high flow, increased sand content has been observed further downstream as the dimensions of the transition zone increase toward the downstream side.

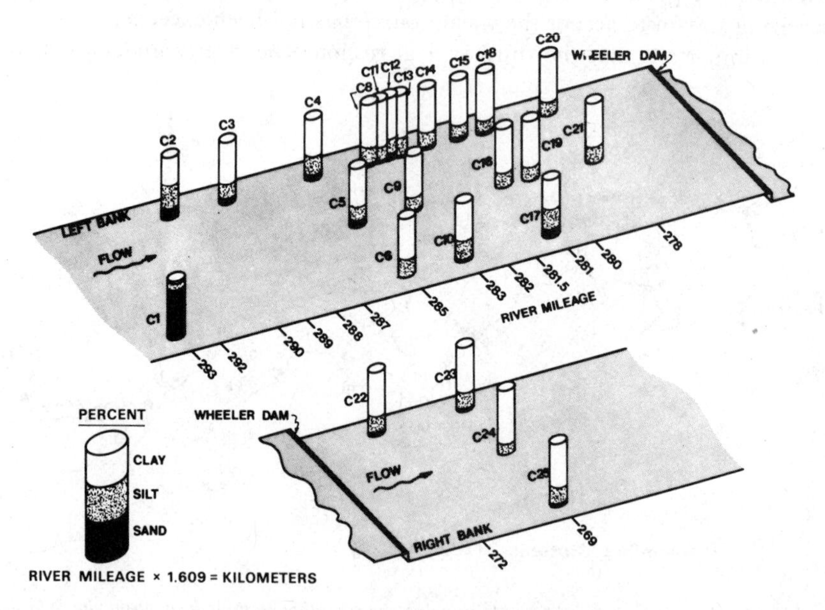

FIG. 2—*An isometric projection of the sediment fractions (percent sand, silt, and clay) for selected sediment samples collected during 1978 from the Tennessee River.*

The macroinvertebrate fauna inhabiting the study area were found to be primarily limnephilic forms. Oligochaeta, Diptera, and Pelecypoda were the most commonly abundant, with several additional orders also being represented. Table 1 contains a species list of the taxa found in the 1978 study. Both the density and diversity of the macroinvertebrate community were moderate, which is typical for such an aquatic habitat [*11*]. The community composition and structure reflected a normal state of environmental well-being.

A gradual change in the macroinvertebrate community structure and total mean density has, however, been observed over the 9-year study period (1970 through 1978). The macroinvertebrate density distribution among the major orders is presented in Figs. 3, 4, and 5 for the study years 1970, 1975, and 1978, respectively. These changes came about through fluctuations in the population densities of several important taxa. The primary cause for a gain in density was due to the significant increase in the Oligochaeta. *Limnodrilus hoffmeisteri,* the predominant oligochaete, increased in mean density by more than sevenfold between 1970 and 1978. The Diptera, Pelecypoda, and other orders demonstrated only minor changes over the study period, while the Ephemeroptera densities, primarily *Hexagenia* specie, fluctuated dramatically. The *Hexagenia* population was rela-

TABLE 1—*Representative list of all taxa collected by Ekman grab from the Wheeler and Wilson Reservoirs of the Tennessee River near Courtland, Ala., during June 1978.*

Turbellaria	Trichoptera
Family Planariidae	*Cyrnellus*
Oligochaeta	*Hydropsyche*
Arcteonais lomondi	*Decetis*
Aulodrilus pigueti	Diptera
Branchiura sowerbyi	*Ablabesmyia*
Dero	Family Ceratopogonidae
Family Enchytraeidae	*Chaoborus*
Ilyodrilus templetoni	*Chironomuc*
Limnodrilus hoffmeisteri	*Coelotanypus*
Nais bretscheri	*Cryptochironomus*
Piguetiella michiganensis	*Epoicocladius*
Stylaria fossularis	*Harnischia*
Unidentified immature Tubificidae with capilliforms	*Paracladopelma*
	Polypedilum
Unidentified immature Tubificidae without capilliforms	*Procladius*
	Tanytarsus
Hirudinea	Gastropoda
Helobdella	Family Planorbidae
Amphipoda	*Pleurocera*
Hyalella	*Viviparus*
Ephemeroptera	Pelecypoda
Hexagenia	*Corbicula*
Megaloptera	*Sphaerium*
Sialis	Unidentified specimen

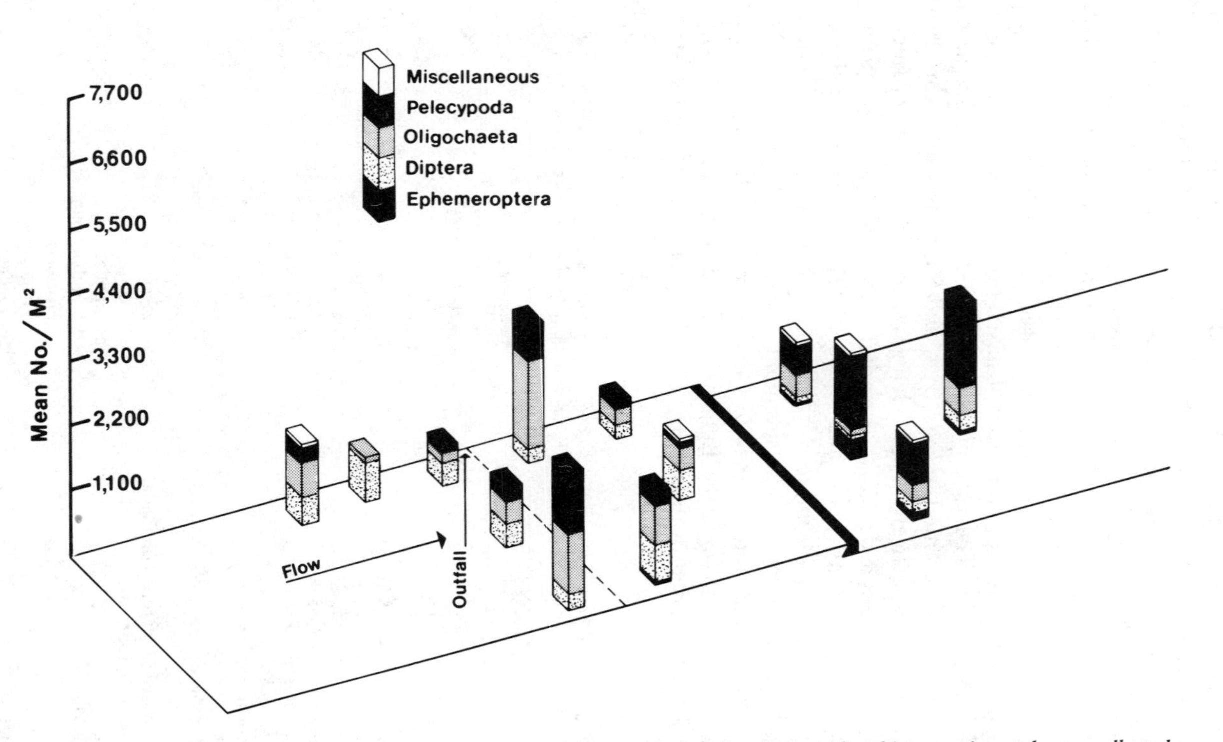

FIG. 3—*An isometric projection of the mean total densities and relative abundance of benthic macroinvertebrates collected by Ekman grab from the Tennessee River during 1970.*

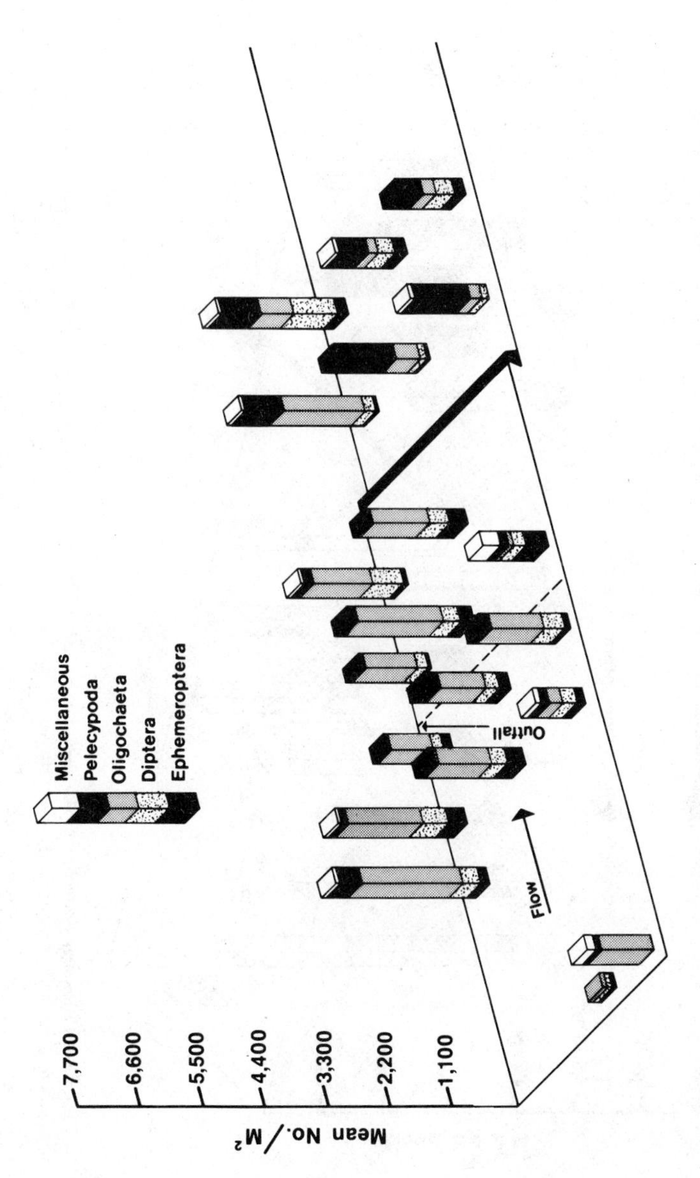

FIG. 4—*An isometric projection of the mean total densities and relative abundance of benthic macroinvertebrates collected by Ekman grab from the Tennessee River during 1975.*

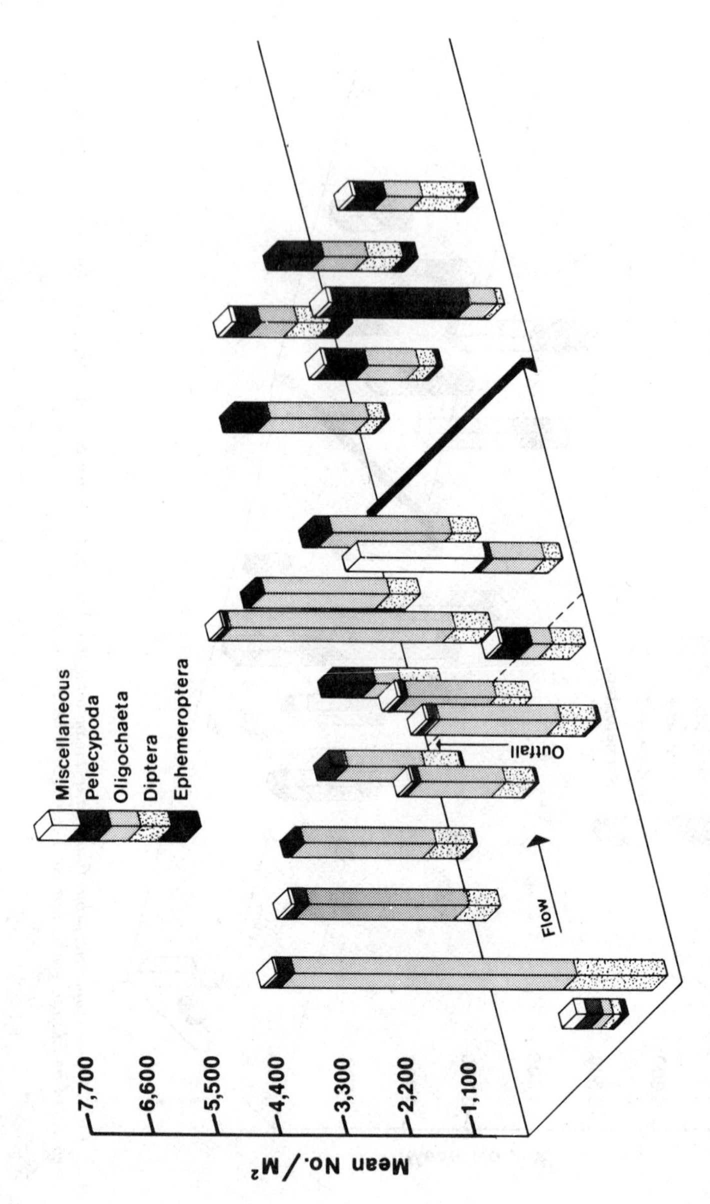

FIG. 5—*An isometric projection of the mean total densities and relative abundance of benthic macroinvertebrates collected by Ekman grab from the Tennessee River during 1978.*

tively low in 1970, progressively increased through 1975, and progressively declined through 1978.

The progressive increase in density, accompanied by fluctuations in the community composition, over the 9-year study period represented the most dramatic changes. Perhaps the most interesting fluctuations involved the Oligochaeta, especially *L. hoffmeisteri* and the mayfly *(Hexagenia)*. These two organisms represent nearly opposite abilities to withstand pollutional inputs. *Limnodrilus hoffmeisteri* has been typically associated with organic pollution and is relatively tolerant of a degraded aquatic environment [*4*]. On the other hand, *Hexagenia* is considered relatively intolerant of pollution and typically disappears from chronically degraded waters [*12*].

The increasing abundance of Oligochaeta in the vicinity of the mill discharge is best illustrated by the data for the years 1970, 1975, and 1978 in Fig. 6. The gradual increase in Oligochaeta took place predominantly between 1975 and 1978 and appeared to be quite similar both upstream and

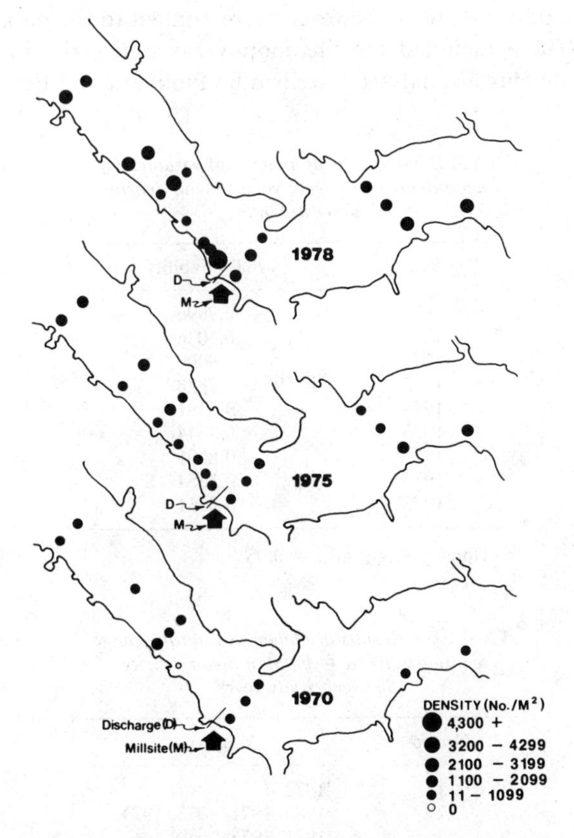

FIG. 6—*Distribution of oligochaete worm densities for three of the nine sampling years (1970, 1975, 1978) collected from selected stations on the Tennessee River.*

downstream from the outfall. An analysis of variance between the mean upstream oligochaete density and the mean downstream oligochaete density for each study year was performed. It was demonstrated that there was no significant difference ($\alpha = 0.05$) between the upstream and downstream oligochaete density in any of the nine study years (Table 2). Significant variation ($\alpha = 0.05$) in the mean oligochaete density was found between the various study years. Four significant year groups were revealed by the Duncan's multiple range test (Table 3).

The density of *Hexagenia* in the vicinity of the discharge for the selected years 1970, 1975, and 1978 is presented in Fig. 7. Again, little difference can be seen in the population distribution in relation to the discharge point. What is strongly evident, however, is the lack of *Hexagenia* in 1970 and 1978, while the 1975 densities were high. Such conditions suggest the possibility of a cyclic nature to the *Hexagenia* population.

Considerable evidence points toward the noninvolvement of the discharge studied in the fluctuations observed within the macroinvertebrate community. Two additional tests, however, were applied to the data for further verification. These included the Shannon-Weaver diversity index [5] and the multivariate cluster analysis described by Pinkham and Pearson [*13*]. A

TABLE 2—*Results of analysis of variance for mean density upstream versus mean density downstream.* [a]

Year	*F* Probability
1970	0.7996
1971	0.1036
1972	0.9009
1973	0.7826
1974	0.5391
1975	0.3844
1976	0.8601
1977	0.3554
1978	0.3251

[a] Nonsignificant at $\alpha = 0.05$

TABLE 3—*Results of Duncan's multiple-range comparisons ($\alpha = 0.05$) of mean density between study years.*

Group	Year(s)
1	1972
2	1970, 1971, 1973, 1974
3	1971, 1975
4	1976, 1977, 1978

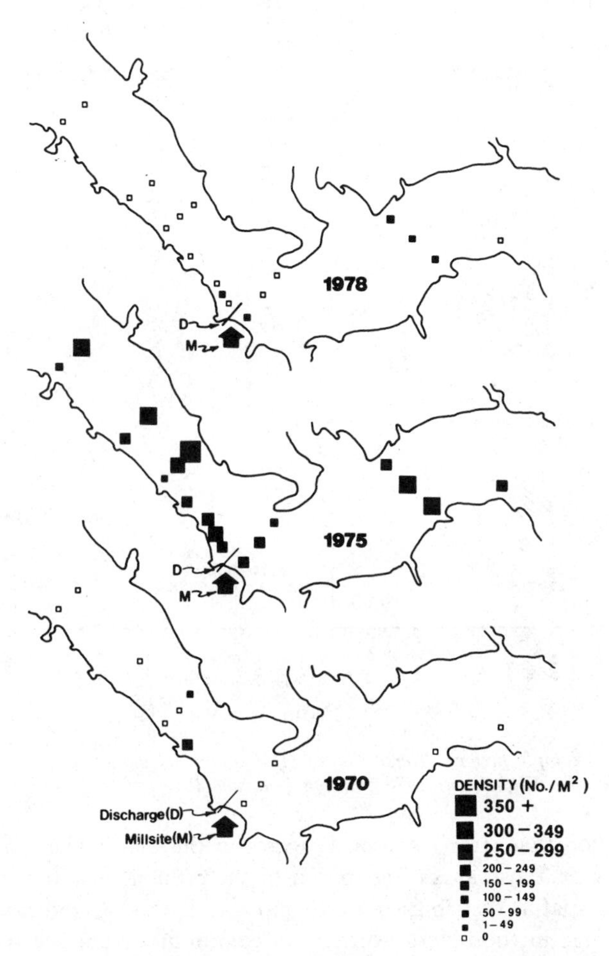

FIG. 7—*Distribution of* Hexagenia *densities for three of the nine sampling years (1970, 1975, 1978) collected from selected stations on the Tennessee River.*

representative plot of the diversity index values (1978 data) is presented in Fig. 8. As with the previous information, no trends in the diversity index values could be detected between upstream and downstream areas.

The dendrogram produced by the cluster analysis of the 1978 data presented in Fig. 9 was also typical of past study years. Major clustering of the sampling station data was completed at a similarity greater than 7.0. Such a response was expected because of the relatively similar habitat throughout the study area. One major clustering group (A) and two small clustering groups (B and C) are noted in Fig. 9. When the stations comprising each of these major groups are displayed spatially (Fig. 10), two interesting patterns result. First, the largest group (A) was composed of stations located both upstream and downstream from the outfall. This was, again, a

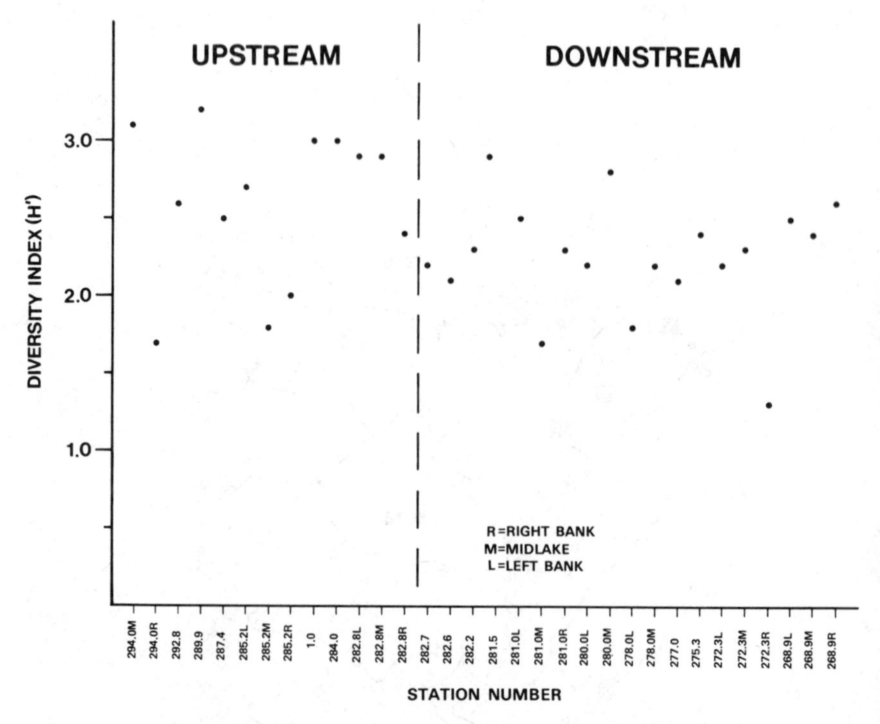

FIG. 8—*An example plot of diversity index (H') values calculated from benthic macroinvertebrate data collected during 1978 from the Tennessee River.*

firm indication that strong similarity exists in the macroinvertebrate community between those areas unexposed to the effluent and those exposed. Second, the spatial distribution of Groups A, B, and C indicated a progressive change in the macroinvertebrate community from the transitional area to the fully lentic pool area, a natural and expected phenomenon.

The influence upon the macroinvertebrate community from the studied discharge was negligible. The data for the 9-year study period indicated that significant changes had occurred in the macroinvertebrate community. However, these changes were primarily temporal, involving the entire study area, rather than spatially related to the discharge. When examined from an ecological standpoint, the observed shifts in population structures and total density suggest that the processes of cultural eutrophication were accelerating.

Conclusions

The macroinvertebrate community is an integral part of the biological resources comprising an aquatic ecosystem. As such, its composition and structure are the product of the prevailing environmental conditions and the perturbations applied to an aquatic ecosystem. These inherent princi-

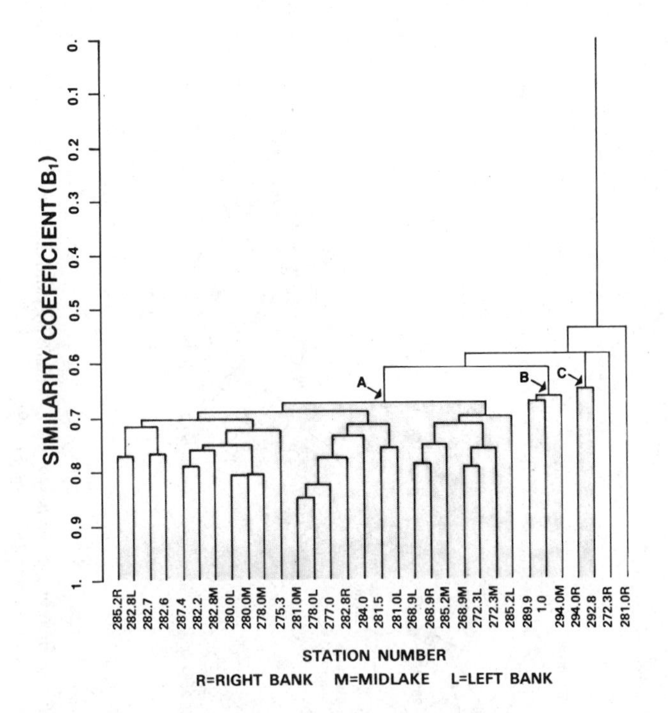

FIG. 9—*Dendrogram of the cluster analysis based on benthic macroinvertebrate data collected during 1978 from the Tennessee River. The shaded areas enclose major clusters forming above the 0.65 level of similarity.*

ples comprise the basis for using the macroinvertebrate community as a monitoring tool.

Certain restrictions must be adhered to if annual macroinvertebrate sampling is to be used successfully:

1. The naturally occurring fluctuations in the macroinvertebrate community under investigation must be identified.
2. The major environmental hazards associated with a pollutant input must be capable of influencing the macroinvertebrate community.
3. The conclusions drawn from the changes in the macroinvertebrate community can be associated with the general impact of the pollutant input but do not represent a definitive impact assessment.

Annual macroinvertebrate sampling is relatively low in cost. A greater amount of information is obtained per research dollar for macroinvertebrate sampling than for the study of other representative biological groups.

Acknowledgment

This research was supported by the Champion International Corporation and The Institute of Paper Chemistry.

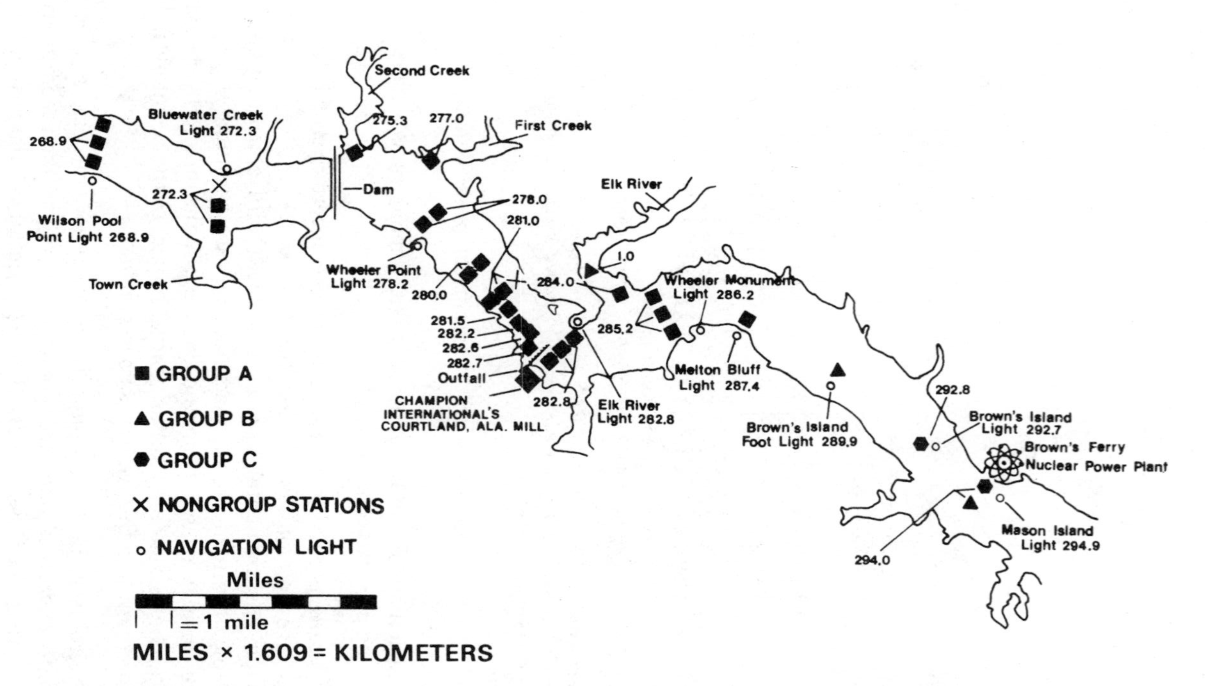

FIG. 10—*Spatial distribution of major biological similarity cluster analysis groups based on the results of comparisons made using benthic macroinvertebrate data collected from the Tennessee River during 1978 (see Fig. 9).*

References

[1] Mackenthun, K. M. and Ingram, W. M., *Pollution and the Life in Water.* Special Publication No. 4, Pymatuning Laboratory of Ecology, University of Pittsburgh, Pittsburgh, Pa., 1964, pp. 136-145.

[2] Mount, D. I. in *Biological Monitoring of Water and Effluent Quality. ASTM STP 607.* American Society for Testing and Materials, Philadelphia, 1977, pp. 20-23.

[3] Patrick, R. in *Biological Monitoring of Water and Effluent Quality. ASTM STP 607.* American Society for Testing and Materials, Philadelphia, 1977, pp. 157-189.

[4] Hart, C. W., Jr., and Fuller, S. L. H., Eds., *Pollution Ecology of Freshwater Invertebrates.* Academic Press, New York, 1974.

[5] Weber, C. I., Ed., *Biological Field and Laboratory Methods for Measuring the Quality of Surface Waters and Effluents.* EPA-670/4-73-001, U.S. Environmental Protection Agency, 1973.

[6] Rydholm, S. A., *Pulping Processes.* Wiley, New York, 1965.

[7] Gaufin, A. R. and Tarzwell, C. M., *Sewage and Industrial Wastes.* Vol. 28, No. 7, pp. 906-924.

[8] Macan, T. T., *Freshwater Ecology.* Camelot, London, 1963.

[9] Mackenthun, K. M., *The Practice of Water Pollution Biology.* U.S. Department of the Interior, Washington, D.C., 1969.

[10] Mason, W. T., Jr., Lewis, P. A., and Anderson, J. B., *Macroinvertebrate Collections and Water Quality Monitoring in the Ohio River Basin 1963-1967.* Cooperative Report, Office of Technical Programs, Ohio Basin Region and Analytical Quality Control Laboratory, WQO, U.S. Environmental Protection Agency, 1971.

[11] Hynes, H. B. N., *The Ecology of Running Waters.* Liverpool University Press, Liverpool, England, 1970.

[12] Mackenthun, K. M. and Ingram, W. M., *Biological Associated Problems in Freshwater Environments.* U.S. Department of the Interior, Washington, D.C., 1967.

[13] Pinkham, C. F. A. and Pearson, J. G., "A New Measure of Biotic Similarity Between Samples and Its Applications with a Cluster Analysis Program," Edgewood Arsenal Technical Report EB-TR-74062, Department of the Army, Washington, D.C., 1974.

R. B. Foster[1]

Use of Asiatic Clam Larvae in Aquatic Hazard Evaluations

REFERENCE: Foster, R. B., **"Use of Asiatic Clam Larvae in Aquatic Hazard Evaluations,"** *Ecological Assessments of Effluent Impacts on Communities of Indigenous Aquatic Organisms, ASTM STP 730,* J. M. Bates and C. I. Weber, Eds., American Society for Testing and Materials, 1981, pp. 280-288.

ABSTRACT: Investigative efforts to predict the potential impact of single chemical species or mixed effluents on natural waterways emphasize laboratory studies of sensitive nonindigenous aquatic organisms reared under controlled laboratory conditions. Integrated aquatic hazard evaluations appropriately require species representative of a number of trophic levels—algae, invertebrates, and fish—and of different ecological niches—benthic versus pelagic, sessile versus mobile. The oyster embryo toxicity test is often applied in hazard evaluations and is a useful measure of water quality in marine and estuarine environments. Parallel methods have been developed in which freshwater Asiatic clam larvae are used to measure the relative toxicity of industrial chemicals. Two applications, the benthic acute lethality test and the larval transformation toxicity test, are presented. The results compared with other representative aquatic species of fish and invertebrates show that these applications are appropriately sensitive to industrial chemicals. Based on the simplicity and utility of these methods, the author recommends that these and other Asiatic clam monitoring techniques be adopted as tools for assessing the impact of effluents on freshwater aquatic environments.

KEY WORDS: water quality, mollusks, Asiatic clams, toxicity, ecology, effluents, aquatic organisms

The water quality criteria proposed by the U.S. Environmental Protection Agency (EPA) [*1*][2] and mandated by Section 304 of the 1977 Clean Water Act are in part based upon the response of sensitive aquatic organisms to toxic pollutants following acute and chronic exposures in standardized laboratory toxicity tests. From the integration of many of these standardized tests into an overall scheme for evaluating the potential of a chemical to cause adverse effects in aquatic environments has evolved the discipline of aquatic hazard assessment [*2,3*].

[1]Director of Marketing, E G & G Bionomics, Wareham, Mass. 02571.
[2]The italic numbers in brackets refer to the list of references appended to this paper.

Commensurate with the development of predictive toxicology laboratory tests is the need to develop and apply field studies that are suitable for monitoring the biological integrity of surface waters. Appropriate techniques have been applied by Dickson et al [4] and Foster and Bates [5] with the use of caged animals within the receiving water body. The use of aquatic organisms caged *in situ* for monitoring receiving water quality and effluent effects has accurately reflected the influence of certain pollutants, particularly those with a tendency to bioconcentrate, and has become fairly widespread in marine environments [6].

Certain benefits are derived by selecting a single representative species to function in both aquatic hazard assessments and in-stream biological monitoring programs. By selecting a species which, due to habitat or behavior, is more indicative of bioaccumulative substances in the receiving water and simultaneously acutely sensitive to a variety of toxic substances, an appropriate reduction in testing costs can be achieved. Freshwater mussels, particularly the Asiatic clam, *Corbicula fluminea,* are suitable test organisms for either purpose, and the added advantage of the ease with which they are maintained in the laboratory makes them model test organisms.

Most freshwater hazard assessment methods have parallels in saline environments, for example, the freshwater fathead minnow (*Pimephales promelas*) and saltwater sheepshead minnow (*Cyprinodon variegatus*) early life stage toxicity tests, but the oyster embryo toxicity test developed by Davis and Hidu [7] and perfected by Woelke [8] has no counterpart. This is due to the lack of appropriate culture techniques for adult mussel culture and reproductive studies in the laboratory, which would ensure an adequate supply of test animals. Adult *Corbicula* are, however, easily maintained under laboratory conditions, and the larval forms spawned are useful in toxicity tests. The purpose of this report is to discuss the methods used for rearing and testing larval *Corbicula* and to present results of comparative toxicant exposures with other representative aquatic fauna.

Life-Cycle Considerations

Unlike freshwater unionid mussels, which propagate by means of a parasitic glochidial larva and, with few exceptions, are heterosexual, the Asiatic clam spawns a free-living larva and is monoecious [9]. The precocious adults are naturally gravid from April through October and can be induced to spawn in the laboratory by a variety of means. The simplest of these is to excise the inner pair of marsupial gills from a reproductively mature adult prior to the natural spawn. The fertilized embryos develop within the adult over a period of approximately 3 weeks before they are expelled into the water column. After their deposition in the gill water tubes, the embryos can be "prematurely aborted" at any time during marsupial development. Premature abortion consists of removing the inner gills and gently teasing the embryos out of the water tubes into culture water. Employing this technique can result in the availability of several thousand organisms for toxicity tests from a single adult.

Procedure

Adult *Corbicula* collected from natural populations and maintained under laboratory conditions retain their normal cycle of reproduction—peak spring spawns followed by declining numbers of larvae until fall—for periods of up to 2 years. Longer periods may be possible but have not been investigated. By taking advantage of the mollusk's natural cycle, substantial numbers of embryo and larval forms can be isolated from reproductively mature adults during a 7-month period, April through October. Alternatively, adult mollusks possessing recrudescent gonads can be collected from field populations several weeks before their use in toxicity tests and conditioned to spawn.

The immature forms of *Corbicula* are removed from the marsupium in four stages of development—fertilized embryo, embryo-trochophore, veliger, and benthic larva. All forms can be cultured in the laboratory for up to 1 month, provided that the culture water is periodically renewed and an adequate source of food (*Scenedesmus* and trout chow at 1 mg/ml is acceptable) is made available. The morphology of individual larvae within the marsupial gills is not synchronous; all forms can be present simultaneously, but, within a 48 to 72-h period following their removal from the gills, most embryos will have transformed into the veliger and benthic larval forms.

Figures 1 and 2 depict the early stages of bivalve embryo development, the embryo and trochophore stages, in which the body organs are visible. Ideally, the time of transformation from embryonic trochophore to larval veliger (Fig. 3) is the time during which exposure to the test substance of interest is initiated. This transformation occurs unaided in culture water even after the embryos have been prematurely aborted. Mortality (acute lethality) tests are conducted on the shelled veliger and benthic larval forms (Fig. 4), because the end point of lethality is more easily distinguished.

Larval Culture

The following techniques were employed over a 7-month period in conducting acute lethality tests on a variety of pesticidal compounds. The marsupial gills of several gravid *Corbicula* were removed and placed in reconstituted or natural dilution water of a quality similar to that of the collection site (Delaware River). Water temperatures of 20°C proved suitable for culture and test purposes, but, in order to delay development and transformation, embryos were occasionally cultured at 4°C with no lethal effects being observed. Glass or plastic petri dishes were used as culture chambers

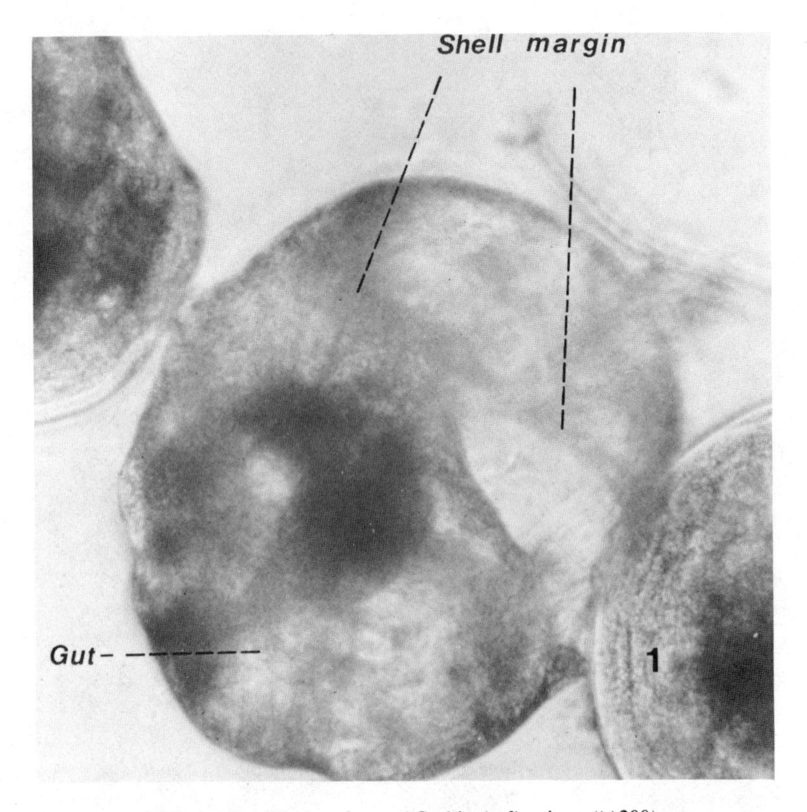

FIG. 1—*Fertilized embryo of* Corbicula fluminea *(×200).*

and operating chambers in removing the embryos and larvae from the excised gills. A small probe was used to push the organisms out of the water tubes and into the culture (dilution) water. Preferably, the organisms should be removed from the open, excised base of the gill and not the outer edge, which is intact.

Following their removal, the organisms are pooled in approximately 25 ml of dilution water in a shallow circular container and gently swirled in order to separate the heavier veliger and benthic larval forms from the embryo-trochophores. The heavier, shelled larvae will congregate in the center of the container and can be transferred to another container for acclimation to test conditions. The remaining embryo-trochophores require a number of simple transfer steps to clean the dilution water in order to remove gill fragments and other debris associated with the removal operation. Subsequent to these transfers, two distinct populations of test organisms, the veliger/benthic and embryo-trochophore populations, are available for acute lethality and larval transformation tests, respectively.

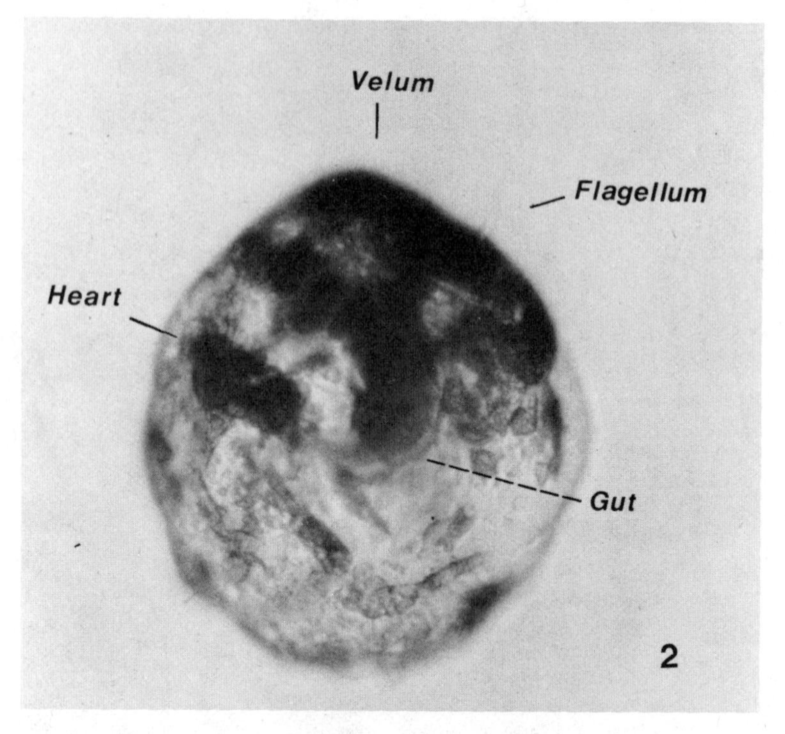

FIG. 2—*Embryo-trochophore of* Corbicula fluminea *(×200).*

Acute Lethality Tests

Although the larval transformation test shows a degree of promise for evaluating the potential effects of toxic substances in a program of hazard evaluation, this test was not extensively evaluated. In a test of acrolein, the embryo-trochophore forms were more sensitive than the veliger/benthic larva by a factor of approximately 0.25 (24-h EC_{50}, 70 and 300 μg/litre, respectively), but a comparative data base was not established. Instead, research efforts focused on determining the relative sensitivity of the veliger/benthic larva of *Corbicula* in acute lethality tests and comparing these results with those from studies of the more common laboratory test species.

For acute lethality tests, larvae were removed from two adults, pooled, and isolated, as previously described, in toxicity test dilution water. For each concentration of the toxicant to be tested and the controls, two 100-ml glass containers were used as the exposure chambers. Following the preparation and distribution of toxicant concentrations into each exposure chamber, between 60 and 100 larvae were transferred to each chamber and exposed for a period of 24 h. At the completion of this exposure, the surviving larvae in the controls and in all the test concentrations were enumerated. Larval mortality was

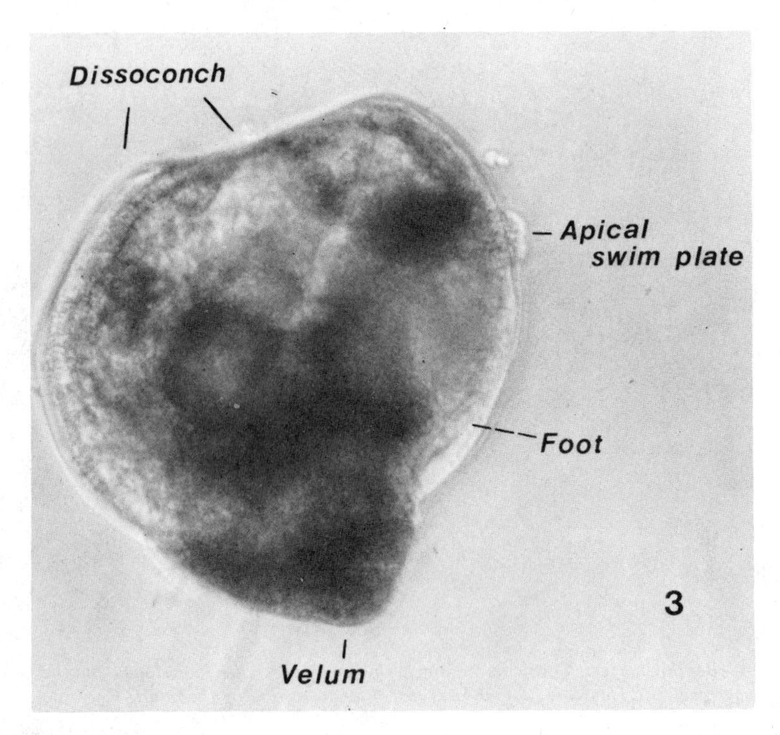

FIG. 3—*Veliger larva of* Corbicula fluminea: *the foot is partially developed and the receding velum is encircled by shell margins (×200).*

based on the number of individuals surviving in relation to the mean number originally introduced and was used to calculate the mortality in the control and treatment populations according to the formula [*10*]

$$\text{Percent larval mortality} = 1 - \frac{\text{number of surviving larvae}}{\text{number of larvae introduced}} \times 100$$

Mortality in the control populations did not exceed 2 percent in any of the acute lethality tests; therefore, no corrections were necessary to adjust the observed results.

Results

The results of seven acute lethality tests with *Corbicula* larvae are shown in Table 1. Comparative results with other freshwater fish and invertebrate species are also shown. These tests were performed in accordance with the "Methods for Acute Toxicity Tests with Fish, Macroinvertebrates, and Amphibians" [*11*].

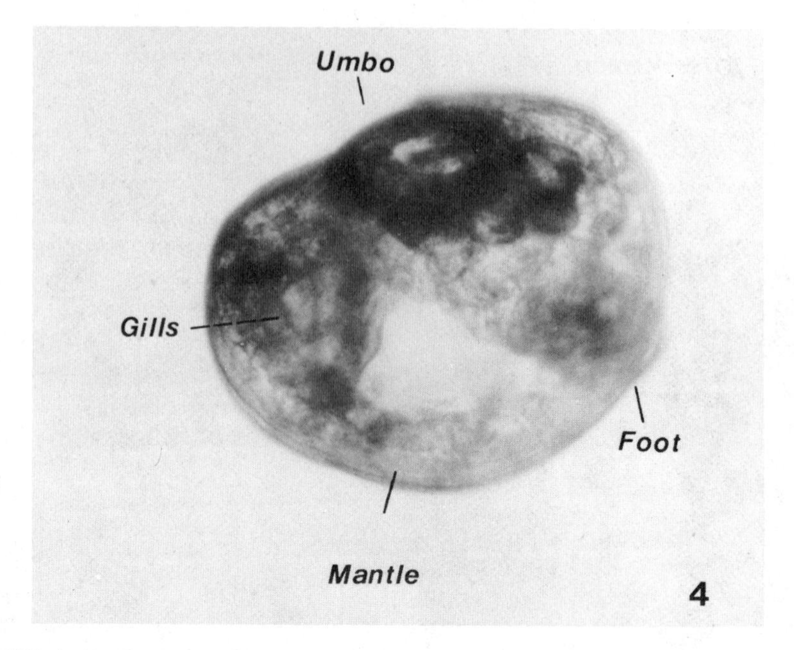

FIG. 4—*Benthic larva of* Corbicula fluminea: *the foot is completely developed and the velum is no longer present. This is the larval form just prior to release by the adult.*

TABLE 1—*Results of acute lethality tests with* Corbicula *larvae and other common test species.*

	Concentration, mg/litre nominal			
Test substance	*Corbicula* 24-h EC_{50}	Rainbow trout 96-h LC_{50}	Bluegill sunfish 96-h LC_{50}	*Daphnia magna* 48-h EC_{50}
Acrolein	0.3	0.08	0.07	0.03
Giv Guard BNS	2.8	0.11	0.11	...
Dichloroisocyanurate, sodium salt	0.6	0.29	...	0.15
Cytox 2410	1.2	0.57	...	...
Dow EC-7	1.6	0.20	0.86	0.08
Tributyltin oxide	2.1	...	0.24	0.07

The most sensitive of the test animals compared is *Daphnia magna,* a freshwater cladoceran, and the least sensitive, the *Corbicula* larva. In only two test cases, however, the tests with Giv Guard β-bromo-β-nitrostyrene (BNS) and tributyltin oxide, are these differences an order of magnitude apart. The results of the tributyltin oxide exposure are rather surprising in themselves in that this compound is frequently used as a component of antifouling paints and is a recognized molluscicide. Undoubtedly, portions of the differences in observed sensitivity can be resolved by conducting the *Corbicula* tests under exposure periods equal to those used for the *D. magna*, that is, 48-h acute lethality tests, but it is unlikely even under these conditions that the *Corbicula* larvae would be equally sensitive. The sensitivity of the larval transformation test ($\times 0.25$) in relation to the acute lethality test for acrolein exposures is, however, indicative of a level of response equal to that of the more common test species and suggests that the transformation period is a more ideal exposure time for measuring the effects of toxic substances than any other.

Summary and Conclusions

Acute lethality tests with a wide variety of representative aquatic fauna are necessary to characterize the acute hazards of potential environmental contaminants. The use of *Corbicula* larvae in these tests provides an alternative test species that is not commonly tested and is representative of a habitat usually ignored in hazard evaluations, the benthos. Although not as sensitive as *D. magna*, the simplicity of adult conditioning, larval culture, and testing—and the fact that biological problems such as age, size, sex, food, and precondition of test animals, which tend to confound the results of acute lethality tests with other organisms, are not present because all larvae have the same parents and are exactly the same age and size [*7*]—indicates that this test has utility in aquatic hazard evaluations. The application of this test to studies of the potential environmental effects of industrial effluents is a logical extension of this methodology and can be accomplished, on site, by modification of the exposure apparatus to accommodate the small-sized (250 μm) larvae.

Adult *Corbicula* are not conducive to acute, short-term evaluations of single compounds or mixed effluents because of their response behavior to toxic conditions. Use of the adult forms can, however, be advantageous in both field [*12*] and laboratory [*13*] programs intended to assess the propensity of effluent constituents to bioconcentrate in animal tissue. *Corbicula* is an important source of food to freshwater fish (for example, bluegills and catfish) and is similar to oysters (*Crassostrea*, used in saltwater bioconcentration tests) in that it is a filter feeder and amenable to effluent monitoring studies in which the organisms are caged *in situ* in the receiving water. Although the ability of this mollusk to bioconcentrate chemical residues has not been fully evaluated, there are preliminary indications that this species is as suitable for freshwater hazard evaluations as the oyster has become in marine environments. The ubiquitous nature of *Corbicula* in fresh water, the simplicity of larval toxicity tests, and the potential of adult bioconcentration studies suggest that these applications are suitable parallels for the oyster embryo toxicity and bioconcentration tests and should be given careful consideration for evaluating the impacts of industrial effluents on communities of indigenous freshwater aquatic organisms.

Acknowledgments

The author wishes to express his gratitude to Betz Laboratories, Inc., Trevose, Pa., for sponsoring this research and to Stephen Box for his support during its conduct.

References

[*1*] *Water Quality Criteria*, Office of Water and Hazardous Materials, U.S. Environmental Protection Agency, Washington, D.C., 1976.

[*2*] Duthie, J. R. in *Aquatic Toxicology and Hazard Evaluation, ASTM STP 634*, American Society for Testing and Materials, Philadelphia, 1977, pp. 17-35.

[*3*] Macek, K. J. and Sleight, B. H. III in *Aquatic Toxicology and Hazard Evaluation, ASTM STP 634*, American Society for Testing and Materials, Philadelphia, 1977, pp. 137-146.

[*4*] Dickson, K. L., Hendricks, A. C., Crossman, J. S., and Cairns, J. C., *Environmental Science and Technology*, Vol. 8, 1974, p. 845.

[*5*] Foster, R. B. and Bates, J. M., *Environmental Science and Technology*, Vol. 12, No. 8, 1978, p. 958.

[*6*] Arimoto, R. and Feng, S. Y., "Changes in the Levels of PCBs in *Mytilus edulis* Associated with Dredge Spoil Disposal," DAMOS Contribution No. 10, Marine Sciences Institute, University of Connecticut, Groton, Conn. 06340.

[*7*] Davis, H. C. and Hidu, H., *Bureau of Comm. Fish. Laboratory, Fisheries Bulletin*, Vol. 67, No. 2, 1969, pp. 393-404.

[*8*] Woelke, C. E. in *Water Quality Criteria, ASTM STP 416*, American Society for Testing and Materials, Philadelphia, 1967, pp. 112-120.

[*9*] Sinclair, R. M. and Isom, B. G., "Further Studies on the Introduced Asiatic Clam *Corbicula* in Tennessee," Tennessee Stream Pollution Control Board, Nashville, Tenn., 1963.

[*10*] "ASTM Standard Practice for Conducting Acute Toxicity Tests with Larvae of Four Species of Bivalve Mollusk (E 724-80)." American Society for Testing and Materials, Philadelphia, 1980.

[*11*] "Methods for Acute Toxicity Tests with Fish Macroinvertebrates and Amphibians," *Ecological Research Series*, EPA 600-13/75/009, U.S. Environmental Protection Agency, Washington, D.C., 1975.

[*12*] Shuba, D. J., Tatem, H. E., and Carroll, J. H., Technical Report D-78-50, U.S. Army Engineers Waterways Experimental Station, Vicksburg, Miss., 1978

[*13*] Sanborn, J. R. and Yu, C. C., *Bulletin of Environmental Contamination and Toxicology*, Vol. 10, No. 6, 1973, pp. 340-347.

I. W. Duedall,[1] *F. J. Roethel,*[1] *J. D. Seligman,*[1] *S. A. Oakley,*[1] *J. H. Parker,*[1] *H. B. O'Connors,*[1] *P. M. J. Woodhead,*[1] *B. K. Roberts,*[2] *H. Mullen,*[2] *and B. Chezar*[3]

Physical and Chemical Behavior and Environmental Acceptability of Stabilized Scrubber Sludge and Fly Ash in Seawater

REFERENCE: Duedall, I. W., Roethel, F. J., Seligman, J. D., Oakley, S. A., Parker, J. H., O'Connors, H. B., Woodhead, P. M. J., Roberts, B. K., Mullen, H., and Chezar, B., **"Physical and Chemical Behavior and Environmental Acceptability of Stabilized Scrubber Sludge and Fly Ash in Seawater,"** *Ecological Assessments of Effluent Impacts on Communities of Indigenous Aquatic Organisms, ASTM STP 730,* J. M. Bates and C. I. Weber, Eds., American Society for Testing and Materials, 1981, pp. 289-306.

ABSTRACT: Flue gas desulfurization sludges and fly ash pose a major disposal problem for coal-burning power plants. Disposal of stabilized, blocklike forms of these wastes in the ocean in the form of a reef is an alternative being studied at the Marine Sciences Research Center, Stony Brook, N.Y.

In laboratory experiments, physical integrity studies and components species leaching rates indicate that, in general, stabilization was found to maintain the strength of the blocks and minimize trace metal leachings in aerobic seawater over prolonged exposure.

In field studies, which are in progress, nine blocks (0.028 m^3) of stabilized scrubber sludge and fly ash were placed in approximately 6 m of water in Conscience Bay near Long Island Sound. Within 3 weeks after placement of the coal waste reef and the control structure, a heavy crop of attached plants and animals were observed growing on both reefs. The organisms attached to the coal waste reef showed no elevation in heavy metals in relation to the heavy metal composition of organisms attached to the concrete control.

KEY WORDS: ecology, effluents, aquatic organisms, coal, fly ash, energy, coal-fired power plant, ocean disposal, heavy metals leaching, quicklime, fishing reefs, reefs, ettr-

[1]Associate professor, doctoral candidate, analytical chemist, analytical chemist, staff oceanographer, assistant professor, and professor, respectively, Marine Sciences Research Center, State University of New York, Stony Brook, N.Y. 11794.

[2]Supervisor of customer services and director of government/industry relations, respectively, IU Conversion Systems, Inc., Horsham, Pa. 19044.

[3]Research and development engineer, Power Authority of the State of New York, New York, N.Y. 10019.

ingite, stabilization of energy wastes, desulfurization sludges, calcium sulfate/sulfite sludges, sulfate/sulfite sludges

Definition of the Problem

The present national energy policy encourages rapid increase in the production and utilization of coal. To achieve this policy, however, it is necessary for new coal-burning power plants to be fitted with flue-gas desulfurization scrubbers to remove harmful sulfur oxides in order that they comply with the standards of the Clean Air Act.

According to the U.S. Environmental Protection Agency, the use of flue-gas scrubbers is a reliable technology [*1*][4]. The combustion of coal in furnaces fitted with scrubbers, however, results in the production of very large amounts of scrubber sludge, principally calcium sulfate and calcium sulfite. Fly ash, another waste product of coal combustion, is also produced in large volumes. The ratios of scrubber sludge to fly ash produced vary with the type of coal burned and the cleanup equipment used.

In areas with minimal land available for disposal, a major unsolved problem in the widespread use of scrubbers is the disposal of the calcium sulfate–sulfite sludge and the fly ash. This problem must be overcome in order to develop the full potential for coal combustion using flue-gas scrubbers.

Several methods for the disposal of these coal combustion wastes have been suggested and discussed (for example, Lunt et al [*2*]). They include disposal in mine shafts, landfill applications, and ocean disposal. The present report considers ocean disposal of solid blocks formed by stabilizing scrubber sludge and fly ash.

Ocean Disposal of Stabilized Scrubber Sludge and Fly Ash

Lunt et al [*2*] suggest that the disposal of untreated sludges (either as liquid slurries or as dewatered sludges) on the seabed is probably environmentally unacceptable. One of the major potential problems is the depletion of dissolved oxygen in the water column and the toxicity of sulfite to some organisms. Other potentially serious problems include increased turbidity in the water column, smothering of benthic communities on the seabed, and the release of trace metals that might be harmful to organisms. Lunt et al [*2*] do suggest, however, that the disposal of stabilized scrubber sludge in solid, bricklike forms in the ocean appears promising.

For two years, we have been investigating the composition, physical and chemical behavior, and biological acceptability of blocks of stabilized scrubber sludge and fly ash (SSFA) in seawater, as well as their acceptability as

[4]The italic numbers in brackets refer to the list of references appended to this paper.

bases for biological colonization in the sea. The blocks of SSFA were supplied by IU Conversion Systems (IUCS), Inc. The investigation of the blocks included laboratory testing and experiments in the marine environment at a tidal site in Conscience Bay, Long Island Sound. Specifically, we sought to measure:

1. The composition of SSFA, with respect to selected major and minor components, and its physical properties (density, porosity, permeability).
2. Laboratory studies of the leaching behavior of the material in seawater.
3. The compressive strength of SSFA exposed to seawater in the laboratory and in Conscience Bay.
4. The biological colonization upon blocks 0.028 m^3 of SSFA, compared with colonization upon control blocks of concrete placed in an estuarine environment.

The aim of the studies reported here was to make preliminary measurements of some aspects of the physical, chemical, and biological behavior and characteristics of SSFA in seawater. These first findings were used in the design of a larger comprehensive experimental and demonstration program to determine the environmental acceptability of blocks of SSFA for marine disposal and particularly to explore the possibilities of using such blocks for the construction of artificial fishing reefs in the ocean (see Addendum).

This latter project is now underway at the Marine Sciences Research Center. Some of the results in the present paper are given in greater detail in available reports [*3-6*].

Materials and Methods

Test Samples

The test sample blocks of stabilized scrubber sludge and fly ash used in our work had a fly ash to sludge filter cake ratio of about 5:1 (by weight, based on 100 percent solids). The basic SSFA mix design (Poz-O-Tec, IUCS, Inc.) was 16 percent scrubber sludge, 80 percent fly ash, 4 percent quicklime. Depending on the source of the scrubber sludge, four variations (Mixes 4, 5, 6, and 7) of the basic mix design were tested. In Mixes 4 and 5, the scrubber sludge came from the Conevilles power plant at Columbus and Southern Ohio Electric; in Mixes 6 and 7, the scrubber sludge was from the Elrama power plant at Duquesne Light. The fly ash used in all the mixes came from the Elrama power plant. In Mix 6, 20 percent of the fly ash was substituted by bottom ash.

In the laboratory work, which was started before the field studies began, the test samples were in the form of proctors 7.6 cm by 7.6 cm [*3*].

In the field work, which was started in May 1977, nine blocks of Mix 7, each 0.028 m^3 in volume, were placed in Conscience Bay, Long Island

Sound, New York (Fig. 1). The blocks were placed about 6 m below the water surface. The blocks were arranged to expose surfaces to seawater and at the same time to provide crevices that might diversify niches available for the biological colonization (Fig. 2). A duplicate set of concrete blocks was placed in this arrangement nearby. In addition, smaller blocks of SSFA and concrete were placed adjacent to the large blocks. The Conscience Bay site was periodically visited by scuba divers to make photographic surveys, to remove

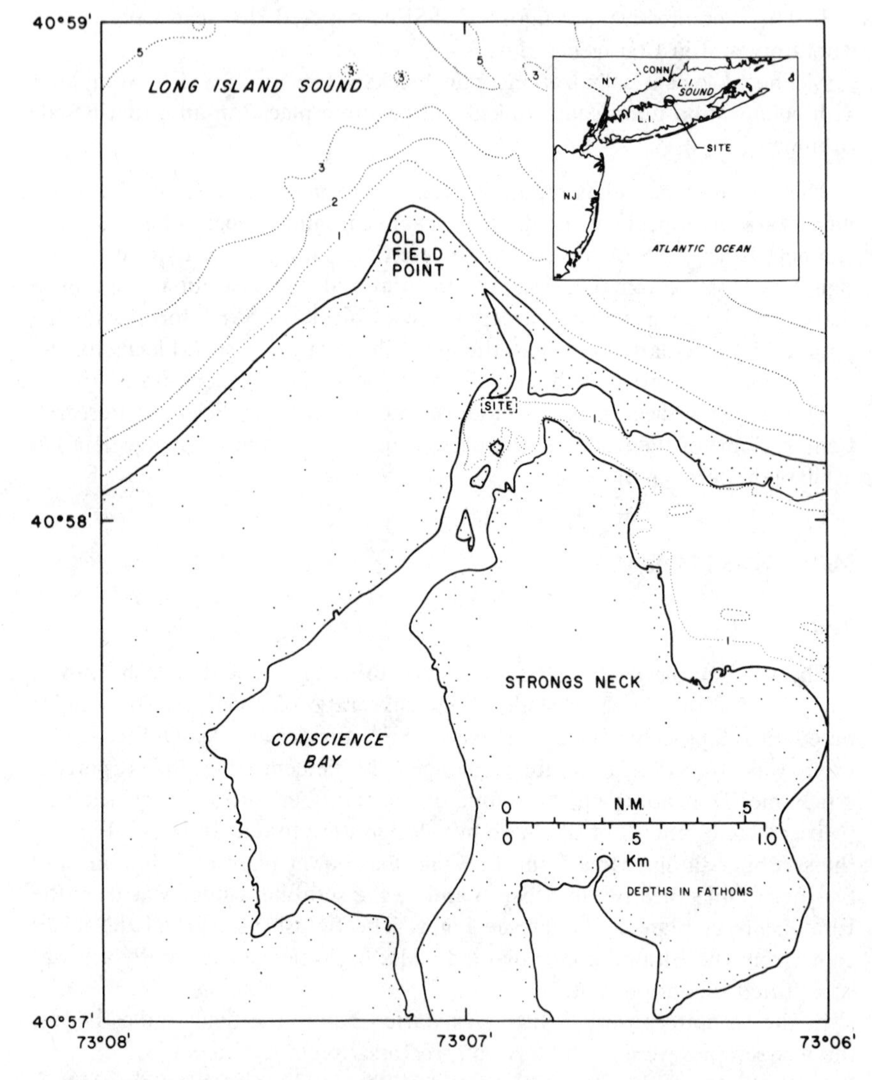

FIG. 1—*Conscience Bay Test Site: latitude 40°58'11"N, longitude 73°06'5"W, adjacent to Port Jefferson Harbor on Long Island Sound. Sounding in feet at mean low water (MLW).*

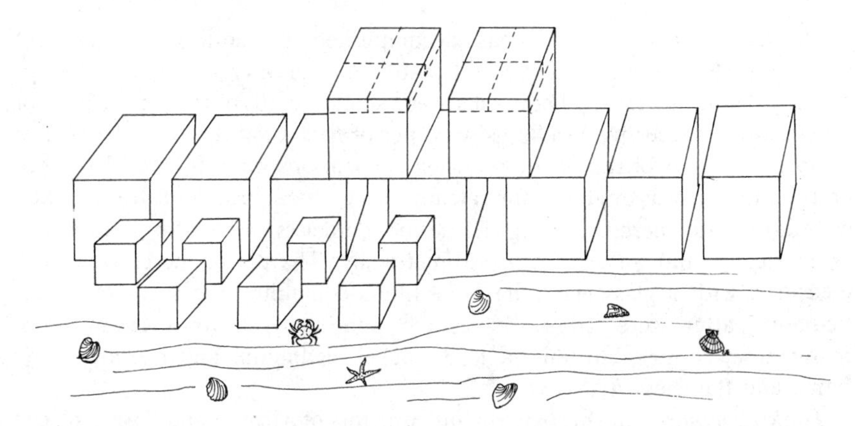

FIG. 2—*Arrangement of SSFA and concrete blocks submerged in Conscience Bay. Small blocks are used for compressive measurements.*

small test blocks for compressive strength testing, and to remove test slabs for species identification and heavy metals analysis of colonizing organisms.

For purposes of comparison of analytical techniques and interpretations, samples of construction-grade concrete blocks (0.028 m^3) and proctors (7.6 cm by 7.6 cm) were prepared using methods given by the American Society for Testing and Materials [*7*].

Analytical Methods

The analytical and physical methods used to characterize the test material are described in detail elsewhere [*3-6,8*]. Briefly, they were as follows:

Bulk Composition—Flame and flameless atomic absorption spectrophotometry (AAS) for calcium and the trace metals iron, manganese, zinc, copper, chromium, lead, nickel, cadmium, and mercury; carbonate by gas buret; total carbon by carbon-hydrogen-nitrogen analyzer; organic carbon by difference; total sulfate by gravimetric analysis of barium sulfate; and sulfite from sulfate-sulfite ratio provided by IUCS, Inc.

Physical Properties—Bulk density was determined by measuring the volume displacement of dried, powdered samples of the SSFAs. Porosity was determined from the change in weight of SSFA samples due to absorption of water; permeability was determined using the falling head technique. Compressive strength was determined by measuring the load needed to achieve total failure of the sample when the load was applied to the sample's vertical axis; a Riehle Universal Testing Apparatus with a loading rate of 0.064 cm/min was used. In the compressive strength tests, the SSFAs had been submerged in seawater in the laboratory and also in the field for different periods of time.

Elutriate Leaching—This was accomplished by adding 150 cm^3 of powdered SSFA samples to 600 ml of filtered seawater in a 1000-ml stoppered Erlenmeyer flask. The flasks were shaken or bubbled for 24 h. (For Mix 7, three successive leachings were performed using the same SSFA test sample.) Portions of the SSFA-seawater suspensions were filtered (0.45 μm), and the filtrate was analyzed for calcium, iron, zinc, chromium, lead, nickel, cadmium and mercury using flame or flameless AAS, for sulfite colorimetrically, and for sulfate turbidimetrically. The pH of the elutriate was measured with a glass electrode. The filters containing the suspended particulate matter were dried, reweighed, and analyzed for calcium, iron, manganese, copper, chromium, lead, nickel, cadmium, and mercury using flame and flameless AAS.

Tank Leaching—In this experiment, proctors of Mixes 5 and 7 were placed in tanks containing 3 litres of seawater (unstirred) for 30 days. The tanks were sampled by syringe at the intervals 1, 3, 5, 10, 20, and 30 days and analyzed for iron, copper, nickel, and calcium using either flame or flameless AAS.

Analysis of Trace Metals in Colonizers—In the field studies, samples of encrusting organisms were removed from the SSFA and the concrete blocks, washed, and then freeze-dried for 48 h; the dehydrated biomass was finely ground. Three l-g portions from each sample were digested in nitric acid and analyzed for cadmium, lead, copper, chromium, zinc, mercury, and silver using AAS.

Results

Chemical Composition

For the concentrations of major and minor components, no significant variation existed between samples of the same mix type for any of the components measured (Table 1). The high-sulfite mixes, Mixes 4 and 5, had relatively high calcium carbonate concentrations (Table 1). In contrast, the high-sulfate mixes (6 and 7) were low in calcium carbonate. The high-sulfate mixes had a high organic carbon residue due to incomplete combustion of coal and a high sulfate-sulfite ratio, probably due to inefficient furnace operation (S. Taub, IUCS, Inc., personal communication). The mole ratios of calcium: sulfite + sulfate were highest in the high-sulfite mixes, 5:1 as opposed to 2:1 in the high-sulfate mixes. These ratios were a function of the different scrubber reaction efficiencies and of the amount of stabilizers added, which were high in calcium.

The high-sulfate mixes had lower concentrations of trace metals, except for mercury, than those observed in the high-sulfite mixes. Cadmium, copper, and mercury were present in SSFAs in appreciable higher concentrations than in concrete. The remaining heavy metals measured had concentrations similar to or lower than those found in the control concrete test samples.

Physical Properties

The bulk densities (Table 2) of all the SSFA mixes were lower than that of concrete due to the addition of light-weight fly ash to the SSFA mixes and the absence of high-density aggregate materials. Mix 7 was found to have greater porosity and permeability than Mixes 4 and 5 (Table 2). Additionally, the SSFA mixes were found to be three to four times more porous and, significantly, more permeable than the concrete test samples. The relationship between porosity and permeability was found to be a direct function of the amount of calcium available for calcium aluminosilicate precipitation in the cementation process. Precipitation of such phases results in a general reduction in porosity and permeability.

Compressive Strength

Compressive strength values for the SSFA mixes and for concrete increased significantly with curing time (Table 3). After 30 days curing in air, the SSFA had a compressive strength of 25 to 75 percent of that of concrete. After 5 months of curing, the compressive strength of the SSFA mixes were from 1.2 to 3 times greater than their strengths observed at 30 days, while that of concrete increased fivefold. The observed differences between mix types and concrete are due to the variations in the composition of the samples. Compressive strength of concrete and similar cementitious compounds varies as a function of curing time, humidity, temperature, water content, and additive content.

Blocks of Mix 7 and of concrete were also subjected to 120 days immersion in seawater; Mix 7 lost no strength, but concrete lost 22 percent of its optimum strength. The loss in strength in concrete exposed to seawater is well documented [*9*]. The process involves sulfate ions in seawater, which disrupt the cementitious bonds in cement. The loss in strength occurs as a result of lattice expansion due to the formation of calcium sulfoaluminate precipitate. This precipitate was formed by sulfate ions reacting with the tricalcium aluminate in cement (one of the bonding compounds) to form solid ettringite ($3CaO \cdot Al_2O_3 \cdot 3CaSO_4 \cdot 31H_2O$) *in situ*. Ettringite occupies a 14 percent greater volume than the tricalcium aluminate originally present. As this and other solution reactions occur, the concrete expands and loses its strength due to internal pressures breaking cementitious bonds and lattice structures. The SSFA test samples did not continue to lose strength upon continued exposure to seawater because they already contain minimal lime available to react with sulfate ions in the seawater. They also have high porosities, which permit lattice expansion with less internal pressure.

TABLE 1—*Concentrations of selected*

Sample Mix Type	Bulk Content, mg/g				Acid-		
	$CaCO_3$	Total Carbon	CO_3 Carbon	Organic Carbon[c]	Ca	SO_3	SO_4[d]
4	118	15	14	1	52 200	23 600	2 600
5	118	17	14	3	66 600	26 700	3 000
6	15	20	2	18	33 200	16 300	24 500
7	19	25	2	23	37 400	16 300	24 500
Concrete[e]	...	...	...	...	111 300	...	6 400
Detection limit	5	1	0.5	1	400	100	100
C.V.,[g] %	9.5	15.7	9.5	15.7	10.5	10.7	10.7

[a]Concentrations are presented on a dry weight basis and are average values of sample and subsample replicates.

[b]Nitric acid was used for heavy metal leaching and hydrochloric acid was used for calcium and sulfur oxide leaching.

[c]Organic carbon is determined by subtracting carbonate carbon from total carbon.

[d]Total SO_x was measured analytically; reported values of SO_3 and SO_4 are based on $SO_3:SO_4$ ratios supplied by I. U. Conversions (S. Taub, personal communication).

TABLE 2—*Selected physical properties of test samples.*[a]

Sample Mix Type	Bulk Density, g/cm^3	Porosity Volume of Water Permeable Voids, %	Coefficient of Permeability K, cm/s $\times 10^{-7}$
4	2.17	34	3.9
5	2.18	37	1.0
6	1.99	58	...
7	2.03	48	65.0
Concrete	2.70	13	0.1

[a]All samples were cured 90 days.

TABLE 3—*Physical stability of test samples.*

Days Cured	Days in Seawater	Age of Sample	Compressive Strength of Mix Types,[a] psi				
			Mix 4	Mix 5	Mix 6	Mix 7	Concrete
30	0	30	110	320	160	200	425
150	0	150	215	920	205	355	2160
120	30[b]	152	180	840	...	320	1780
30	120[b]	152	...	...	...	310	1520
120	0[c]	150	...	...	...	245	1650

[a]The coefficient of variation for compressive strength values, based on three replicates of one mix type, was 7 percent. One psi = 6.8948 KPa.

[b]Samples were allowed to dry for 2 days after removal from seawater before testing.

[c]These samples underwent 20 cycles of rapid freezing and thawing in air, requiring 30 days.

major and minor components.[a]

Leached Components,[b] μg/g								
Fe	Mn	Zn	Cu	Cr	Pb	Ni	Cd	Hg
3660	58.6	23.2	19.1	16	11	7	0.6	0.048
4750	53.0	30.5	22.7	18	19	7	0.7	0.059
4930	35.1	13.7	12.3	7	12	6	0.4	0.272
3340	33.1	14.1	10.9	10	10	3	0.2	0.299
5780	312.5	18.4	5.2	24	6	5	0.2[f]	0.010
30	0.3	0.1	0.4	2	3	2	0.2	0.002
4.9	4.1	21.4	7.2	9.4	20.8	18.9	17.3	12.1

[e]Concrete component concentrations are based upon total weight, including aggregates. The component concentrations presented would be approximately two times higher if the aggregate materials were not included.

[f]Value below the detection limit.

[g]C.V. is the weighted average coefficient of variation (s.d./y × 100) for the replicate samples and subsamples of each mix type for each component measured.

Laboratory Leaching Studies

Elutriate Test—Powdered samples of each SSFA mix were added to filtered seawater and agitated for 24 h then allowed to settle for 36 h. The major and minor components in the elutriate were analyzed for in the dissolved and particulate phases (Tables 4 and 5). Dissolved concentrations of the major components examined—calcium, sulfite, and sulfate—reached the saturation concentrations of calcium sulfite and calcium sulfate in seawater following the reaction of the seawater with the powdered SSFAs in this closed system (Table 4). For the minor components (heavy metals), however, concentrations of the dissolved species (except for iron and zinc in a few instances) decreased when the seawater reacted with the test samples (Table 4). This depletion is probably due to adsorption processes associated with the particulate materials in suspension during mixing. Goldberg [*10*] and Krauskopf [*11*] have shown that adsorption of dissolved metals by suspended particulates and colloidal precipitates can be a primary mechanism for controlling the concentrations of heavy metals in seawater.

The major portion of unstabilized scrubber sludges has a size distribution in the range of 5 to 50 μm [*12*]. Because of this relatively large grain size, a very small fraction of the test samples used in the elutriate test remained suspended as particulates following the 36-h settling period as outlined in the procedure (Table 5). The composition of these remaining suspended particulates was found to be similar to the composition of the bulk parent material (Table 1 and 5), except for mercury, cadmium, zinc, and nickel, which were found to be more concentrated in this lighter fraction. This observed enrichment may be due to the association of these metals with the fly ash present in the SSFAs or to the fact that the suspended particulates

TABLE 4—*Concentrations of dissolved components in the elutriate.*[a]

Seawater and Sample Mix Type	Dissolved Component Concentrations, μg/litre												
	pH	Ca	SO_3	SO_4	Fe	Mn	Zn	Cu	Cr^{6+}	Pb	Ni	Cd	Hg
Seawater (SW)[b]	8.13	630	0.5[c]	2770	3	42	5.8	20.2	2.0	3.1	17.4	8.5	0.7
SW+4	9.08	2900	68.4	3270	7	33	22.4	7.4	0.5[c]	0.4	7.5	1.9	0.3
SW+5	10.72	3020	68.6	3130	3	24	2.6	2.9	0.5[c]	0.1	10.0	0.6	0.6
SW+6	9.12	3430	50.0	3190	11	42	3.0	3.3	0.5[c]	0.2	6.2	3.8	0.5
SW+7[d]													
1	9.49	2980	0.5[c]	2590	19	19	7.8	2.9	0.5[c]	2.1	6.0	5.5	0.6
2	8.83	2180	0.5[c]	3700	3	29	0.6[c]	2.7	0.5[c]	0.1[c]	6.5	1.7	...
3	8.53	2150	0.5[c]	4190	7	34	0.6[c]	2.0	0.5[c]	0.1[c]	8.7	1.9	...
Concrete	12.33	3890	2.6	2340	5	33	9.5	2.2	95.0	1.9	13.3	0.8	0.6
Detection limit	...	40	0.5	6	3	2	0.6	0.4	0.5	0.1	0.2	0.1	0.2
C.V. (%)[c]	0.73	7.5	11	3.1	55.9	11.8	43.5	48.2	22.3	46.8	25.3	31.9	10.0

[a]Mechanical agitation was the mode of agitation used for the elutriate test results presented in this table.
[b]Seawater was obtained 1.6 km (1 mi) east of Montauk Point, Long Island, (salinity 34 parts per thousand)
[c]Value below the detection limit.
[d]Values 1, 2, and 3 represent the concentrations of components in three successive runs of the test using the same sample of SSW but new portions of seawater each time.
[e]C.V. = average coefficient of variation (s.d./$\bar{y}$ × 100) based upon replicate subsamples of each powdered sample.

TABLE 5—*Concentrations of suspended metals in the elutriate.*[a]

Sample Mix Type	Percent Suspended Particulates[b]	Suspended Particulate-Acid-Leached Components, μg/g									
		Ca	Fe	Mn	Zn	Cu	Cr	Pb	Ni	Cd	Hg
4	0.04	87 700	5160	50	100	5.3	28	16	48	2.7	0.74
5	0.03	161 100	4490	20	200	4.8	20	7	23	2.3	1.26
6	0.06	49 600	6330	30	40	3.8	14	7	9	2.2	1.40
7 (1)[c]	0.06	84 100	4060	20	600	2.8	14	16	10	2.6	1.55
7 (2)[c]	0.02	52 100	4910	30[d]	10[d]	6.1	13	3	15	3.7	0.58
7 (3)[c]	0.01	57 800	4580	70[d]	20[d]	9.0	7[d]	4[d]	37	6.7	2.15
Concrete	0.06	197 400	5950	250	110	1.7	39	5	29	1.2	0.17
Detection limit	0.001	200	40	8	5	0.2	3	1	4.5	0.2	0.03
C.V., %[e]	13.0	11.5	34.0	53.3	39.7	32.1	28.6	42.3	58.2	38.2	40.5

[a]Mechanical shaking was the mode of agitation used for the elutriate test results presented in this table.
[b]Percentage of original crushed sample that remained suspended following agitation and 36 h settling.
[c]Values 1, 2, and 3 represent the concentrations of components in three successive runs of the test using the same sample of SSW but new portions of seawater each time.
[d]Value is below this concentration. Here the detection limit is the concentration indicated, owing to the small sample size of suspended particulates.
[e]C.V. = average coefficient of variation (s.d./$\bar{y}$ × 100) based upon replicate subsamples.

have larger surface areas than the ground bulk material, thus presenting more potential sites for adsorption of dissolved heavy metal.

Tank Leaching—Figures 3 and 4 show the concentrations of dissolved iron, copper, and nickel in the seawater containing the SSFA samples, as a function of time. There was an initial increase of dissolved iron and nickel in the first days after exposure to seawater, but after about 10 days, their concentrations decreased to levels near the original seawater concentrations. The concentrations of copper, however, decreased during the first 10 days of the experiment. The behavior of dissolved iron, copper, and nickel concentrations is probably due to desorption-adsorption processes or precipitation reactions occurring on the surfaces of the SSFA and the suspended particulates that were released, as discussed in the preceding section.

Concentrations of calcium in the seawater steadily increased with time (Fig. 5) [*3*]. Mix 7 was found to release calcium at a greater rate than Mix 5. This is primarily due to a higher content of the more soluble calcium compounds—calcium sulfate and calcium hydroxide—in Mix 7 compared to Mix 5, which contains greater concentrations of the less soluble calcium salts, calcium sulfite and calcium carbonate [*3*].

Field Studies

Physical Behavior and Compressive Strength—Over the observational period, significant erosion of SSFA block edges was not observed. Some superficial surface softening was detected during dives 3 and 4 months after placement. Compressive strength tests indicated that during the first 100

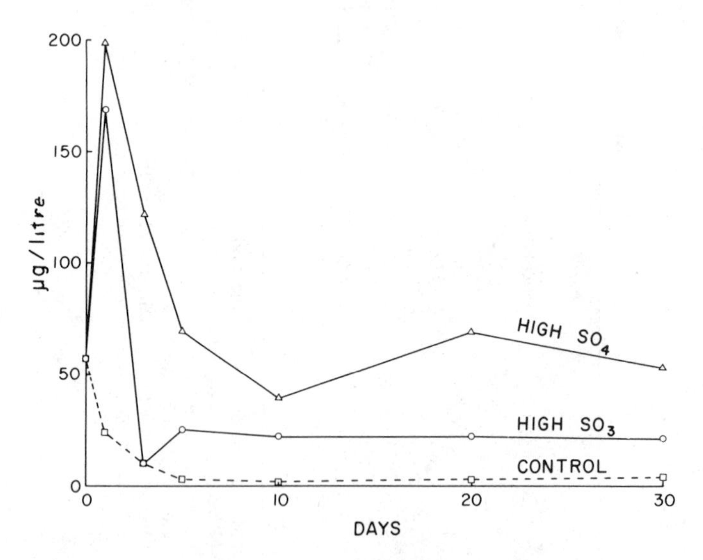

FIG. 3—*Dissolved iron observed in tank experiment.*

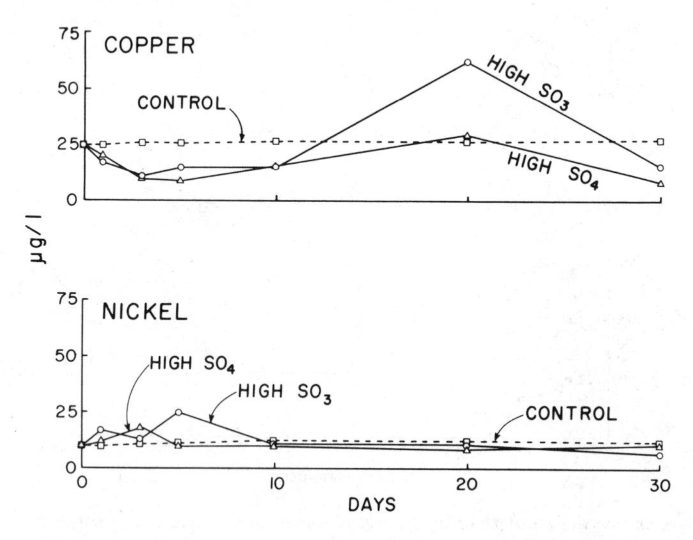

FIG. 4—*Dissolved copper and nickel observed in tank experiment.*

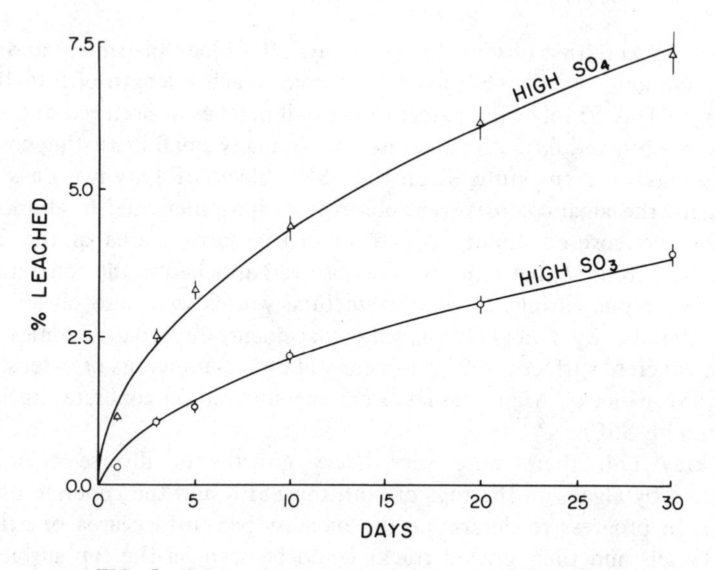

FIG. 5—*Calcium leached from SSFAs in tank experiment.*

days, compressive strength of the coal waste blocks tended to decrease, while that of the concrete initially increased (Fig. 6). After 100 days, however, the compressive strength of coal waste stabilized and later began to increase, while the strength of the concrete demonstrated a dramatic decrease.

Biological Colonization—Of the invertebrates observed growing on the blocks, hydroids were early colonizers on both SSFA and concrete blocks.

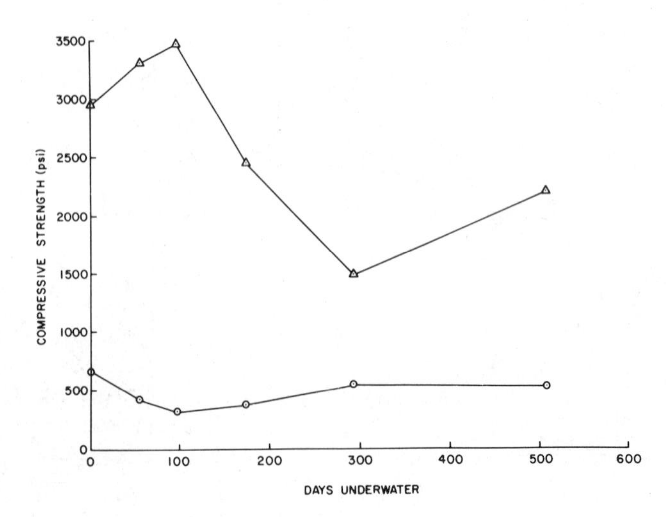

FIG. 6—*Compressive strength of* in situ *test blocks in Conscience Bay [1 psi = 6.89 kN/m^2 (kPa)].*

The colonies were first observed only 20 days after block placement and were about 1 cm long. By Day 82, hydroid colonies reach a length of 5 to 10 cm (Fig. 7), by Day 55 following placement, small patches of both red and green algae were observed on both substrates, while many small incrusting colonies of bryozoans were encountered on the SSFA blocks. Eighty-two days after placement, the algal colonies were observed to have increased in abundance and size and covered about 30 percent of the surface area of the SSFA blocks. Numerous slipper limpets were observed attached to the concrete surfaces. Calcareous casings of polychaete tube worms were also observed on both materials. By Day 111 following placement, bryozoan colonies were seen on concrete surfaces, but they were still not as numerous or extensive as on the SSFA blocks. Many limpets were encountered on concrete, but none were seen on SSFA.

On Day 174, there were very heavy growths of diverse colonizers, dominated by algae, on the tops of both the SSFA and the concrete blocks. Work is in progress to determine the biomass per surface area of exposed block. Snails and their grazed tracks could be seen on the top surfaces of these blocks.

In the second year of observation, more than 460 days after placement, bryozoan, hydroid, and algal growths were very extensive, covering all exposed surfaces. The differences in colonization between SSFA and concrete, which were seen at the earlier stages of colonization in the study, were no longer evident. The bryozoan colonies and tube worms dominated the sides of the blocks, with algae and hydroids attached to the tops, sides, corners, and edges. The colonizers seemed to be forming a stable community. Fish,

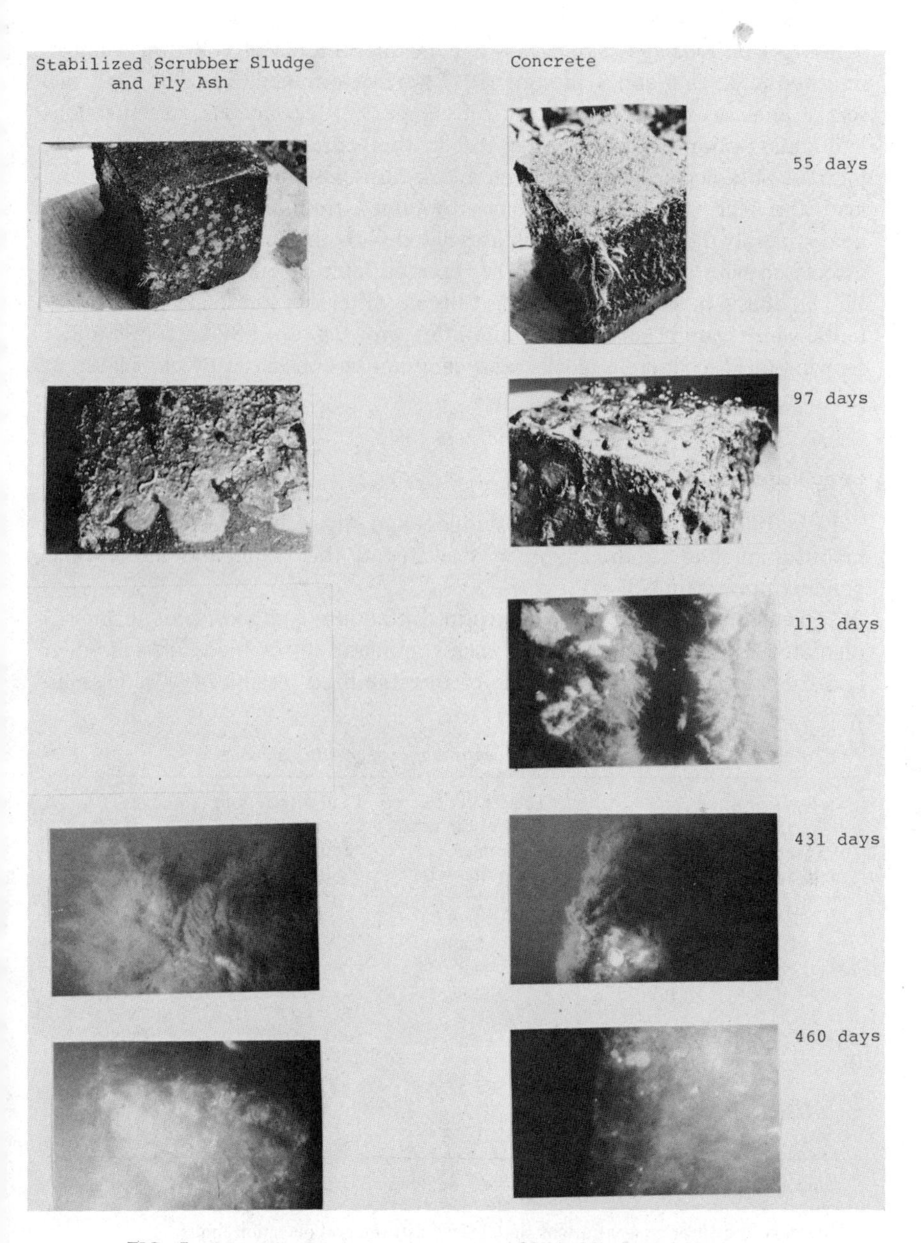

FIG. 7—*Time history of the colonization of SSFA (Mix 7) and concrete.*

crabs, and snails were commonly seen grazing and browsing on the dense growths.

Metal Contents of Colonizers—The metal contents of colonizer biomass removed in August and September 1977 were determined (Table 6). Only two sets of analyses showed significant differences between metals concentrations in biomass obtained from SSFA and in that obtained from concrete. The biomass obtained from the concrete block during September contained more lead and zinc than did the biomass obtained from the SSFA. Additional heavy metals data for organisms associated with the reef and also for those associated with natural rock will be reported later by one of us (F. Roethel, Ph. D. thesis in progress), but basically no differences in heavy metals contents were found between communities growing on SSFA, communities growing on the concrete blocks, and communities on natural rocks placed at the Conscience Bay site.

Conclusions

1. Stabilization of scrubber sludges and fly ash prevents the rapid mobilization and solubilization in seawater of the major and minor components present in SSFA.

2. Leaching of the major calcium-containing compounds is primarily regulated by the concentration of each compound in the block, its solubility, and the surface area/volume ratio of the stabilized test blocks [*3*]. Dissolu-

TABLE 6—*Metals concentration in encrusting biomass.*[a]

Metal[b]	Collection Date	Concentration in SCPW Biomass Sample ($\overline{X}_s$), μg/g dry weight	Concentration in Concrete Biomass Sample ($\overline{X}_{con}$), μg/g dry weight	*t* Test Result[c]
Cd	8/16	0.29	0.60	N.S.
	9/13	0.43	0.55	N.S.
Pb	8/16	40.2	37.7	N.S.
	9/13	33.4	40.4	...[d]
Cu	8/16	46.2	49.7	N.S.
	9/13	59.8	63.5	N.S.
Cr	8/16	22.8	26.8	N.S.
	9/13	22.4	23.1	N.S.
Zn	8/16	287	232	N.S.
	9/13	201	239	...[d]

[a]Data are the means of replicate ($n = 3$) analyses of samples collected on 16 Aug. and 13 Sept. 1977.

[b]Mercury and silver concentrations were below instrumental detection limits.

[c]The hypotheses H_0:$\overline{X}_s = \overline{X}_{con}$, H_a:$\overline{X}_s > \overline{X}_{con}$ or $\overline{X}_s < \overline{X}_s$ were tested for each date, using Student's *t* distribution $\alpha = 0.10$ [*13*], where $\overline{X}_{con}$ and $\overline{X}_s$ are mean metals concentrations measured on replicate analyses of samples obtained from concrete and SSFA surfaces, respectively. N.S. indicates no significant difference observed, that is, H_0 could not be rejected.

[d]The rejection of H_0:$\overline{X}_s = \overline{X}_{con}$; in these cases, $\overline{X}_{con}$ was significantly larger than $\overline{X}_s$.

tion of calcium sulfite is the only major component of environmental concern, but in an aerobic environment, the sulfite is probably rapidly oxidized to sulfate.

3. Trace metal concentrations in blocks of SSFA and leaching experiments [*3*] suggest that heavy metal contamination by SSFA is not likely to be environmentally significant in aerobic environments.

4. SSFA maintains its structural integrity in the sea for extended periods of time.

5. Biological colonization of *in situ* SSFA and concrete surfaces began soon after placement. Differences in the occurrence of species of colonizing macroinvertebrates on SSFA and concrete surfaces were notable after about 50 days. These initial differences *may* be related to different surface hardness and differential selection by the settling colonizers. By Day 111 after placement, clear differences could no longer be seen. After 1½ years, the SSFA blocks had been completely covered by heavy growths of organisms.

6. The concentrations of selected heavy metals in colonizer biomass removed from *in situ* SSFA surfaces were not greater than concentrations measured in biomass removed from concrete control blocks or from hard rock controls at the site.

Acknowledgment

The authors are grateful to the Link Foundation, the New York Sea Grant Institute, and the New York Energy Research and Development Authority for the financial support for this work. We are indebted to Jacqueline Restivo, Annette May, and Mary Ann Lau for assistance in the manuscript preparation and to Ed Smith for fabrication of the SSFA test blocks. This report is Marine Sciences Research Center's Contribution No. 249.

Addendum

Since the preparation of this manuscript, two reports relevant to this work has been published by the Stony Brook group [*14,15*]. In addition, the project has expanded to include a field study in Chesapeake Bay which began in 1979 and was completed in 1980. On 12 Sept. 1980, a 500-ton demonstration reef was constructed in the Atlantic south of Long Island, from 18 000 8 by 8 by 16-in. solid blocks of stabilized fly ash and scrubber obtained from coal burning power plants located in Ohio and Indiana. The reef blocks were fabricated at a commercial concrete block plant in Pennsylvania using automatic block making equipment. The blocks were made using two mix designs whose ratios (dry weight) of fly ash to scrubber sludge were near 1.5:1 and 3:1. The demonstration reef will be monitored for 3 to 4 years to assess environmental impacts that may occur and to measure the development of biological communities that will be associated with the reef. Support for these recent studies comes from the State of Maryland, NYSERDA, the U.S. Department of Energy, the U.S. Environmental Protection Agency, the Electric Power Research Institute, and the Power Authority of the State of New York.

References

[1] Herlihy, J., "Flue Gas Desulfurization in Power Plants," Status Report, Division of Stationary Source Enforcement, Office of Enforcement, U.S. Environmental Protection Agency, Washington, D.C., April 1977.

[2] Lunt, R. R., Cooper, C. B., Johnson, S. L., Oberholtzer, J. E., Schimke, G. R., and Watson, W. I., "An Evaluation of the Disposal of Flue Gas Desulfurization Wastes in Mines and the Ocean, Initial Assessment," EPA 600/7-77-051, U.S. Environmental Protection Agency, Research Triangle Park, N.C., 1977.

[3] Seligman, J. D., "Chemical and Physical Behavior of Stabilized Scrubber Wastes and Fly Ash in Seawater," M.S. thesis, Marine Sciences Research Center, State University of New York, Stony Brook, N.Y. 1978.

[4] Seligman, J. D. and Duedall, I. W., *Environmental Science and Technology*, Vol. 13, 1979, pp. 1082–1087.

[5] Duedall, I. W., O'Connors, H. B., Parker, J. H., Roethel, F. J., and Seligman, J. D., "A Preliminary Investigation of the Composition, Physical and Chemical Behavior, and Biological Effects of Stabilized Coal Fired Power Plant Wastes (SCPW) in the Marine Environment," Final Report, New York Energy Research and Development Authority, Albany, N.Y., 1978.

[6] Duedall, I. W., Seligman, J. D., Roethel, F. J., O'Connors, H. B., Parker, J. H., Woodhead, P. M. J., Dayal, R., Chezar, B., Roberts, B. K., and Mullen, H. in *Ocean Dumping of Industrial Wastes*, B. H. Ketchum, D. R. Kester, and P. K. Park, Eds., Plenum, N.Y., 1981.

[7] *1974 Annual Book of ASTM Standards*, American Society for Testing and Materials, Philadelphia, 1974.

[8] *Aquatic Disposal Field Investigations: Eatons Neck Disposal Site, Long Island Sound*, Appendix B: *Final Report*, Contract No. DACW51-75-C0016, Work Unit 1A06B, Waterways Experiment Station, Marine Sciences Research Center, Army Corps of Engineers, Vicksburg, Miss., 1978.

[9] Swenson, E. G., *Performance of Concrete: Resistance of Concrete to Sulfate and Other Environmental Conditions*, University of Toronto Press, Toronto, 1968.

[10] Goldberg, E. D., *Journal of Geology*, Vol. 62, 1954, pp. 249–269.

[11] Krauskopf, K. B., *Geochimica et Cosmochimica Acta*, Vol. 9, 1956, pp. 1–32.

[12] Mahloch, J. L., "Chemical Fixation of FGD Sludges—Physical and Chemical Properties," presented at U.S. Environmental Protection Agency Symposium on Flue Gas Desulfurization," New Orleans, La., 1976.

[13] Sokal, R. R. and Rohlf, F. J., *Biometry: The Principles and Practice of Statistics in Biological Research*, Freeman, San Francisco, 1969.

[14] Woodhead, P. M. J. and Duedall, I. W., "Coal Waste Artificial Reef Program, Phase I," FP-1252, Research Project 134-1, Electric Power Research Institute, Palo Alto, Calif., 1979.

[15] Roethel, F. J., Duedall, I. W., O'Connors, H. B., Parker, J. H. and Woodhead, P. M. J., *Journal of Testing and Evaluation*, Vol. 8, 1980, pp. 250–254.

J. R. Gammon,[1] *A. Spacie,*[2] *J. L. Hamelink,*[3] *and R. L. Kaesler*[4]

Role of Electrofishing in Assessing Environmental Quality of the Wabash River

---

REFERENCE: Gammon, J. R., Spacie, A., Hamelink, J. L., and Kaesler, R. L., **"Role of Electrofishing in Assessing Environmental Quality of the Wabash River,"** *Ecological Assessments of Effluent Impacts on Communities of Indigenous Aquatic Organisms, ASTM STP 730,* J. M. Bates and C. I. Weber, Eds., American Society for Testing and Materials, 1981, pp. 307-324.

ABSTRACT: The fish community of the middle Wabash River has been studied since 1973 from the standpoint of its capacity for measuring the biological impact of various kinds of point influences, such as thermal, municipal, and industrial effluents. Of particular interest has been the potential value of various community parameters as indicators of environmental quality, since individual species populations tend to vary markedly from year to year.

The primary sampling method consisted of repeated direct-current electrofishing through a series of 0.5-km-long zones located strategically throughout 274 km (170 miles) of river, divided into twelve reaches. The use of cluster analysis and community indexes, including a composite index, has been of value in isolating problem areas prior to the development of lethal environmental conditions and in gaging the degree of environmental benefit resulting from improved waste treatment.

KEY WORDS: ecology, environments, aquatic ecology, diversity, hierarchical diversity, stream surveys, clustering, effluents, aquatic organisms

There is an undeniable need for methods of measuring the effect of various human activities on the aquatic communities of large rivers. Negative effects that lead to the elimination of entire species or to their replacement by predominantly pollution-resistant species may be detected with little difficulty. Likewise, changes resulting from substantially improved environmen-

[1]Professor, Department of Zoology, DePauw University, Greencastle, Ind. 46135.

[2]Assistant professor, Department of Forestry and Natural Resources, Purdue University, Lafayette, Ind. 47907.

[3]Research scientist, Lilly Research Laboratories, Greenfield, Ind. 46104.

[4]Professor, Department of Geology, University of Kansas, Lawrence, Kans. 66045.

tal conditions are relatively easy to detect and evaluate. Techniques are now needed to provide an appropriate information base for decision making in situation wherein cause and effect may be more subtle.

Perhaps the most common product of an environmental investigation is a list of species that, in itself, is supposed to reflect the "health" of the biological community. Such a list has certain value, but it also has severe limitations, because it does not provide a quantitative measure of population size. Larimore and Smith [*1*][5] found that increasing levels of organic pollution in streams first caused a reduction in the number of species present, then a reduction in the total weight or biomass, and finally a reduction in the total number of individual fish. Consequently, some measure of relative numeric density is frequently used to evaluate the status of fish populations in rivers.

Recently, some measurement of diversity has been included in the analysis of field surveys. The value of diversity indexes applied to samples of fish has met with mixed success. Bechtel and Copeland [*2*] found that diversity indexes based on otter trawl tows were useful indicators of pollution in Galveston Bay, Texas. McErlean and Mihursky [*3*] employed various measurements of species diversity to a variety of fish collections in estuaries to demonstrate seasonal changes and annual trends [*4*]. Richards [*5*] found diversity indexes of value in interpreting faunal changes in the Au Sable River over a 50-year period. However, Denoncourt and Stambaugh [*6*] found relatively high species diversities at several sites that had received moderate to heavy pollution and recommended against their use when evaluating fish populations.

The impact of two electrical generating stations on fish communities in the middle Wabash River has been studied for several years. Species shifts consistent with the altered thermal conditions have been documented [*7,8*], but other changes noted in the fish community structure could not be solely attributed to the electrical generating station. Consequently, a program was initiated to extend the investigation into flanking reaches of the river.

During 1973 and 1974, the fish communities over 161 km (100 miles) of the middle Wabash River were examined by electrofishing at 24 similar sites, including the electric generating station. A majority of the species of fish were found to be nonrandomly distributed throughout the river [*9*]. For example, shortnose gar and bowfin were much more abundant in the lower section of river near bayous, which these fish used for reproduction. On the other hand, redhorse (*Moxostoma*) were more common in the upper section of the river, probably indicating a preference for the cooler water upriver from the electrical generating stations. The species composition was found to shift considerably from place to place in response to both natural and man-

[5]The italic numbers in brackets refer to the list of references appended to this paper.

induced factors. Nevertheless, it was also evident that, while species associations were variable from place to place, the various community parameters used to assess abundance and diversity remained relatively unaffected. The only statistically significant findings in the analyses of community structure were a lower diversity below Terre Haute, Indiana, and a greater numeric catch rate of fish upriver from the Cayuga, Indiana, electrical generating station.

Refinements in selecting the collecting stations were added in 1975, and an additional 161 km (100 miles) of upper river was included for study [*10*]. In 1977 and 1978, a coordinated effort was launched which concentrated upon 233 km (145 miles) of river, including not only the two electrical generating stations, but also major centers of heavy industry and chemical manufacturing. This summary focuses upon the 1977 and 1978 research programs and relates the findings to those of prior studies.

The Wabash River

The Wabash River in western Indiana is the largest free-flowing tributary of the Ohio River, although there are six major reservoirs that influence flow patterns somewhat on the upper and middle Wabash. It has a drainage basin area of 31 600 km^2 (12 200 mile^2) and an average discharge of 288.8 m^3 (10 200 ft^3) [*11*], draining most of Indiana and parts of eastern Illinois. The present river valley is a low-gradient vestigial glacial sluiceway now filled with sand and gravel deposits. The upland areas adjacent to the middle Wabash River are mostly glacial till plains.

The basin is predominantly agricultural and contains some of the most productive soils in the Midwest. Corn, soybeans, cattle, and hogs form the backbone of the rural economy and influence the water quality throughout the entire basin. Approximately half of the human population lives in urban areas. The small cities of Danville, Illinois, and Terre Haute and Lafayette, Indiana, are centers of heavy industry and chemical manufacturing (Fig. 1). Industrial growth away from population centers along the main stem of the Wabash River has been rapidly expanding and is expected to continue to grow rapidly.

Two large coal-fired electrical generating stations discharge waste heat into the Wabash River, the Wabash electrical generating station (970 MW), located just north of Terre Haute, and the Cayuga electrical generating station (1000 MW), located about 56 km (35 miles) upriver. Both utilize once-through cooling, although topping cooling towers are operational at the Cayuga electrical generating station during the summer. Strip mining for coal, much of which is conveyed to the two electrical generating stations, is common from the level of Big Vermillion River southward.

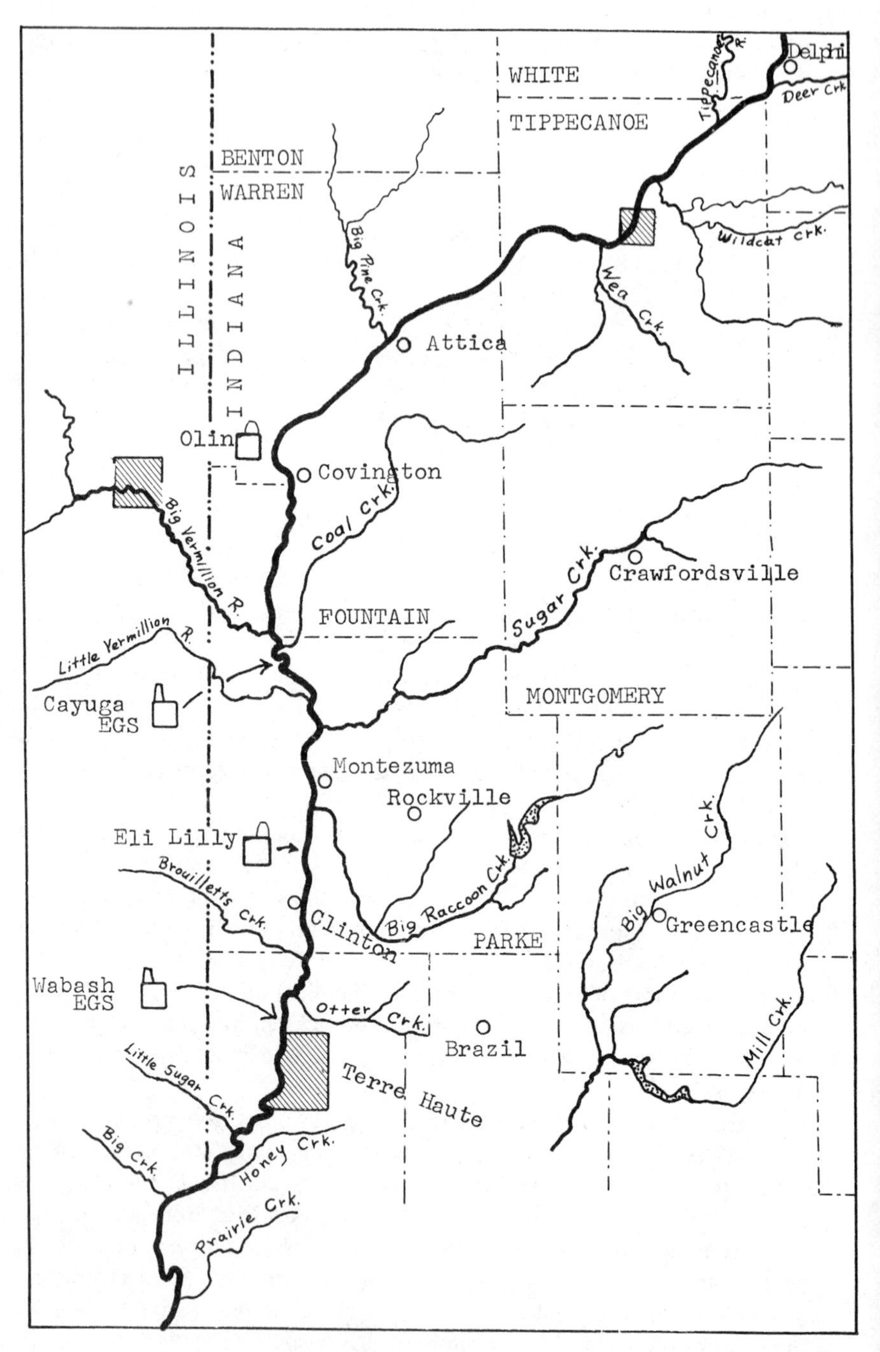

FIG. 1—*The middle Wabash River and important tributaries, towns, and industries.*

Methods

The study area consisted primarily of the middle Wabash River from Delphi, Indiana, [River Mile (RM) 330] to Darwin, Illinois, (RM 185) and was subdivided into eleven reaches of varying lengths (Fig. 1, Table 1). In 1977, four collecting stations were scattered throughout each reach and sampled four times each. An additional 40-km (25-mile) reach was added in 1978 from Darwin, Illinois, to Merom, Indiana, (RM 185 to RM 160) and five collecting stations were situated in each reach and visited three times each. Each collecting station consisted of a 0.5-km section of shoreline, measured with a Leitz rangefinder. Generally, the collecting stations were positioned in relatively swift water areas having good cover and an average depth of 1.5 m or less.

The collecting stations were sampled by electrofishing from upstream to downstream near the shore and near any existing cover. The electrofishing apparatus consisted of a Smith-Root Type VI electrofisher powered by a 3500-W generator. The pulse configuration consisted of a triangular wave interrupted into 60 pulses per second and producing 400 to 600 V dc at approximately 5.5 A. The pulsed direct current was fed into an electrode system consisting of two circlets of short stainless steel anodes extended by booms about 2.4 m (8 ft) off the bow of a 4.9-m (16-ft) boat and two gangs of long cathodes suspended from port and starboard gunwales near the bow.

Upon capture, all the fish were placed in an on-board holding tank. At the conclusion of each sampling run, the fish were identified as to species, weighed, measured, and then returned to the river unharmed. The data were then subsequently recorded on magnetic tape and computer disk for analysis.

TABLE 1—*Twelve reaches of the Wabash River studied.*

Reach	River Mile	(km)	Location Description
1	330 to 312	(531 to 502)	Delphi to Lafayette
2	311 to 302	(502 to 486)	within Lafayette area (effluents)
3	301 to 285	(486 to 459)	below Lafayette (river mile 305) to Big Pine Creek
4	284 to 269	(459 to 433)	Attica to Covington
5	268 to 251	(433 to 404)	Covington to above Cayuga electrical generating station
6	249.6	(402)	within Cayuga electrical generating station area
7	247 to 233	(397 to 375)	below Cayuga electrical generating station to above Eli Lilly plant at Clinton
8	232 to 218	(375 to 351)	Clinton to Wabash electrical generating station
9	215.4	(347)	within Wabash electrical generating station area
10	213 to 203	(343 to 327)	below Wabash electrical generating station to Terre Haute sewage treatment plant
11	202 to 186	(325 to 299)	Terre Haute sewage treatment plant to Darwin, Ill.
12	185 to 160	(299 to 257)	Darwin, Ill., to Merom, Ind.

The distributions of individual species throughout the middle Wabash River were examined. In addition, we examined several quantitative measures of community structure, including (1) relative density (number per kilometre), (2) relative biomass (kilogram per kilometre), (3) number of species per station, (4) Shannon index, H, of diversity based on numbers, (5) Shannon index, H, of diversity based on weights, and (6) evenness indexes associated with the Shannon diversity indexes. The diversity indexes were calculated using natural logarithms. The evenness indexes consisted of $H{:}H_{max}$ ratios.

In addition, a composite index of "well-being" was calculated as

$$I_{wb} = 0.5 \ln N + 0.5 \ln W + \text{Shannon}_{no} + \text{Shannon}_{wt}$$

where

N = the number captured per kilometre,
W = the weight in kilograms captured per kilometre
Shannon_{no} = Shannon index based on numbers, and
Shannon_{wt} = Shannon index based on weights.

The rationale and validation of this approach is detailed elsewhere [*10*]. Basically, the composite index incorporates two widely used indexes of diversity and two widely used indexes of abundance in approximately equal quantities to form a single value reflective of both the diversity and the abundance of fish in a collection. This composite value also appears to reflect environmental quality somewhat better than any single community index or indicator species.

Seven other methods of analysis were also explored, among which the following proved to be the most informative when 1977 data were used:

1. The number of species found in subsamples of 120 fish drawn at random from the aggregate collections of each reach of the river.
2. Comparison of the fish communities in each reach with those of the reach immediately upstream, using correlation coefficients and a reduced data set from which the 26 rarest species had been discarded.
3. *Q*-mode cluster analysis to compare the eleven reaches of river with each other using the reduced data set.
4. *R*-mode cluster analysis grouping the 24 most common species of fish on the basis of their abundance and distribution in the eleven reaches of the river.

Results and Discussion

Composite Index

From 1968 through 1978, a total of more than 40 000 fish of 71 species were collected by electrofishing. The results and analyses emphasized in this paper are based on approximately 3800 fish taken during 1977 and 4700 fish taken during 1978.

An examination of the patterns of change of the composite index over space and time (Fig. 2, Table 2) indicate that during years of relatively great summer discharge (for example, 1973, 1974, 1975, and 1978), there was little difference between reaches, while during 1977, a year of very low summer flows, a substantially altered pattern was evident with much higher values upriver from Lafayette than in the lower section of river. These areas of depression are believed to be caused by combinations of effluents, some effects of which are fairly simple to detect (for example, thermal effects from the Cayuga and Wabash electrical generating stations) while others are more complex (for example, in the Clinton area and below Terre Haute).

Overall, there was a positive correlation between river discharge and I_{WB} in impacted reaches of the river and little influence in relatively unimpacted reaches (Fig. 3). Both the absolute values of the composite index and the slope of the discharge/composite index regression appear to have value in assessing pollution effects.

The trends exhibited by the composite index were also followed generally by all of the other community parameters except for the evenness indexes, which tended to remain fairly constant (0.7 to 0.9) throughout the study section.

The variability of the community parameters for the reaches based on coefficient of variation calculations was high for the relative density and biomass (40 to 80 percent), intermediate for the number of species and the two Shannon diversity indexes (20 to 50 percent), and low for the composite index (10 to 20 percent). This characteristic makes the composite index attractive for statistical anaylsis.

The responsiveness of community indexes is well demonstrated by the annual changes in the composite index of Reaches 7 and 8 (Fig. 4). The values were relatively depressed in these reaches during 1973 through 1975, and it was postulated that this was an environmentally stressed section, although the cause of the stress was not identified [*10*]. The values were even more depressed in early summer 1977 prior to the development of a pronounced oxygen sag, which developed in late July and early August. This dissolved oxygen sag extended from the Cayuga region (RM 250) downstream approximately 40 km (25 mi) and caused a partial fish kill [*12*]. The diversity indexes declined still further at the time of the dissolved oxygen sag, but the composite index did not because of increased catches.

An explanation for the dissolved oxygen sag was not found. Heavy deposits of organic matter were noted in Reach 7 in early summer. These deposits had accumulated during several months of exceptionally low river flow in late winter and early spring. Very dense populations of chironomid adults flanked the afflicted section of river during much of the summer. These organic deposits might well have caused the depressed fish community in early sum-

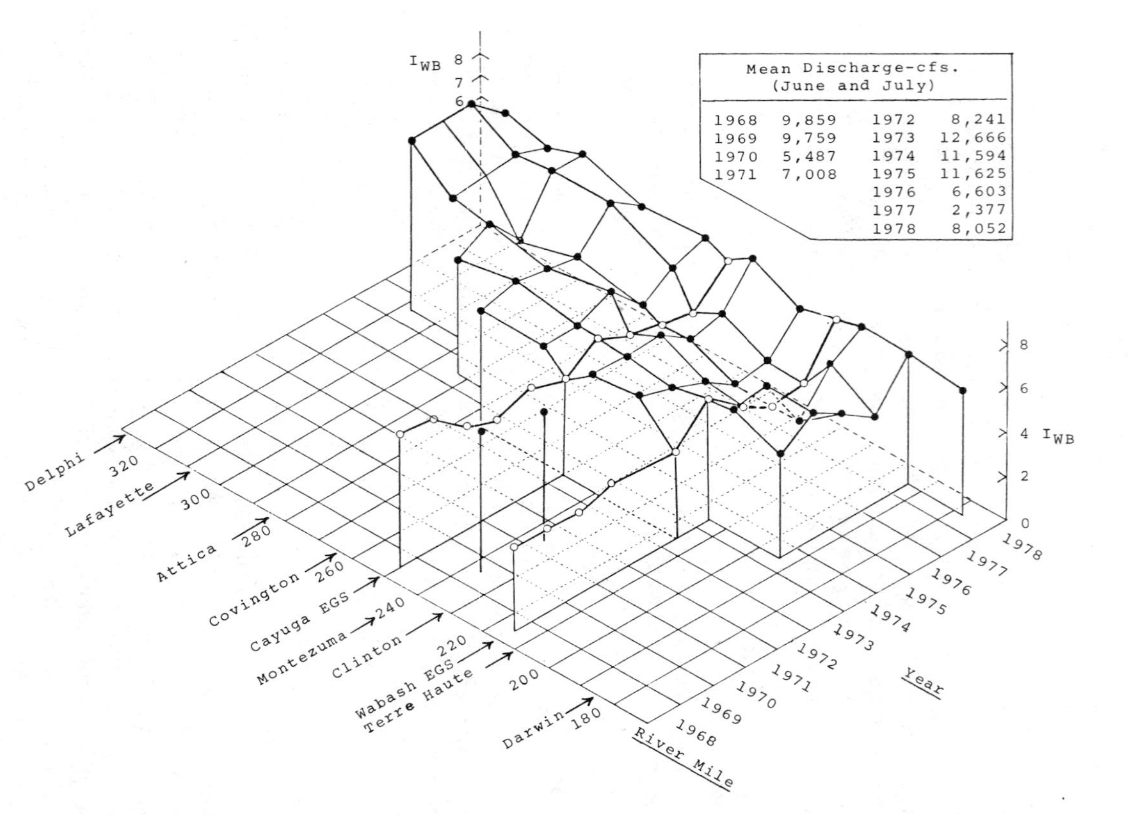

FIG. 2—*Spatial and temporal patterns of the mean annual composite index for 1968 through 1978. The 1969 and 1971 values at RM 238 are based on a single collection each. One river mile × 1.6093 = the distance from the mouth in kilometres.*

TABLE 2—*Mean annual composite index values in eleven reaches of the Wabash River—1968 Through 1978.*

	Mean Annual Composite Index Value										
						(Cayuga)			(Wabash)		
Year	1	2	3	4	5	6	7	8	9	10	11
1968	...	...	...	...	...	6.030	...	...	3.801	...	...
1969	...	...	...	...	...	5.927	7.262[a]	...	3.809	...	...
1970	...	...	...	...	...	4.770	...	...	3.672	...	...
1971	...	...	...	...	...	4.089	6.349[a]	...	4.101	...	...
1972	...	...	...	...	...	4.757	...	...	...	...	...
1973	...	...	...	5.246	5.272	4.331	5.266	5.266	5.432	...	...
1974	...	...	5.270	5.783	5.270	5.298	5.179	4.869	5.527	5.792	4.875
1975	7.793	6.236	6.032	5.506	6.069	4.607	5.395	4.331	4.163	5.795	4.711
1976	...	...	4.464	5.076	4.610	4.068	4.313	3.401	3.394	3.763	4.831
1977	7.950	6.761	6.893	6.844	5.488	3.952	4.619	3.687	3.469	5.212	3.924
1978	6.673	6.030	6.807	5.826	6.253	5.469	5.970	5.114	5.657	6.015	5.856

[a]based on a single collection at RM 238.

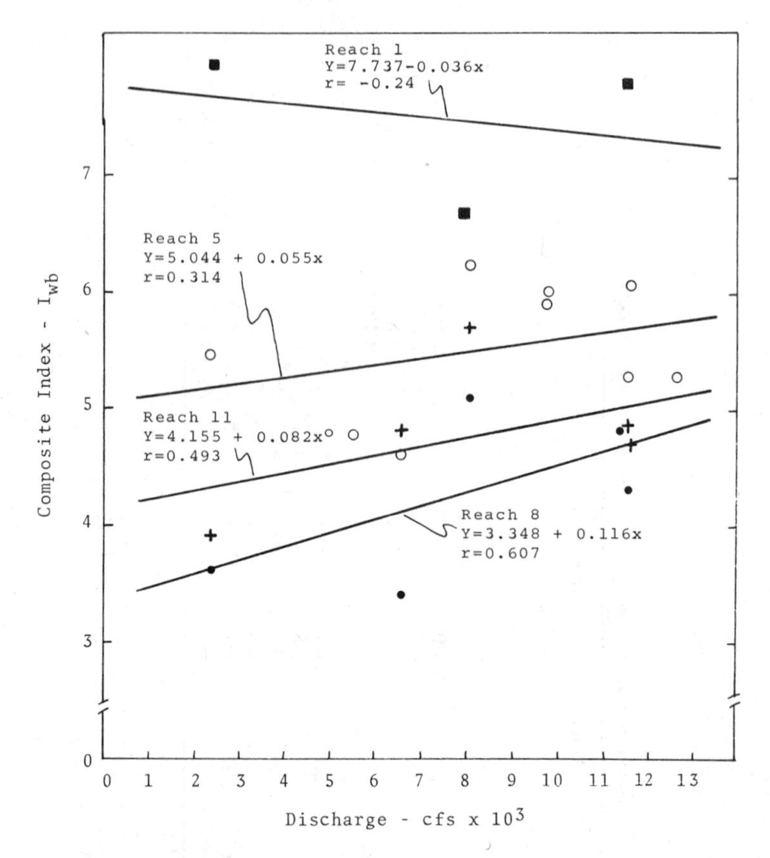

FIG. 3—*Relationship of the composite index to river flow in control and impacted reaches of the river.*

mer and the oxygen depletion later on. None of these phenomena were repeated in 1978, and the composite index values were correspondingly high then in Reaches 7 and 8.

Number of Species

The unequal number of fish from each reach made direct comparison of the number of species found in each reach impossible. The likelihood of species being missed by chance alone increases as the total number of fish decreases. The equation describing the relationship is

$$Y = (1 - p)^N$$

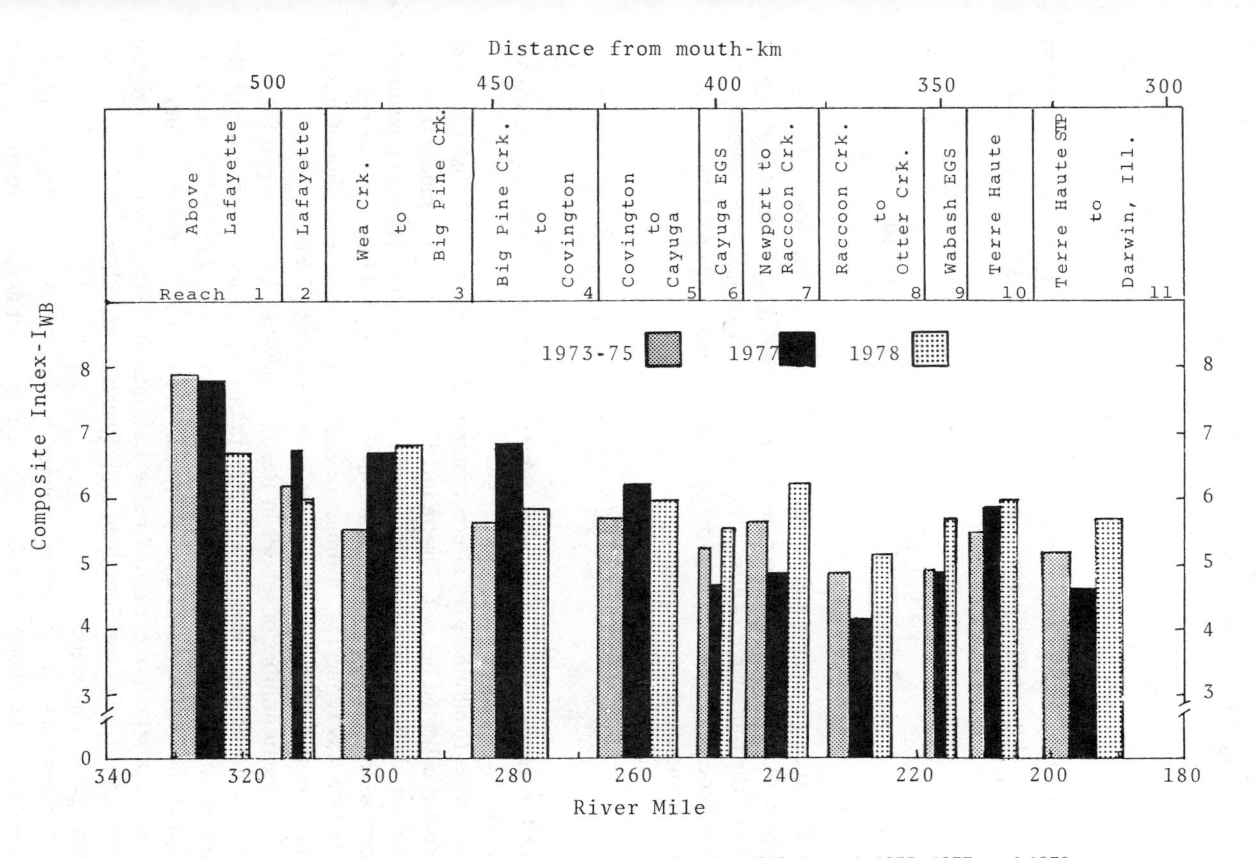

FIG. 4—*Profiles of mean annual composite index values for 1973 through 1975, 1977, and 1978.*

where p is the proportion a species of fish comprises in the community, N is the sample size, and Y is the probability of failing to detect the species. The equation may be solved for N

$$N = \frac{\ln Y}{\ln(1 - p)}$$

For example, how many fish (N) must be collected at random to detect species that make up 1 percent ($p = 0.01$) of the community with a chance of failure of 5 percent ($Y = 0.05$)?

$$N = \frac{\ln(0.05)}{\ln(1 - 0.01)} = \frac{-2.99573}{-0.01005} = 300 \text{ fish}$$

The average sample size in Run 1 (1977) was 22 fish. Nearly half of the species of fish captured were rare, each contributing only about 1 percent or less of the total catch. Thus, with the very small sample sizes collected, the likelihood of failing to collect the species that comprise only 1 percent of the community is about 80 percent

$$Y = (1 - 0.01)^{22} = 0.80$$

For this reason, the samples were combined in eleven reaches of the stream (Table 1).

In order to eliminate the effects of unequal sample sizes, 120 fish (the total number collected below Lafayette in 1975) were selected at random from the total number collected in each reach in 1977, thus standardizing the sample sizes on the basis of the minimum number collected. The process of random selection was replicated 25 times for each reach, and the number of species thus selected was averaged over the 25 replications (Table 3).

The mean number of species in Reach 3 (between Lafayette and Big Pine Creek) during 1977 was 24.0, higher than that for any other reach of the stream, and the mean number found in Reach 2 in the Lafayette area was as high as the number in the upstream areas of Reach 1. All of these values were higher than the number found below Lafayette in 1975 (17 species). A large number of species was found through Reach 5, just above the Cayuga electrical generating station.

A major drop of 13 species occurred in the vicinity of the Cayuga plant. Below Cayuga and above Clinton, the mean number of species dropped again to about 15 and continued at about this value until the area below the Terre Haute sewage treatment plant.

TABLE 3—*Standardization of sample sizes based on the minimum number of fish collected.*[a]

Reach of Stream (year)	Sample Size Reduced to 120		Total of Samples		
	Number of Species	Number of Fish	Number of Fish	Fish per km	Stations
3 (1975)[b]	17[b]	120	120	34.3	5[c]
1 (1977)	20.6	120	411	74.7	11
2 (1977)	20.1	120	252	56.0	9
3 (1977)	24.0	120	409	54.5	15
4 (1977)	23.6	120	197	43.8	9
5 (1977)	23.2	120	181	32.9	11
6 (1977)	13.2	120	416	34.7	24
7 (1977)	17.6	120	368	46.0	18
8 (1977)	14.9	120	341	22.7	30
9 (1977)	16.1	120	258	22.4	23
10 (1977)	15.36	120	343	38.0	18
11 (1977)	11.00	120	301	22.3	25

[a]In 1975 in Reach 3 of the Wabash River (below Lafayette), 120 fish were collected in 5 samples occupying 3.5 km of river. To make the 1977 samples comparable, the number of fish in each sample was reduced to 120 by selecting fish at random. For each of 11 reaches of the stream, this random subsampling was repeated 25 times, and the mean number of species in the 25 subsamples was determined. The number of fish per kilometre in the total of all samples collected in 1977 in each reach of the stream was also determined, assuming that each sample occupied 0.5 km of the river. Note that the five samples collected in Reach 3 in 1975 occupied 3.5 km.

[b]Total number of fish collected = 120.

[c]Five stations occupied 3.5 km.

Comparison of Sequential Correlation Coefficients

Results of preliminary cluster analysis of data from the Wabash River were difficult to interpret, because meaningful patterns were confounded by the seemingly random presence or absence of rare species. The numerical code used to identify the species studied is presented in Table 4 according to the criteria shown in Table 5. The species included in the next three approaches were those that had mean proportions greater than 0.01 in the eleven reaches or, if rarer on average, those that had standard deviations greater than or equal to 0.005.

Figure 5 shows correlation coefficients computed between successive pairs of reaches. No statistical test of the level of significance was possible, because the data failed to meet the assumptions of the statistical test. Nevertheless, the value of the coefficient may be taken to indicate the degree of similarity among reaches. Reaches 2 and 3 are more closely similar to each other than Reaches 1 and 2, suggesting limited effluent effects in the Lafayette area. Reaches 4 and 5 are the least similar of the three pairs of reaches. From the Cayuga plant downstream, all the reaches were very highly correlated with each other. The correlation coefficients, it should be noted, compare reaches

TABLE 4—*Species of fish studied.*

2	Silvery lamprey	117	Shorthead redhorse
4	Shovelnose sturgeon	118	Golfish-carp hybrid
6	Paddlefish	119	River redhorse
8	Bowfin	120	Northern river carpsucker
10	Longnose gar	121	Highfin carpsucker
11	Shortnose gar	122	Blue sucker
13	Gizzard shad	123	Smallmouth buffalo
14	Skipjack herring	124	Black buffalo
18	Goldeye	134	Emerald shiner
19	Mooneye	139	Spotfin shiner
30	Yellow bullhead	148	Creek chub
32	Channel catfish	154	Silvery minnow
36	Flathead catfish	157	River chub
41	American eel	165	Rockbass
57	White bass	169	Bluegill
61	Freshwater drum	170	Longear sunfish
104	Carp	171	Smallmouth bass
105	Quillback carpsucker	172	Largemouth bass
107	White sucker	173	White crappie
109	Northern hogsucker	174	Black crappie
110	Bigmouth buffalo	175	Spotted bass
111	Spotted sucker	177	Redear sunfish
112	Silver redhorse	183	Greenside darter
113	Black redhorse	188	Logperch
114	Golden redhorse	199	Sauger

TABLE 5—*Scheme to eliminate rare species from cluster analysis.*[a]

Proportion (species omitted if proportion < value given below)	Species Remaining[b]
0.05	13, 104
0.04	13, 104, 117, 120
0.03	13, 104, 117, 120, 36, 114, 170
0.02	13, 104, 117, 120, 36, 114, 170
0.01	13, 104, 117, 120, 36, 114, 170, 10, 11, 18, 32, 57, 61, 109, 112, 199
0.01 unless species ≤ 0.005	13, 104, 117, 120, 36, 114, 170, 10, 11, 18, 32, 57, 61, 109, 112, 199 19, 119, 121, 122, 123, 139, 171, 175

[a]The average proportions of all species in all samples, grouped by reach of stream, were computed. Species rarer than certain average proportions were omitted. Finally, species rarer, on average, than 0.01 were omitted unless the standard deviation of their proportions in each reach of the stream was greater than or equal to 0.005. The rationale behind this was that some rare species ought to be included if their distribution was sufficiently variable so that they were possible indicators of particular environments or reaches of the stream.

[b]The numbers correspond to those in Table 4.

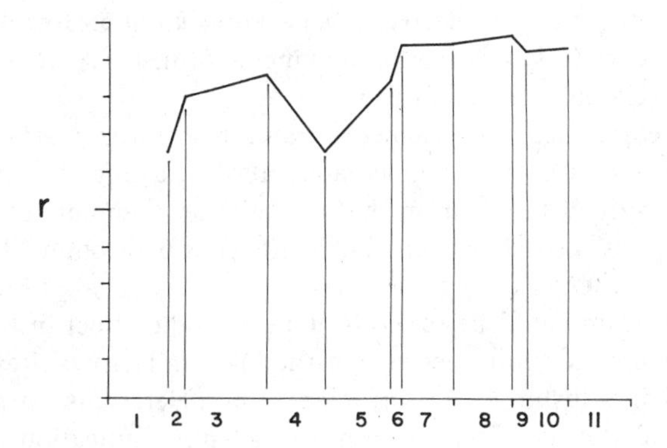

FIG. 5—*Product-moment correlation coefficients comparing the relative abundances of fish in sequential pairs of reaches downstream. No statistical tests were warranted. The reduced data set described in Table 4 was used.*

on the basis of their relative proportions of fish, not on the absolute numbers collected.

Q-*Mode Cluster Analysis*

Figure 6 shows the results of *Q*-mode cluster analysis. The reaches are divisible into two clear-cut clusters marked A and B. Cluster A comprises the

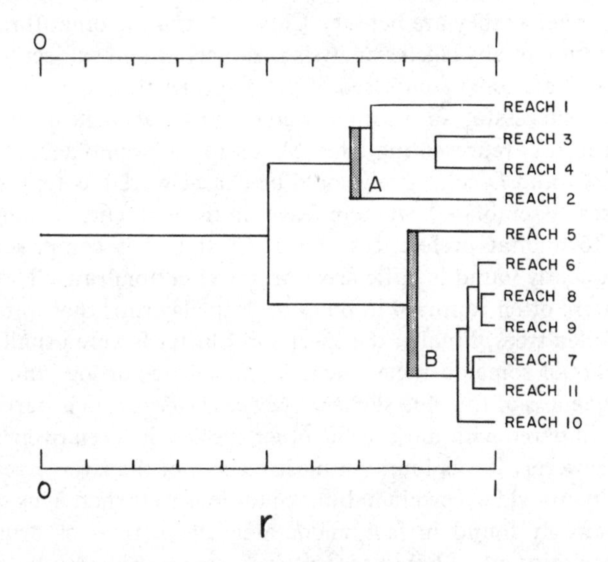

FIG. 6—*Results of Q-mode cluster analysis of the reduced fish data from eleven reaches of the Wabash River using a matrix of product-moment coefficients and the unweighted pair group method of clustering with arithmetic averages (UPGMA). The cophenetic correlation coefficient (r_{cc} = 0.901) indicates little distortion introduced during clustering.*

reaches above the Cayuga electrical generating station. Reach 2 has the lowest average correlation with other members of Cluster A. Reaches 3 and 4 are highly correlated with each other.

Cluster B comprises the reaches at and below the Cayuga electrical generating station. Reach 5, immediately above Cayuga, has a fish fauna that is appreciably different from that of the other members of the cluster. Otherwise, as was also shown in Fig. 5, Reaches 6 through 11 are closely similar to each other.

This dendrogram could be taken to indicate some effect in the Lafayette area on the fish community because of the low similarity of Reach 2 to the other reaches in Cluster A. In any event, any deleterious effect was soon eliminated, and Reaches 3 and 4 seem to shown no indication of it.

R-*Mode Cluster Analysis*

The 24 most common species of fish in the Wabash River are grouped into four clusters by cluster analysis, with one species, the blue sucker, left unclustered (Fig. 7). Cluster A is quite distinct from the other three clusters, having a *negative* average correlation of greater than -0.3, which indicates pronounced dissimilarity of distribution and abundance. The other three clusters are also quite distinct from each other, although average correlations are not negative, as they are between Cluster A and the other three clusters.

Interpretation of any cluster analysis is subjective and subject to criticism. Cluster A appears to be comprised of those species that may be regarded as being more successful in warmer waters. In that regard, it might be characterized as representing the Mississippi assemblage. Conversely, Clusters B, C, and D collectively could be characterized as representing the Great Lakes assemblage, wherein each individual cluster might be attributable to habitat preferences. Thus, Cluster D is comprised of those species frequently found in riffle areas or gravel bottom runs. The species in Cluster C were often captured in pools or deep clay runs that were generally devoid of fallen trees. Finally, the species in Cluster B were usually captured in the vicinity of some obstacle, such as a fallen tree or log jam.

The uniqueness of the blue sucker (*Cycleptus elongatus*) is particularly intriguing. Compared with most of the other species, it is relatively rare in the Wabash. However, it was found in most reaches of the river, irrespective of water quality, provided certain habitat conditions were met. This species was almost invariably found in fast, moderately deep, tree- or brush-covered chutes. Since this type of habitat is relatively scarce in the Wabash, it seems reasonable to postulate that the amount of habitat available for this species limits its abundance, rather than water quality or other man-caused factors.

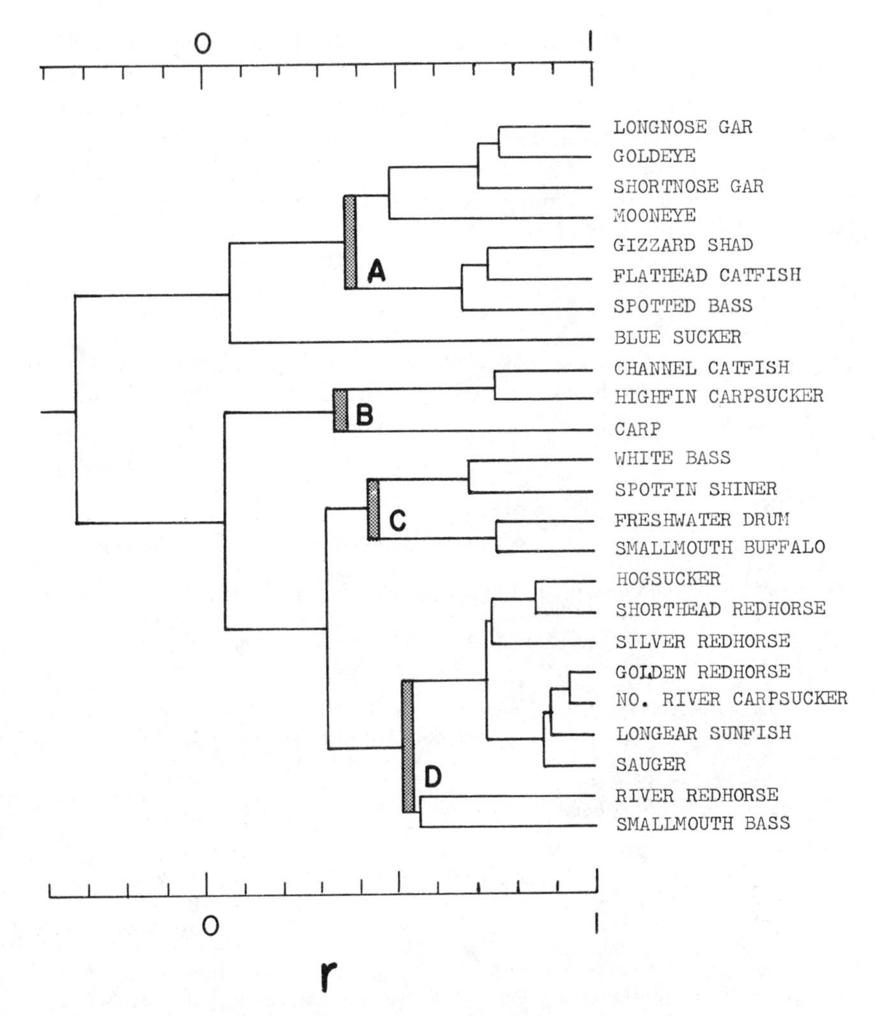

FIG. 7—*Results of* R-*mode cluster analysis (UPGMA) of a matrix of product-moment correlation coefficients computed from the reduced fish data* (r_{cc} = *0.820).*

Conclusions

1. Environmental quality in a large, warm-water river can be monitored by electrofishing.

2. A composite index, reflective of both species diversity and abundance, appears to provide a more precise measure of environmental quality than any single community index or indicator species of large-sized fish.

3. Large sample sizes and repetitive surveys are needed to make reliable measurements of species diversity within each reach when a substantial pro-

portion of the population is made up of species having a relatively low frequency of occurrence.

4. Cluster analysis appears to be a powerful technique for identifying similar fish communities and species associations.

5. River flow may strongly influence the distribution of fish communities. Impact assessment studies must include surveys during years of exceptionally low flow as well as years of normal flow.

Acknowledgments

We wish to thank the many students who have assisted in the collection of this data. The research was funded by grants from Eli Lilly and Co. and Public Service Indiana.

References

[*1*] Larimore, R. W. and Smith, P. W., *Illinois Natural History Survey Bulletin,* Vol. 28, No. 2, 1963, pp. 299-382.
[*2*] Bechtel, T. J. and Copeland, B. J., *Contributions to Marine Science,* Vol. 15, 1970, pp. 103-132.
[*3*] McErlean, A. J. and Mihursky, J. A., *Proceedings,* 22nd Annual Conference of the Southern Coast Association, U.S. Fish and Game Commission, 1969, pp. 367-372.
[*4*] McErlean, A. J., O'Connor, S. G., Mihursky, J. A., and Gibson, C. I., *Estuarine and Coastal Marine Science,* Vol. 1, No. 1, 1973, pp. 19-36.
[*5*] Richards, J. S., *Transactions of the American Fisheries Society,* No. 1, 1976, pp. 32-40.
[*6*] Denoncourt, R. F. and Stambaugh, J. W., *Proceedings,* Pennsylvania Academy of Science, Vol. 48, 1974, pp. 71-78.
[*7*] Gammon, J. R., *Proceedings,* 3rd National Symposium on Radioecology, Vol. 1, 1971, pp. 513-523.
[*8*] Gammon, J. R., "The Effect of Thermal Inputs in the Population of Fish and Macroinvertebrates in the Wabash River," Technical Report No. 32, Water Reservoir Research Center, Purdue University, Lafayette, Indiana, 1973, pp 1-105.
[*9*] Teppen, T. C. and Gammon, J. R., *Thermal Ecology II,* ERDA Symposium Series, CONF-75-425, 1976, pp. 272-283.
[*10*] Gammon, J. R., "The Fish Populations of the Middle 340 km of the Wabash River," Technical Report No. 86, Water Reservoir Research Center, Purdue University, Lafayette, Ind., 1976, pp. 1-73.
[*11*] Todd, D. K., *The Water Encyclopedia,* Water Information Center, Port Washington, N.Y. 1970.
[*12*] Gammon, J. R. and Reidy, J. M., "The Role of Tributaries During an Episode of Low Dissolved Oxygen in the Wabash River, Indiana," Symposium on Warmwater Streams, in press.

Summary

The papers included in this volume were selected as representative of present work in the field of environmental impact assessment. While no single paper meets all of the criteria of the symposium at which they were presented, these papers are probably typical of present environmental assessment study reporting.

The principal shortcoming of all these papers is failure to address adequately the validity and feasibility of application of the methods presented—that is, the spectrum of relevance or range of application, comparison with the alternatives (other approaches), cost-effectiveness, and the level of competence required. Consideration of these factors is essential for priority ordering of community and parameter protocols.

The following brief comments on the papers, while the sole responsibility of this editor, incorporate the responses of symposium attendees and manuscript referees, as compiled by Sally Dennis.

The introductory overview presented by C. Weber gives perspective in understanding the kinds of assessments associated with the aquatic environment and provides a working definition of the term "biological integrity." His discussion of the basic properties of communities provides the essential framework for the current prioritization efforts of the Ecosystem Section of Subcommittee D19.23 on Biological Field Methods of ASTM Committee D-19 on Water.

Herricks et al provide a rationale for the use of a computer-based predictive ecosystems model. While this approach has potentially broad application, it presently suffers from the constraints imposed by the limited availability of an adequate data base. Furthermore, it assumes a level of interpretive expertise not always available. The applicability of this kind of program will be greatly enhanced as more ecological data become available.

The use of scale model streams is proposed by Russell et al. The approach taken in this paper would at present seem more applicable to mobile bioassays than to in-stream assessments. As pointed out by the authors, field validation of results is essential.

Power-plant-related studies are presented in two papers. Bogardus discusses the problem of species selection for monitoring studies and presents a useful rationale for the selection of target species, using examples from studies of demonstrations under Section 316(a) of the Federal Water Pollution Control Act. Sage and Olson present a detailed case history of an estuarine zooplankton study, discussing at length the evolution of a sampling program. This paper has particular merit in that it points out some of the pitfalls encountered in sampling design.

The use of a uni-algal culture to measure uptake of radionuclides is proposed by Gertz and Suffet. While this approach may be useful in monitoring effluents, it, as yet, has little applicability to assessment of in-stream impact.

Four papers that deal with the use of algal periphyton in stream assessments address various aspects of application. These papers point up problems associated with periphyton methods and use of the data generated. Sullivan et al attempt to arrive at a cost-effective periphyton analysis method. Weber and McFarland (Effects of Copper . . .) caution against the use of density and diversity data without analysis of the community structure (species identification). Weitzel and Bates address the relationship between the counting effort and the information yield for periphytic diatoms. An additional paper by Weber and McFarland (Effects of Exposure . . .) points up potential effects of exposure time, season, and substrate type on periphyton colonization.

Putnam et al present a case history of analysis of impacts of TNT waste without evaluation of the method used or recommendations for its future application.

A paper by Guthrie et al presents a study of biological succession in an ash settling basin. While not dealing directly with in-stream assessment, this paper suggests a method for reducing impacts of power plant effluents on receiving waters.

Kovalak provides a discussion of the application and limitations of two diversity indexes (Shannon-Weaver and Brillouin) in benthic analysis, a discussion which will be particularly useful to biologists not familiar with these indexes. Further, a mechanistic model based on available oxygen and used for predicting species richness is presented.

In a paper by Tesmer and Wefring, it is suggested that annual macroinvertebrate sampling may in some instances suffice as a low-cost tool for assessing effluent impacts. While this approach has obvious shortcomings, it may have limited application.

Foster proposes an alternative test organism (*Corbicula*) for use in hazard evaluation studies and suggests the possible extension of this application to in-stream assessment.

Gammon et al discuss the potential for using data resulting from electrofishing in assessing environmental quality. While presenting a good case for such as application, this paper does not discuss the advantages of this method over other, less labor intensive, sampling programs. The range of application is, likewise, not addressed.

While the paper by Duedall et al does not have application to in-stream assessment, it was included because it provides an interesting potential solution to the problems associated with fly ash disposal.

These symposium papers are suggestive of the range of sampling protocols

TABLE 1—*Recommended priorities in sample parameters for ecological assessment of effluents and receiving waters.*

Parameters	Biological Communities Present						
	Detritus (Tripton)	Periphyton	Macrophyton	Phyto-plankton	Zooplankton	Benthos	Fish
Community characteristics							
Taxonomic composition	...[a]	...	...	...	...	...	...
Density	...	...	...	...	...	...	...
Biomass	...	...	...	...	...	...	...
Age/size distribution	...	...	...	...	...	...	...
Drift	...	...	...	...	...	...	...
Condition factor	...	...	...	...	...	...	...
Diseases and parasites	...	...	...	...	...	...	...
Tissue characteristics							
Organic compounds	...	...	...	...	...	...	...
Metals content	...	...	...	...	...	...	...
Radionuclides	...	...	...	...	...	...	...
Flesh tainting	...	...	...	...	...	...	...
Physiological characteristics							
Productivity	...	...	...	...	...	...	...
Respiration	...	...	...	...	...	...	...

[a] Sampling priority to be designated.

available for assessing the aquatic environment. The selection of the community or communities to be sampled and the relevant community characteristics to be investigated is the most difficult and also the most important decision to be made in the development of an ecological assessment protocol. Somewhere between the one-man (have Hach Kit and dip net—will travel) approach and the interstellar consulting consortium (base price, one million dollars) approach is a rational yet affordable protocol for assessing environmental impacts.

The Ecosystems Section of ASTM Subcommittee D19.23 on Biological Field Methods is attempting to resolve this apparent dilemma through consensus prioritization of sample parameters and adoption of effluent/receiving water classifications. Table 1, arrived at through consensus of the Ecosystems Section, presents an attempt to develop this priority ordering. An initial and much more detailed document was deemed overly ambitious for present purposes. The physical and chemical parameters included in this initial document will be reintroduced at a later date.

It is hoped that this effort will lead us away from both the "laundry-list" approach, which often results in oversampling (to be on the safe side), and the "rationalization" approach, which results in overemphasis of those parameters with which we feel most comfortable. While a standardized sampling design may not be for everyone, it would provide a sound, defensible program having wide application in environmental monitoring.

J. M. Bates

Ecological Consultants, Inc., Ann Arbor, Mich. 48103; symposium chairman and editor.

Index

D

E

F

N

O

P

R

S